Electroactive Polymers
(EAP)

MATERIALS RESEARCH SOCIETY
SYMPOSIUM PROCEEDINGS VOLUME 600

Electroactive Polymers (EAP)

Symposium held November 29–December 1, 1999, Boston, Massachusetts, U.S.A.

EDITORS:

Q.M. Zhang
The Pennsylvania State University
University Park, Pennsylvania, U.S.A.

Takeo Furukawa
Science University of Tokyo
Tokyo, Japan

Yoseph Bar-Cohen
Jet Propulsion Laboratory
Pasadena, California, U.S.A.

J. Scheinbeim
Rutgers University
Piscataway, New Jersey, U.S.A.

Materials Research Society
Warrendale, Pennsylvania

CAMBRIDGE UNIVERSITY PRESS
Cambridge, New York, Melbourne, Madrid, Cape Town,
Singapore, São Paulo, Delhi, Mexico City

Cambridge University Press
32 Avenue of the Americas, New York NY 10013-2473, USA

Published in the United States of America by Cambridge University Press, New York

www.cambridge.org
Information on this title: www.cambridge.org/9781107413269

Materials Research Society
506 Keystone Drive, Warrendale, PA 15086
http://www.mrs.org

First published 2000
First paperback edition 2013

Single article reprints from this publication are available through
University Microfilms Inc., 300 North Zeeb Road, Ann Arbor, MI 48106

CODEN: MRSPDH

ISBN 978-1-107-41326-9 Paperback

This work was supported in part by the Office of Naval Research under Grant Number
ONR: N00014-00-0166. The United States Government has a royalty-free license
throughout the world in all copyrightable material contained herein.

CONTENTS

APPLICATIONS

FERROELECTRIC POLYMERS

*Invited Paper

v

PIEZOELECTRIC, ELECTROSTRICTIVE, AND DIELECTRIC ELASTOMER

*Invited Paper

*Invited Paper

*Invited Paper

COMPOSITES AND OTHERS

PREFACE

For many years, electroactive ceramic, magnetostrictive material and shape memory alloys have been the primary source of actuation materials for manipulation and mobility systems. Electroactive polymers (EAP) received relatively little attention due to their limited capability. In recent years, effective EAP materials have emerged changing the paradigm of these materials' capability and potential. Their main attractive characteristic is the operation similarity to biological muscles where under electrical excitation a large displacement is induced. The potential to operate biologically inspired mechanisms using EAP as artificial muscles and organs is offering exciting applications that are currently considered science fiction.

This MRS symposium, "Electroactive Polymers," held November 29–December 1 at the 1999 MRS Fall Meeting in Boston, Massachusetts, was initiated for the first time this year in an effort to promote technical exchange of EAP research and development as well as provide a forum for progress reports. The symposium has the input from the international mix in participation. Eminent EAP researchers from the U.S.A., Japan, and Europe presented Invited Papers covering the cutting edge of their material state of the art capabilities and limitations. Generally, two groups of materials were covered: Dry - including electrostrictive, electrostatic, piezoelectric, and ferroelectric; as well as Wet - including IPMC, nanotubes, conductive polymers, gels, etc. While overall the dry types require high voltage for their operation, they provide larger mechanical energy density and they can hold a displacement under a DC voltage, and some of the DRY EAP are capable of operating to high frequencies (>10 kHz). On the other hand, the WET EAP are superior in requiring low actuation voltage (~ a few volts) with high strain generation capabilities, but are sensitive to drying, and some have difficulties holding a displacement under DC activation.

This proceedings volume brings together many of the oral and poster presentations at the symposium and includes papers on the following topics:

- ▼ EAP applications and their characterizations
- ▼ Ferroelectric based polymers
- ▼ Piezoelectric, electrostrictive polymers, and high strain dielectric elastomers
- ▼ Polymer gels and biological muscles
- ▼ Conductive polymers
- ▼ Polymer composites
- ▼ EAP synthesis and processing
- ▼ Models and analysis of EAP

We hope this volume can provide a resource for the future development in this exciting interdisciplinary field. Clearly, the EAP developed offer great potential in many applications, and also challenge to improve them further to meet the great expectations.

Q.M. Zhang
Takeo Furukawa
Yoseph Bar-Cohen
J. Scheinbeim

January 2000

ACKNOWLEDGMENTS

We would like to extend special thanks to all the speakers, poster presenters, session chairs, and symposium attendees who made this symposium and the publication of this proceedings possible. We would also like to thank the MRS staff and the meeting chairs whose efforts made our tasks much easier. We acknowledge and would like to thank Ms. Kim Sterndale for her assistance in the preparation of the symposium program and compiling the proceedings. The generous financial support from the following organizations is greatly appreciated and acknowledged:

> Office of Naval Research
> Daikin US Corporation
> Kureha Chemical Industry Co., Ltd.
> Toray Techno Co., Ltd.

MATERIALS RESEARCH SOCIETY SYMPOSIUM PROCEEDINGS

MATERIALS RESEARCH SOCIETY SYMPOSIUM PROCEEDINGS

Prior Materials Research Society Symposium Proceedings available by contacting Materials Research Society

Applications

COMPLIANT ACTUATORS BASED ON ELECTROACTIVE POLYMERS

S G. Wax*, R.R. Sands**, L. J. Buckley***
*Defense Science Office, DARPA, Arlington, VA
**Technology Consultant, Arlington, VA
***Naval Research Laboratory, Washington, DC

ABSTRACT

The field of Electroactive Polymers has experienced a considerable amount of expansion over the last decade. Much of this work has been concentrated on developing polymeric materials that mimic biological systems or that exhibit electronic and optical properties similar to inorganic materials. This paper briefly reviews some of the nearer term applications that electroactive polymers might impact: image processing and sonar. In addition, a review of compliant actuators based on the unique properties inherent in electroactive polymers is provided. Emphasis will be placed on the mechanisms responsible for actuation and on the limited mechanical, electrical and chemical data current available. A comparison between mammalian muscle properties and electroactive polymer actuator properties is provided.

INTRODUCTION

The interest in electroactive polymers (EAP) has grown significantly over the last several years, as indicated by the growing number of conferences that now address this topic. This interest seems to be due in part to the potential for EAPs to provide compliant actuation for use in biological or biomimetic applications. (Here the term "compliant" is used to describe the flexibility that emulates that of real muscles) The actuation promise of these materials is one of the major reasons the Defense Advanced Research Projects Agency (DARPA) has been sponsoring this technology. In addition, other applications that involve the unique electro-optic response of the materials are being pursued. Table I, modified from a previous paper [1], outlines the many possible applications being examined.

The primary focus of this paper is to discuss how the unique advantages of EAP for actuation drive the requirements for the performance of these materials and therefore the research needs. However, it is also illuminating to examine another biologically inspired application for EAPs in order to note how even simple control of the chemistry can make huge differences in capability.

There is a critical defense need for rapid signal processing of images, particularly important as a weapon attempts to identify and follow its target amid clutter of all spatial frequencies. Traditional sensors are pixel based and require a great deal of digital processing, costly in time and power. Conversely, our eyes and early visual system sort through scenes in an analog fashion, blending "pixels" as needed and sending only the critical

Table I: Defense Applications for Electroactive Polymers
Actuation and Sensing
• Artificial Muscles
• Smart Skins
• Acoustic (Sonar)
• Biomimetic Devices
• Bio-sensors
Electro-Optical Response
• Analog Processing
• Large Area, Flexible Displays
• Polymer FET's
• Sensor Protection

features of an image on to the brain for "processing." A program with Uniax, Raytheon Infrared Center of Excellence (RIRCOE) and Computational Sensors Corporation is using EAPs to provide the analog processing by using the anisotropic conductivity property of polyaniline.

Figure 1 shows how the conducting polymer is used to form a resistive layer which, upon closing the switches, connects capacitive elements holding charges Q(i). These charges represent the intensity of pixels in an image passed to the processor from a charge coupled device (CCD) in an imaging sensor. The amount of charge leakage between the elements of the sensor is determined by the resistivity of the polymer layer and the time the switches are held closed. The longer the closed time, the greater the blurring of the image, removing more and more of the high spatial frequencies and thus creating an analog low pass filter. It is also possible to form simple high pass and band pass filters by simple modifications. Figure 2 illustrates the power of this

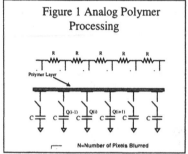

Figure 1 Analog Polymer Processing

N=Number of Pixels Blurred

processing approach. In the figure the image was enhanced through the use of a set EAP image processing chips, which are performing local contrast control. The enhanced image allows for information to be pulled from the dark areas without loss of information from the remaining regions. This approach can be more than 10 times faster than digital processing and has the scaling ability to process pixel arrays on

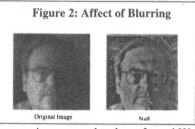

Figure 2: Affect of Blurring

Original Image   N=R

the order of 2000x2000, and larger, with minimum increase in power and no loss of speed [2]. The next step is to learn how to produce controllable anisotropic properties in the polymer films to allow for selection of the directionality of the filtering, and ultimately to control the conductivity of the polymer in response to the target being sought. However, the simple near term capability is more than sufficient to begin to apply this technology in defense applications [3].

PROMISE OF COMPLIANT POLYMER ACTUATORS

In this paper, the primary focus of discussion will be on three classes of compliant actuators: polymer gels, ionic polymer and conducting polymers. Given the number of actuators that exist, it is always important to ask why there is a need for another. From the authors' perspective, it is not simply the actuation potential of the materials, which arguably might be achievable by a number of alternative actuator designs, but rather that potential in concert with the compliant nature of polymers. A general view of the requirements of such actuators is to compare them with biological muscles as shown in Figure 3. From this graph it appears that the compliant actuators have the potential to match muscles, but have not yet the frequency. Other applications include microactuation in which the simplicity and flexibility of these actuators also show promise, but are not yet in use.

Before undertaking a discussion of the concepts and the limitations that have to be overcome, it is constructive to examine an EAP actuator that is moving rapidly toward an application. Specifically, the development of electrostrictive actuators, poly(vinylidene difluoride-co-trifluoroethylene (P(VDF-TrFE)) co-polymers, that are already under consideration for Navy sonar. The actuation properties of these co-polymers are based on a phase transition (ferroelectric to paraelectric) that these materials undergo when subjected to an electric field. This phase transition leads to a large lattice change within the material that generates large volume changes. Initial data [4] on P(VDF-TrFE)) actuators shows a 4% strain capability with <15Mpa stress generation. P(VDF-TrFE)) exhibits improved strain capability over existing electrostrictive actuators that are being used for Navy sonar. In this case, the polymers have low density, an excellent acoustic match with water, and can be placed conformally along the length of the submarine. As shown in Table II, work at Penn State University [5] has shown remarkable progress in increasing the performance of irradiated PVDF-TrFE to the point where they may well be competitive with the new PZN-PT relaxor ferroelectrics. Current work is focused on lowering the actuation voltage of these electrostrictive polymers and to demonstrate these co-polymer systems in a sonar test bed at Raytheon.

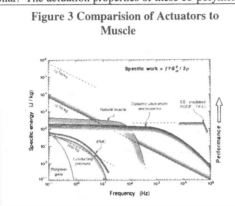

Figure 3 Comparision of Actuators to Muscle

The rapid movement to applications of these materials is based on two factors. First, the application domain is well understood for the sonar transducer insertions. Secondly, there is an excellent understanding of the mechanisms of actuation and what it will take to positively affect them. This is in sharp contrast with the compliant "biomimetic"actuators for which the targets of opportunities are less clear and the ability to control the mechanisms of actuation are not well established. In the next sections the characteristics of each of the compliant actuator concepts discussed along with their limitations will be in this context.

Table II Properties of Selected Actuators for Sonar Applications

Materials		Y (GPa)	S_m (%)	$YS_m^2/2$ (J/cm³)	$YS_m^2/2\rho$ (J/kg)
Electrostrictive	S_3	0.50	5	0.6	300
PVDF-TrFE	S_1	**1.3**	**3.5**	**0.8**	**400**
Piezoceramics		64	0.2	0.13	17
Magnetostrictor		100	0.2	0.2	21.6
PZN-PT Crystal		**7.7**	**1.7**	**1.0**	**131**
Polyurethane		0.017	11	0.1	83

CHEMISTRY AND CHARACTERISTICS OF COMPLIANT ACTUATORS

Polymer Gels

Many researchers [6,7,8] have described polymer gels as analogous to biological gels such as the vitreous humor of the eye. The resemblance between natural gels and artificial gels has inspired researchers to attempt to mimic the functionality found in biological systems with artificial polymer gels. These activities have created the area of artificial muscles/actuators based on polymer gels. Earlier work by Wasserman [9], and Katchalski [10], demonstrated that polymer gels could be engineered to produce large dimensional changes based on phase changes within the gel. Today many researchers are investigating polymer gels for actuation (locomotion, drug delivery, membrane separation, MEMs), sensing (physical, chemical, electrical and biological), and energy conversion. A very active area is in drug delivery systems where drugs are incorporated into the polymer network and allowed to release through diffusion or due to environmental conditions within the body such as pH, and temperature.

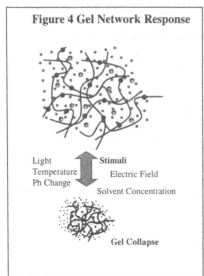

Figure 4 Gel Network Response

Light
Temperature
Ph Change
Stimuli
Electric Field
Solvent Concentration

Gel Collapse

As shown in Figure 4, polymer gels are comprised of macromolecules crosslinked to form a three-dimensional network, where the interstitial space in the network is occupied with a fluid. The properties of the gel are directly dependent on the interactions between the polymer network and the fluid, and these interactions or forces depend on the elastic properties of the polymer network, movement of ions in and out of the gel, and electrochemical forces due to ionic interactions. The volume of the gel is contingent upon the equilibrium of these forces (internal and external). Depending on the environmental conditions, a stable phase for the gel will be present. For example, applying an electric field to the gel can cause the movement of ions in or out of the polymer network causing swelling or shrinking.

Investigators [11] have reported that Gels can swell or shrink as much as 1000 times in response to external stimuli (pH, temperature, light, solvents, etc.) DeRossi [12] has reported mechanical data of a hydrogel of polyvinyl alcohol (PVA) and polyacrylic acid (PAA) that showed a power density of 0.1kW/kg (for a 10 micron thick film) that is similar to the power density for human muscle, however human muscle has a peak power density close to 1.0kW/kg. Furthermore, DeRossi [13] has reported that the PVA-PA system exhibits a force-velocity relationship similar to natural muscle.

The amount of volume change that can be generated in a gel is directly related to the elastic properties of the polymer network and the interaction of the network and fluid with the surrounding environment. There is a trade-off between modulus, displacement, and capacity for work where networks that high elastic modulus display small volume changes, and networks possessing lower modulus have large volume changes.

Gel actuators have several major issues. First, the actuation mechanism is based on ion diffusion, its speed is transport limited and tends to be very slow. Second, these systems require an aqueous solution for operation. The need for an aqueous environment causes severe constraints on the packaging of these actuators. Finally, the low modulus of the gel might make it difficult to design an actuator with reasonable stress capability.

Ionic Conducting Polymer Actuators

Ionic conducting polymers have been used in many industrial applications ranging from fuel cells to specialty chemical sensors (oxygen, humidity, glucose, etc.). There exists a large body of knowledge that covers the chemical and physical properties of these materials. However, the use of these for actuators is relatively new and data concerning the mechanical and chemical behavior of the materials systems is limited and incomplete. Several research groups in the

United States, Japan and Europe have been developing and characterizing actuators based on the unique properties exhibited by ionic conducting polymer membranes. One popular system is the Ionic Polymer Metal Composite (IPMC) actuator [14,15]. Typical construction of these actuators involves a perfluorinated cation-exchange membrane (Nafion® or Flemion ®) plated with a noble metal (typically Pt or Au) for use as electrodes.(See Figure 5). Further processing, (details found elsewhere [15,16,17]) allow for the ion exchange of selected ions (eg., Na+, Li+, H+, Ca++) with constituents found in the membrane structure, and the incorporation of water. A unique feature of these perfluorinated membranes is their ability to selectively allow water molecules and cations to move through its porous structure while blocking the motion of anions. This feature, in conjunction with the noble metal electrodes, allows for the control of cation movement via an e. :ctric field. Applying an electric potential across the membrane, (typically 1.5 volts or less to avoid electrolysis) causes cations to move toward the cathode inducing internal charge imbalance that leads to internal stresses.

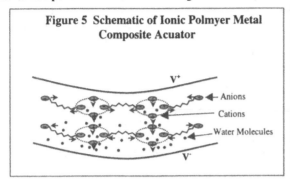

Figure 5 Schematic of Ionic Polmyer Metal Composite Acuator

V+
— Anions
— Cations
— Water Molecules
V-

While the actual mechanisms responsible for actuation (shape change) are not perfectly understood, it appears that two processes may be active, perhaps simultaneously. In one mechanism the water molecules moving with the cations create an osmotic pressure effect. These stresses are reduced through movement of the membrane that reduces the charge imbalance. The observation of water being expelled at the cathode during operation tends to support this mechanism. The movement of water molecules produces a pressure gradient within the membrane that supports further movement of the structure to reduce osmotic pressure. In another mechanism, which may occurs simultaneously, it is the redistribution of the cations that cause the motion. This certainly is the case in those instances where a non-aqueous cation has been used. Preliminary work at the University of California, San Diego seems to indicate that the observed actuation resulting from each of these mechanisms appears to be dramatically different [18].

In addition, consistent and complete mechanical data for the behavior of IMPC type actuators is limited. For example, investigators have published force versus displacement curves that provide little insight on the active mechanism responsible for actuation. Figure 6 is typical data found in the literature for IMPC actuators (based on Nafion) [14]. The test setup is based on a cantilever approach where the electrodes are attached to one end of the IMPC bar and the deflection measurements are made at the opposite end. This data set demonstrates the complex behavior of this polymer composite and the

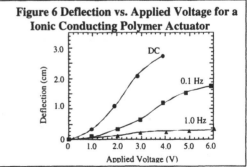

Figure 6 Deflection vs. Applied Voltage for a Ionic Conducting Polymer Actuator

Deflection (cm)

3.0 — DC
2.0 — 0.1 Hz
1.0 — 1.0 Hz

0   1.0   2.0   3.0   4.0   5.0   6.0
Applied Voltage (V)

dependence on frequency and applied voltage. However, it provides little information on force generation, work capability, bandwidth or other information needed for designers of actuators to seriously consider these materials. Such work is underway, but not yet available.

These material systems have the same limitations as with the polymer gels in that they need an aqueous medium within the membrane and that the available data indicates that they have a limited work and force capability [19]. Researchers [15] are developing packaging approaches to seal the IPMC actuators to retain the aqueous electrolyte within the membrane. Another important issue with IPMC is their bandwidth (response time), which can be improved by controlling the diffusion distance necessary for actuation and through better design and fabrication of electrodes.

To address these limitations is will be necessary to understand well the mechanism(s) of actuation and how these may be affected by the complex processing of the materials. Work underway at Environmental Robotics and the University of California, San Diego is beginning this task. Figure 7 shows work looking at how the noble metal is infiltrated into the material, the first step in understanding how processing affects the field strength across the actuator.

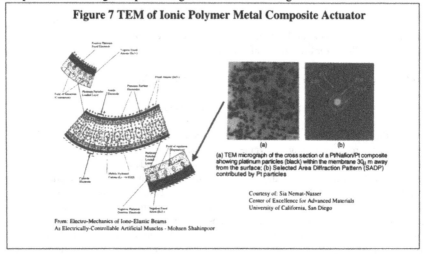

Figure 7 TEM of Ionic Polymer Metal Composite Actuator

(a) TEM micrograph of the cross section of a Pt/Nafion/Pt composite showing platinum particles (black) within the membrane 30µm away from the surface; (b) Selected Area Diffraction Pattern (SADP) contributed by Pt particles

Courtesy of: Sia Nemat-Nasser
Center of Excellence for Advanced Materials
University of California, San Diego

From: Electro-Mechanics of Iono-Elastic Beams
As Electrically-Controllable Artificial Muscles - Mohsen Shahinpoor

Conducting Polymers

Electrically conductive polymers have been an intense area of research for about 25 years and have led to some interesting phenomena in sensing and actuation [20]. The application of electrically conductive polymers specifically as actuators has been investigated for the past 10 years [21]. The mechanism of actuation via a volume change involves dopant ion insertion and deinsertion with an electrolyte and electrodes. This is a rather slow process when compared to other forms of polymer actuators such as electrostrictive and piezoelectric polymers [4]. For thin samples (less than 0.5 microns), where the dopant ions have smaller distances to travel, the response time may be less of a problem. Many issues arise when these actuators are compared to other systems. For example, the process of doping and dedoping is not completely reversible so the total number of cycles has intrinsic limitations. This will depend upon the polymer, dopant, electrolyte and morphology of the system and may not be an issue for low cycle applications. Atomic Force Microscopy (AFM) studies on polyaniline have shown that an extensive amount of change in the surface structure occurs during the first several cycles of doping and dedoping [22]. This was dependent upon the dopant and electrolyte conditions (i.e. pH).

Outstanding issues with conductive polymer actuators are limited force generation at relatively large strains and the need for an electrolyte (that is liquid, or at best, a soft solid or gel) for operation. The need for the electrolyte in these systems poses some very challenging packaging problems and will require some creative approaches in order to allow for long term stability of these actuators. The type of packaging will also determine the environmental robustness of the actuator. For example, environmental exposure to ambient moisture, oxygen and common pollutants can directly influence the mechanistic change responsible for actuation. The process of doping polyaniline involves exposure of the emeraldine base form of the polymer to an acid. Environmental exposure to acidic or basic conditions over time will affect the overall lifetime of the material. Packaging will delay but not completely eliminate these external effects. Encapsulated polypyrrole actuators have recently been studied using a gel electrolyte but have limited lifetimes of several hours due to the gel drying out (i.e. packaging) and gas bubble formation at the electrode/electrolyte interface [23]. Stability studies that included the parameters of temperature, air or nitrogen exposure have revealed a stability ranking of conductive polymers that place aryl sulfonate-doped polypyrroles better than protonic acid-doped polyanilines that, in turn, are better than alkyl-substituted polythiophenes [24,25,26].

As in the case of ionic actuators, steps are being taken to overcome these limitations. Specifically, work at Sante Fe Science and Technology, Inc.(SFST) is using cyclic voltammetry to look at the mechanisms and stability of the conducting polyaniline polymers. In this technique voltage is cycled at a given rate and current and each redox peak at each potential is observed. An example of this is shown in Figure 8. Each peak in the curve corresponds to an electrochemical reaction. When there are pairs of peaks of approximately equal intensity, the reaction is reversible. In addition, the peak heights, widths and positions provide information on the mechanisms, efficiency and time behavior of the polymers.

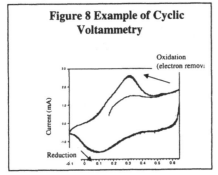

Figure 8 Example of Cyclic Voltammetry

As discussed above, other critical issues in using these materials as actuators is the need to have an electrolyte present and to increase the actuation response time. This will require clever design of the actuators. SFST will attempt to overcome these by using hollow conducting fiber actuators or controlled porosity filled with a solid electrolyte. This will allow an internal conducting fiber to slide inside the hollow fiber emulating the movement of natural muscle. To accomplish this, improvements in the spinning of polyanaline with controllable porosity must be made. Though the work is still in the early stages, Figure 9 shows progress on this account.

Figure 9 Improvements in Controlling Polyaniline Fiber Porosity

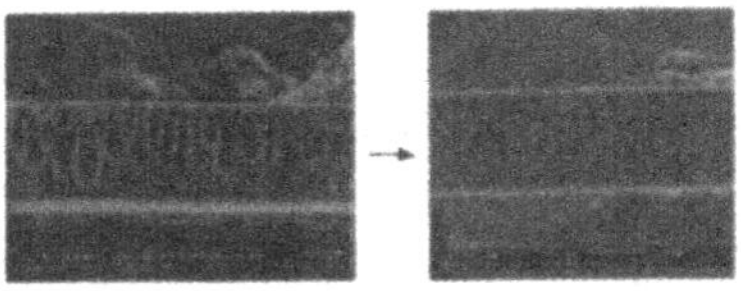

SUMMARY

While the opportunity for compliant actuators to have a significant impact on actuator technologies remains, it is becoming clear that the qualitative, early demonstrations were perhaps overly optimistic and the barriers to finding practical applications were perhaps underestimated. These barriers are mainly due to lack of complete and conclusive mechanical, chemical and electrical data necessary for designers to confidently work with. When comparing EAP based actuator with existing technologies, it becomes obvious that the polymer systems need substantial improvement. As described above, this is now happening. However, there are lessons to be learned in both the EAP image processing chip and the electrostrictive polymer examples presented. In both these cases, there were clearly defined application opportunities and thus a clear path to polymer development and improvement. When considering application for compliant actuators choosing an application that exploits the materials attributes and brings innovative solutions to a difficult problem may well be the first step. Furthermore, these novel materials provide an opportunity to approach actuation problems from a different perspective. Typically, compliance in actuation systems is designed in and can cause a great deal of system overhead. Compliant EAP actuators could (for selective applications) provide for the development of unique actuator designs that function like natural muscle, having dynamic mechanical properties as well as intrinsic sensing for feedback/feedforward control. However success will depend on understanding well these applications and then doing the careful mechanistic research to achieve them.

References

1. S. Wax, and R. Sands, Proceedings of the SPIE Conference on Electroactive Polymer Actuators, 3669, pp. 2-10, Newport Beach, CA, Mar. 1999.

2. D. Scribner, R. Klein, J. Schuler, G. Howard, J. Langan, J. Finch, C. Trautfield, S. Taylor, R. Behm and M. Costolo, IRIS Specialty Group on Passive Sensors, August 1999.

3. J. Langan, private communication.

4. Q. M. Zhang, Vivek Bharti, and X. Zhao, Science 280, pp. 2101-2104, (1998).

5. Q. M. Zhang, Vivek Bharti, Z.Y. Cheng, T.B. Xu, S. Wang, T.S. Ramotowski, F. Tito, and R. Ting, Proceedings of the SPIE Conference on Electroactive Polymer Actuators, 3669, pp. 134-139, Newport Beach, CA, Mar. 1999.

6. D. DeRossi, P. Parrini, P. Chiarelli, and G. Buzzigol, Trans. Am. Soc. Artif. Intern. Organs. 31, pp. 60-65, (1985).

7. P. Chiarelli, and D. DeRossi, Progr. Colloid. Polymer Sci. 78, pp. 4-8, (1988).

8. Y. Osada and J.P. Gong, Proceedings of the SPIE Conference on Electroactive Polymer Actuators, 3669, pp. 12-18, Newport Beach, CA, Mar. 1999.

9. A. Wasserman, A., ed. 1960. *Size Changes of Contractile Polymers*, Pergamon Press, New York, 1960.
10. A. Katchalsky, T. Michaeli, and H. Zwick, *Size Changes of Contractile Polymers*, Pergamon Press, New York, 1960.

11. E.S. Matsuo and T. Tanaka, J. Chem. Phys. 89, pp. 1695-1701, (1988).

12. D. DeRossi, M. Suzuki, Y. Osada, and P. Morasso, J. of Intell. Mater. Syst. And Struct. 3, pp. 75-95, (1992).

13. D. DeRossi, P. Chiarelli, G. Buzzigoli, C. Domenici, and L. Lazzeri, , Trans. Am. Soc. Artif. Intern. Organs. 32, pp. 157-162, (1986).

14. Y. Bar-Cohen, S. Leary, M. Shahinpoor, J. Harrison, and J. Smith, Proceedings of the SPIE Conference on Electroactive Polymer Actuators, 3669, pp. 57-63,Newport Beach, CA, Mar. 1999.

15. K. Oguro,N. Fujiwara, K. Asaka, K. Onishi, and S. Sewa, Proceedings of the SPIE Conference on Electroactive Polymer Actuators, 3669, pp. 64-71,Newport Beach, CA, Mar. 1999.

16. Y. Bar-Cohen, T. Xue, B. Joffe, S.-S. Lih, P. Willis, J. Simpson, J. Smith, M. Shahinpoor, and P. Willis, Proceedings of SPIE, Vol. SPIE 3041, Smart Structures and Materials 1997 Symposium, Enabling Technologies: Smart Structures and Integrated Systems, Marc E. Regelbrugge (Ed.), ISBN 0-8194-2454-4, SPIE, Bellingham, WA, pp. 697-701, (1997).

17. B. K. Kaneto, M. Kaneko, and W. Takashima, Oyo Buturi, vol. 65, no. 1, pp. 803-810, (1996).

18. Sia Nemat-Nasser, private communications.

19. Y. Bar-Cohen, S. Leary, M. Shahinpoor, J. Harrison, and J. Smith, Proceedings of the SPIE Conference on Electroactive Polymer Actuators, 3669, pp. 51-56, Newport Beach, CA, Mar. 1999.

20. Chiang, C, Druy, M., Gau, S., Heeger, A., Louis, E., MacDiarmid, A., Park, Y., Shirakawa, H., *J. Am. Chem. Soc.*, 100, 1013 (1978).

21. Baughman, R., Shacklette, L., Elsenbaumer, R., Plitcha, E., Becht, C., Conjugated Polymeric Materials: Opportunities in Electronics, Optoelectronics, and Molecular Electronics, vol. 182 of NATO ASI Series E: Applied Sciences, Kluwer, Dordrecht, Netherlands, pp. 559-582 (1990).

22. Xie, L., Buckley, L., Josefowicz, J., *Journal of Materials Science*, 29, 4200-4204 (1994).

23. Madden, J., Cush, R., Kanigan, T., Brenan, C., Hunter, I., *Synthetic Metals*, 105, 61-64 (1999).

24. Wang, Y., Rubner, M., *Synthetic Metals*, 39, 153-175 (1990).

25. Wang, Y., Rubner, M., Buckley, L., *Synthetic Metals*, 41, 1103 (1991).

26. Wang Y., Rubner, M., *Synthetic Metals*, 47, 255 (1992).

CHALLENGES TO THE TRANSITION TO THE PRACTICAL APPLICATION OF IPMC AS ARTIFICIAL -MUSCLE ACTUATORS

Y. Bar-Cohen[1], S. Leary[1], A. Yavrouian[1], K. Oguro[2], S. Tadokoro[3], J. Harrison[4], J. Smith[4] and J. Su[4],

1 JPL/Caltech, (MC 82-105), 4800 Oak Grove Drive, Pasadena, CA 91109-8099, yosi@jpl.nasa.gov, website: http://ndeaa.jpl.nasa.gov

2 Osaka National Research Institute, Osaka, Japan;

3 Dept. Computer & Systems Eng., Kobe University, Kobe, Japan;

4 NASA Langley Research Center, Advanced Materials and Processing Branch, MS 226, Hampton, VA 23681-2199

ABSTRACT

In recent years, electroactive polymers (EAP) materials have gained recognition as potential actuators with unique capabilities having the closest performance resemblance to biological muscles. Ion-exchange membrane metallic composites (IPMC) are one of the EAP materials with such a potential. The strong bending that is induced by IPMC offers attractive actuation for the construction of various mechanisms. Examples of applications that were conceived and investigated for planetary tasks include a gripper and wiper. The development of the wiper for dust removal from the window of a miniature rover, planned for launch to an asteroid, is the subject of this reported study. The application of EAP in space conditions is posing great challenge due to the harsh operating conditions that are involved and the critical need for robustness and durability. The various issues that can affect the application of IPMC were examined including operation in vacuum, low temperatures, and the effect of the electromechanical and ionic characteristics of IPMC on its actuation capability. The authors introduced highly efficient IPMC materials, mechanical modeling, unique elements and protective coatings in an effort to enhance the applicability of IPMC as an actuator of a planetary dust-wiper. Results showed that the IPMC technology is not ready yet for practical implementation due to residual deformation that is introduced under DC activation and the difficulty to protect the material ionic content over the needed 3-years durability. Further studies are under way to overcome these obstacles and other EAP materials are also being considered as alternative bending actuators.

INTRODUCTION

Consideration of practical applications for electroactive polymers (EAP) has began only in this decade following the emergence of new materials that induce large displacements [Hunter and Lafontaine, 1992; Kornbluh, et al, 1995; and Bar-Cohen, 1999a]. These materials are highly attractive for their low-density and large strain capability, which can be as high as two orders of magnitude greater than the striction-limited, rigid and fragile electroactive ceramics (EAC) [Bar-Cohen, et al, 1997; and Osada & Gong, 1993]. Also, these materials are superior to shape memory alloys (SMA) in their temporal response, lower density, and resilience. However, EAP materials reach their elastic limit at low stress levels, with actuation stress that falls far shorter than EAC and SMA actuators. The most attractive feature of EAP materials is their ability to emulate biological muscles with high fracture tolerance, large actuation strain and inherent vibration damping. EAP actuation similarity to biological muscles gained them the name "Artificial Muscles" and potentially can be used to develop biologically inspired robots. Such biomimetic robots can be made highly maneuverable, noiseless and agile, with various shapes including insect-like. Effective EAP offers the potential of making science fiction ideas a faster reality than feasible with any other conventional actuation mechanism. Unfortunately, the force actuation and mechanical energy density of EAP materials are relatively low, and therefore limiting the practical applications that can be considered. Further, there are no commercially available effective and robust EAP materials and there is no reliable database that documents the properties of the existing materials. To overcome these limitations there is a need for development in numerous multidisciplinary areas from computational chemistry, comprehensive material science, electromechanical analysis, as well as actuation characterization and improved material processing techniques. Efforts are needed to gain a better understanding of the parameters that control the electromechanical interaction. The processes of synthesizing, fabricating, electroding, shaping and handling will need to be refined to maximize their actuation capability and robustness.

Since EAP can be used as actuators that are light, compact and driven by low power, the authors sought to take advantage of their resilience and fracture toughness to develop space applications. The harsh environment associated with space environment poses great challenges to the application of EAP. Addressing these challenges has been the subject of the NASA task called Low Mass Muscle Actuators (LoMMAs). Under the lead of the principal author several EAP materials and applications were investigated. The emphasis of this paper is on Ion-exchange membrane metallic composites (IPMC), which are bending EAP materials, first reported in 1992 [Oguro et al, 1992; Sadeghipour, et al, 1992 and Shahinpoor, 1992]. The various issues that can affect the application of IPMC were examined including operation in vacuum, low temperatures, and the effect of the electromechanical and ionic characteristics of IPMC on its actuation capability. The finding that IPMC can be activated at low temperatures and vacuum, paved the way for the serious consideration of this class of materials for space applications [Bar-Cohen, et al, 1997]. Its bending characteristics offered the potential to address the critical issue of planetary dust that affects solar cells and imaging instruments on such planets as Mars.

Throughout the authors' studies, several problem areas were identified as needing attention to assure the practicality of IPMC for space applications. The authors addressed these issues and the results of their study are reported in this manuscript.

EAP ACTUATOR DRIVING DUST WIPER

Lessons learned from Viking and Mars Pathfinder missions indicate that operation on Mars involves an environment that causes accumulation of dust on hardware surfaces. The dust accumulation is a critical problem that hampers long-term operation of optical instruments and degrades the efficiency of solar cells to produce power. To remove dust from surfaces one can use a similar mechanism as automobile windshield wipers. Contrary to conventional actuators, bending EAP has the ideal characteristics that are necessary to produce a simple, lightweight, low power wiper mechanism. Specifically, the IPMC responds to activation signals of about 0.3-Hz with a bending angle that can exceed 90 degrees span each way depending on the polarity. For dust cleaning from windows, it is necessary to place the wiper outside the viewing area and move it inward to clean the window. This necessitates the use of two wipers that are placed on opposite sides of the window as shown in Figure 1.

To demonstrate the wiping capability, an IPMC was attached to a blade with a fiberglass brush and it was used to mechanically remove 20-μm dust particles that were sprinkled onto a glass plate (see bottom of Figure 2). Since this mechanism effectively addressed the critical issue of dust, the MUSES-CN mission selected it as a baseline technology for the Nanorover's infrared camera window. This mission is a joint effort of NASA and the NASDA (National Space Development Agency of Japan), which is scheduled to be launched from Kagoshima, Japan, in January 2002, to explore the surface of a small near-Earth asteroid. A photograph of the dust-wiper, a schematic view of the rover and an artist view of the mission, are shown in Figure 2.

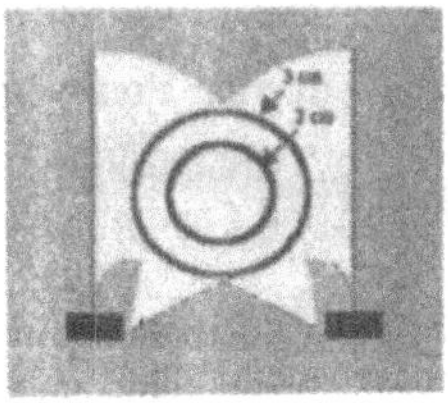

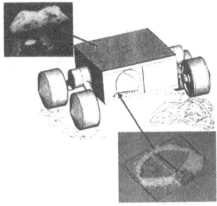

FIGURE 1: A schematic view showing the area that is covered by two bending EAP-actuators sweeping a window from two opposite sides.

A unique ~100-mg wiper blade was constructed by ESLI (San Diego, CA) using a composite (graphite fibers and DuPont Kapton™ resin) beam with fiberglass brush see Figure 3). The blade was processed at temperature of 623K (350°C) and was tested to demonstrate endurance down to 77K. A 15-mm x 6-mm IPMC film was bonded to the ESLI blade using platinum electrode strips bonded on its other end to provide electrical excitation of the EAP wiper. Since mechanical wiping of a surface using a soft brush may not remove minute dust particles, which are smaller than the distance between the whiskers, a high voltage repulsion mechanism was introduced. The blade (beam and brush) was coated with gold and activated by 1-2-KV bias DC voltage, whereas the EAP wiper was induced with 1-3V (Figure 3). In an effort to bring the technology to space flight readiness the critical issues associated with the IPMC material as an actuators were addressed.

FIGURE 2: Schematic view of the EAP dust-wiper on the MUSES-CN's Nanorover (middle) and a photograph of a prototype EAP dust-wiper (right-bottom).

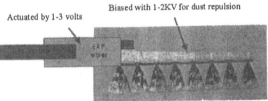

FIGURE 3: A schematic view of the EAP actuated dust-wiper

MODELING THE ACTUATOR
Design and prediction of the response of the IPMC dust wiper requires an effective analytical modeling of the material and its mechanical constraints. For this purpose, the Kanno-Tadokoro model was adopted using a gray box approach relating the experimental input and output data. The voltage applied to an actuator is transformed to current distribution through the membrane. The current generates distributed internal stress, which causes strain in the IPMC material and it is affected by its viscoelastic properties. The resistances of surface layers and RC elements approximate the experimental voltage-current response as the electric property. The stress generation property and the viscoelasticity were expressed by an equation similar to the piezoelectric equation.

$$\sigma = D(s)\varepsilon - ei\frac{\omega_n^2 s}{s^2 + 2\zeta\omega_n s + \omega_n^2}$$

Where: σ - internal stress; $D(s)$ – mechanical characteristics including mass, damping and stiffness; ε - strain vector; e - transformation tensor of stress generation; i – current through the actuator; and ζ and ω_n -delay parameters of the 2nd order.

The 2nd order delay approximates the time delay until ionic distribution reaches equilibrium and the internal stress is generated by swelling and electrostatic force. This equation was used to simulate the response of the IPMC actuators and in Figure 4a the results for 15-mm long 8-mm wide actuator strip in shown. The strain near the electrodes clamp (fixture) is larger than the tip. Analysis, using this model, has shows that current concentration near the electrodes causes imbalance of strain distribution. The response speed is faster near the electrodes because of the RC elements and therefore, it is better to design shorter actuator. The whole membrane deforms to curve in two dimensions, where deformation in the direction of width obstructs the wiper motion. When a 2mm long

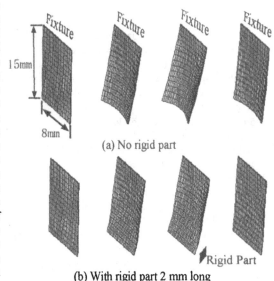

(a) No rigid part

(b) With rigid part 2 mm long

FIGURE 4: Simulation result of actuator motion.

section of the IPMC tip is constrained preventing it from deformation, the bending displacement is improved as shown in Figure 4b. Therefore, crosspiece design is important for efficient actuation.

IPMC AS A BENDING EAP ACTUATOR

Driving a dust wiper using EAP materials requires that material be capable of bending under electro-activation. As can be seen in the first issue of the WW-EAP Newsletter [Bar-Cohen, 1999b], several types of EAP materials can be made to bend under electrical excitation. The authors concentrated on the use of the ion exchange polymer membrane metal composite (IPMC) which has metal electrodes deposited on both sides. Two types of base polymers were used including Nafion® (perfluorosulfonate made by DuPont) and Flemion® (perfluorocaboxylate, made by Asahi Glass, Japan). Prior to using these polymers as EAP base material, they were widely employed in fuel cells and production of hydrogen (hydrolysis). The operation as actuators is the reverse process of the charge storage mechanism associated with fuel cells. In the current study, Nafion® #117 was used with a thickness of 0.18-mm and perfluorocarboxylate films were used having a thickness of 0.14-mm. Initial studies involved the use of Platinum as the metal electrodes however recent studies have shown that gold coating provides superior performance [Yoshiko, et al, 1998]. The gold layer was applied in 7-cycles resulting in a dendritic structure as shown in a cross section view in Figure 5. The counter cation consists of tetra-n-butylammonium or lithium and these two species showed significantly greater bending response than sodium, which was used earlier. Under less than 3-V, such IPMC materials were shown to bend beyond a complete loop and the response follows the electric field polarity.

FIGURE 5: Perfluorocarboxylate membrane with tetra-n-butylammonium cation and 7 cycles of ion exchange and reduction (resulting dendritic growth) of the gold electrodes

When an external voltage is applied on an IPMC film, it causes bending towards the anode at a level that increases with the voltage, up until reaching saturation, as shown in Figure 6. Under AC voltage, the film undergoes swinging movement and the displacement level depends not only on the voltage magnitude but also on the frequency. Generally, activation at lower frequencies (down to 0.1 or 0.01 Hz) induces higher displacement and the displacement diminishes as the frequency rises to several tens of Hz. The drive voltage level at which the bending displacement reaches saturation depends on the frequency and it is smaller at higher frequencies. The applied electrical current controls the movement of the film but the response is strongly affected by the water content of the IPMC serving as an ion transport medium.

FIGURE 6: The response of the bending EAP to various voltage amplitudes at three different frequencies (data obtained for sodium base IPMC).

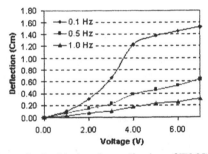

The authors addressed several issues that were determined critical to the application of IPMC (both the Nafion® and Flamion® base):

Film moisture: IPMC is highly sensitive to its moisture content. To maintain the moisture content there is a need for a protective coating that acts as the equivalent of skin, otherwise the material stops to respond after few minutes of activation in dry conditions. Using an etching procedure and silicone coating, an IPMC film was shown to operate for about 4 months. This Dow Corning coating material allows operation in a wide range of temperatures with great flexibility and is durable under UV radiation. However, since the MUSES-CN mission requires operation over 3-years, the 4-month protection period is too short. Analysis of the cause of the degradation indicates that the silicone coating is water permeable and the rate is $3000 \ cm^3 \ x \ 10^{-9} \ per \ sec/cm^2/cm$ at STP and 1 cm•Hg pressure difference [Dow Corning, 1999]. Assuming 2 cm^2 electrode area with 0.1-mm thick silicone coating shows a water loss rate of ~40-50mg/24 hrs. This rate is significantly higher than observed for IMPC and it does not account for the IPMC electrode layers, however it indicates the severity of the issue. To overcome this limitation various alternative coating techniques are being considered including the use of a metallic Self-Assembled Monolayer as an overcoat.

Electrolysis: The wetness of IPMC and the introduction of voltages at levels above 1.03-V introduce electrolysis during electro-activation causing degradation, heat and release of gasses. This issue raises a great concern since the emitted hydrogen accumulates under the protective coating and leads to blistering, which will rupture the coating due to the high vacuum environment of space. The use of tetra-n-butylammonium cations was shown to provide higher actuation efficiency allowing to reduce the needed voltage and to minimizing the electrolysis effect.

Operation in vacuum and low temperatures: In space the temperature can drop to significantly low levels and the ambient pressure is effectively vacuum. The ability to protect IPMC from drying allowed performing tests in vacuum and low temperatures. These tests showed that while the response decreases with temperature, as shown in Figure 7, a sizeable displacement was still observed at ~140°C. This decrease can be compensated by an increase in voltage. It is interesting to point out that, at low temperatures, the response reaches saturation at much higher voltage levels than room temperatures.

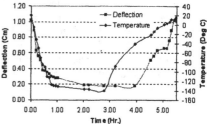

FIGURE 7: Deflection amplitude of sodium-base IPMC as a function of time and temperature.

Besides the need to address the low temperature issue, the material ability to sustain temperatures as high as +125°C is also necessary. For this purpose, several solvents with higher boiling points than water were examined for their potential use as a solvent for the IPMCs. These solvents were examined for their performance equivalence to "antifreeze" in automobile radiators. Various solvents were considered and their effect on the swelling characteristics of Nafion® was investigated. Nafion® strips with an initial size of 5.8mm x 38.1mm were immersed in a series of solvents for a period of 4-days at ambient temperature and the change in mass and size were measure. The results are listed in Table 1 and show both swelling and increase in mass due to water absorption. Examination of IPMC films that were immersed for 24-hours in various solvents, including ethylene glycol, showed a significant reduction in the induced bending amplitude.

Low actuation force: Using thin IPMC with a thickness of 0.14-mm was found to induce a significant bending displacement. However, the induced force was found relatively small making it difficult for the wiper to overcome the electrical forces that are involved with the dust-repelling high-voltage. Further, even though the wiper blade is relatively light, weighing about 104-mg, it is still responsible for significant bending due to gravity pressing the blade onto the window surface and constraining its movement. Alternative 0.18-mm thick film is currently being sought to provide the necessary force.

Permanent deformation under DC activation: Unfortunately, under DC voltage IPMC strips do not maintain the actuation displacement and they retract after several seconds. Further, upon removal of the electric field an overshoot displacement occurs in the opposite direction moving slowly towards the steady state position leaving a permanent deformation. This issue was not resolved yet and would hamper the application of IPMC.

TABLE 1: Changes in mass and size as a result of immersing Nafion® in various liquid media for 4-days at ambient temperature.

Solvent	Initial mass, g	Final mass, g	Change, %	Dimensions after soaking [width x length (mm)]
Water	0.0722	0.0838	16	5.8 x 43.0
N-methyl-2-pyrrolidinone (NMP)	0.0721	0.1124	56	6.4 x 50.8
Ethanol	0.0707	0.1113	57	6.4 x 50.8
Dimethylformamide (DMF)	0.0606	0.0974	61	6.4 x 44.5
Ethylene glycol	0.0720	0.1104	53	6.4 x 50.8
Ammonium hydroxide	0.0719	0. 0795	11	5.8 x 41.4

Challenges and solutions: To allow future design of EAP mechanisms actuated by IPMC, the challenges and solutions were summarized and are listed in Table 2. While most challenges seem to have been addressed, two issues still pose a concern: the introduction of permanent deformation and the need for an effective protective coating. Unless these issues are effectively resolved the use of IPMC for planetary applications will be hampered.

TABLE 2: Challenges and identified solutions for issues regarding the application of IPMC.

Challenge	Solution
Fluorinate base - difficult to bond	Pre-etching
Sensitive to dehydration (~5-min)	Etching and coating
Electroding points cause leakage	Effective compact electroding method was developed
Off-axis bending actuation	Use of load (e.g., wiper) to constrain the free end
Most bending occurs near the poles	Improve the metal layer uniformity
Electrolysis occurs at >1.03-V in Na+/Pt	• Minimize voltage • Use IPMC with gold electrodes and cations based on Li^+ or Perfluorocarboxylate with tetra-n-butylammonium
Survive -155°C to +125°C and operate at -125°C to + 60°C	IPMC was demonstrated to operate at -140°C
Need to remove a spectrum of dust sizes in the range of >3μm	• Use effective wiper-blade design (ESLI, San Diego, CA) • Apply high bias voltage to repel the dust
Reverse bending under DC voltage	Limit application to dynamic/controlled operations
Developed coating is permeable	• Alternative polymeric coating • Metallic Self-Assembled Monolayer overcoat
Residual deformation	Still a challenge
No established quality assurance	• Use short beam/film • Efforts are underway to tackle the critical issues

CONCLUSION

In recent years, electroactive polymers have emerged as actuators with great potential to enable unique mechanisms that can emulate biological systems. A study has taken place to adapt IPMC to planetary applications in an effort to develop a dust wiper for a mission to an asteroid. A series of challenges were identified as obstacles to the transition of such materials to space flight missions. Some of the key issues were effectively addressed while some require further studies. The unresolved issues include the permanent deformation under DC activation and the water permeability of the developed protective coating. Another issue is the limited force that is induced by IPMC, which require compromising between the bending displacement and the actuation force. Further studies are under way to overcome these obstacles and other EAP materials are also being considered as alternative bending actuators.

ACKNOWLEDGEMENT

The research at Jet Propulsion Laboratory (JPL), California Institute of Technology, was carried out under a contract with National Aeronautics and Space Agency (NASA) Code S, as part of the Surface Systems Thrust area of CETDP, Program Manager Dr. Samad Hayati.

REFERENCES

Bar-Cohen, Y., (Ed.), *Proceedings of the Electroactive Polymer Actuators and Devices*, Smart Structures and Materials 1999, Volume 3669, pp. 1-414, (1999a).

Bar-Cohen Y. (Ed.), WorldWide Electroactive (WW-EAP) Newsletter, Vol. 1, No. 1, http//ndeaa.jpl.nasa.gov/nasa-nde/lommas/eap/WW-EAP-Newsletter.html (1999b)

Bar-Cohen, Y., S. Leary, M. Shahinpoor, J. Harrison and J. Smith, "Flexible Low-Mass Devices and Mechanisms Actuated by Electroactive Polymers," ibid, pp. 51-56 (1999).

Dow Corning data for general silicones - based on personal communication, Oct. 1999.

Heitner-Wirguin C., "Recent advances in perfluorinated ionomer membranes: Structure, properties and applications," *J. of Membrane Sci.*, V 120, No. 1, 1996, pp. 1-33.

Hunter I. W., and S. Lafontaine, "A comparison of muscle with artificial actuators," *IEEE Solid-State Sensor and Actuator Workshop*, 1992, pp. 178-165.

Kornbluh R., R. Pelrine and Joseph, J. "Elastomeric dielectric artificial muscle actuators for small robots," *Proceeding of the 3rd IASTED Intern. Conf.*, Concun, Mexico, June, 14-16, 1995.

Oguro, K., Y. Kawami and H. Takenaka, "Bending of an Ion-Conducting Polymer Film-Electrode Composite by an Electric Stimulus at Low Voltage," *Trans. J. of Micromachine Soc.*, 5, (1992) 27-30.

Osada, Y. and J. Gong, "Stimuli-Responsive Polymer Gels and Their Application to Chemomechanical Systems," *Prog. Polym. Sci*, 18 (1993) 187-226.

Sadeghipour, K., R. Salomon, and S. Neogi, "Development of a Novel Electrochemically Active Membrane and 'Smart' Material Based Vibration Sensor/Damper," *Smart Mater. and Struct.*, (1992) 172-179.

Shahinpoor, M., "Conceptual Design, Kinematics and Dynamics of Swimming Robotic Structures using Ionic Polymeric Gel Muscles," *Smart Mater. and Struct.*, Vol. 1, No. 1 (1992) 91-94.

Yoshiko A., A, Mochizuki, T. Kawashima, S. Tamashita, K. Asaka and K. Oguro, "Effect on Bending Behavior of Counter Cation Species in Perfluorinated Sulfonate Memberance-Platinum Composite," *Poly. for Adv. Tech.*, Vol. 9 (1998), pp. 520-526.

Ferroelectric Polymers

STRUCTURE, PROPERTIES, AND APPLICATIONS OF SINGLE CRYSTALLINE FILMS OF VINYLIDENE FLUORIDE AND TRIFLUOROETHYLENE COPOLYMERS

H. OHIGASHI
Department of Materials Science and Engineering, Yamagata University
Yonezawa, Yamagata 992-8510, Japan

ABSTRACT

The single crystalline (SC) film of copolymer of vinylidene fluoride and trifluoroethylene is a highly crystallized, highly double-oriented, and optically transparent ferroelectric film, which can be prepared by annealing a uniaxially drawn film in the paraelectric phase with its ends being clamped and its surfaces free. The SC film is composed of endlessly extended chain crystals whose c-axis aligns along the stretching direction almost perfectly and the polar b-axis is oriented $\pm \pi/6$ off the film normal. It does not contain lamellar crystals nor amorphous phase. Owing to such unique structure, the SC film exhibits the ferroelectricity and related properties inherent to the crystal more clearly than the films composed of lamellar crystals. In this paper, we present the results of our recent studies on the SC films: (1) the structure and morphology studied by POM and X-ray diffractions; (2) the ferroelectric, piezoelectric, and mechanical properties in the ferroelectric phase; (3) molecular chain motions in the paraelectric phase as revealed by dielectric, X-ray diffraction and shear mechanical studies; and (4) applications to shear ultrasonic transducers. The followings are the main results to be emphasized: (A) the Young's modulus in the direction along the stretching axis is much larger (120 GPa at 10K) than that of LC films (15 GPa); (B) the SC film is the most effective piezoelectric polymer film among piezoelectric polymers; (C) the SC film in the paraelectric phase is a liquid crystal of 1D-liquid and 2D-solid, and each chain molecule of $TGTG'$ conformation undergoes rotational and flip-flop motions independently of neighboring chains in a highly regular hexagonal crystal lattice, which leads to anomalous dielectric behaviors in the paraelectric phase; (D) SC film transducers for shear ultrasonic waves are effectively usable for NDE.

INTRODUCTION

Ferroelectric and piezoelectric properties of copolymers of vinylidene fluoride (VDF) and trifluoro-ethylene(TrFE), P(VDF/TrFE), strongly depend on their molar composition ratio, VDF(x)/TrFE($1-x$). The copolymers P(VDF/TrFE) with $0.65 < x < 0.82$ exhibit most strong piezoelectric activity, clear ferroelectric to paraelectric phase transition, and sharp polarization switching phenomena as compared to the copolymers with component ratio outside of this range and to polyvinylidene fluoride (PVDF) as well [1-4]. Large remanent polarization and strong piezoelectric activity of P(VDF/TrFE) ($0.65 < x < 0.82$) are mostly attributed to their crystalline morphology: when their films are annealed in the paraelectric phase, very thick lamellar crystals stacking parallel to each other are highly developed with their lamellar planes perpendicular to the film surface [1-5]. Since the chain molecules are perpendicular to the lamellar plane, the remanent polarization of the film becomes very large after poling treatment. Growth of such large lamellar crystals (or extended chain crystals) are owing to the presence of paraelectric phase (or hexagonal phase), in which trans-gauche trans-guache' ($TGTG'$) chain molecules rotate rather freely around their chain axis [6], and can slide along the chain axis, as in the case of polyethylene crystals in the hexagonal phase appearing at high pressures [7, 8]. Lamellar crystals in uniaxially stretched films have preferential orientation: the orthorhombic c-axis orients along the stretching axis, and the [110] or [200] axis orients parallel to the film normal [1, 5]. However, the orientation distributions of these axes are rather broad.

Mechanical, dielectric and piezoelectric properties of polymer crystals are essentially highly anisotropic. Properties of lamellar crystalline films, however, do not necessarily represent the properties of crystals themselves, because inherent properties of the crystal are screened by weak interlamellar interaction. Recently, we found a highly double-oriented, highly crystallized film of P(VDF/TrFE), in which endlessly extended chain crystals are oriented with their crystal c-axis parallel to the stretching axis, and the b-axis parallel to the directions $\pm \pi/6$ from the film normal. Any lamellar crystals do not exist in the film. We call this film the "single crystalline (SC) film" to differentiate it from double-oriented lamellar crystalline (LC) film [9]. The SC film is the most suitable material to investigate crystal properties of P(VDF/TrFE)

23

by macroscopic measurement methods.

This paper describes our research results obtained with SC films of P(VDF/TrFE); the structure of the SC film; the mechanical, dielectric and piezoelectric properties measured in a wide temperature range; the molecular chain motions in the paraelectric phase revealed by dielectric and X-ray scattering studies; and the application of SC films to shear ultrasonic wave transducers.

STRUCTURE OF SINGLE CRYSTALLINE FILMS OF P(VDF/TrFE)

The SC film (5-100µm thick) can be prepared by uniaxial drawing of a solution-cast P(VDF/TrFE) film, followed by crystallization in the paraelectric (hexagonal) phase with the film ends being clamped at fixed length. It is essential to leave, during crystallization, the film surfaces free from contacting with solid materials which would induce the nucleation of lamellar crystals at contacting points. The film after crystallization increases in length along the stretching direction by about 5% of the length before crystalliza- tion, which is due to disentanglement of chain molecules during crystallization. The SC film can be easily cleaved straight along the stretching axis, but hardly fractured across it even in liquid nitrogen.

Figure 1 (a) shows an optical micrograph taken with crossed polarizers for a fragment of a SC film of P(VDF/TrFE) (75/25 molar ratio). For comparison, a SEM image of a uniaxially oriented LC film prepared by annealing in the paraelectric phase with its surfaces clamped, is shown in Fig. 1 (b). The SC film is optically transparent, and has an essentially uniaxial optic axis along the stretching direction. No microcrystal can be observed. The density of the poled SC film measured by a floating method is 1.935 g/cm^3 at 20 °C. This value implies that the SC film has high degree of crystallinity (almost 100%). (The density of a poled LC film (75/25) is 1.880 g/cm^3 at 20° C.)

The SC film has highly double-oriented and highly ordered structure. Figure 2 shows the X-ray diffraction profiles of a poled SC film observed by transmission method for (a) the 200/110 and 400/220, (b) 310/310 and (c) 001 scattering vectors. To obtain profile (a), for example, the X-ray beams must be scanned on a plane perpendicular to the stretching axis, making the angle between the beams and the normal of the film to be 30°+θ, where θ is the Bragg diffraction angle parameter. These profiles indicate that the orthorhombic crystal axes orient to the specific directions in the film. The orthorhombic $hk0$ reciprocal lattice for the poled SC film is shown in Fig. 3. Reciprocal lattice point distributions at fixed θ's (loking curves) for (a) 001 and (b) 200/110 diffractions are shown in Fig. 4. Orientation distribution of the c-axis is axially symmetric with respect ro the stretching axis. The distribution is very sharp [Fig. 4 (a)]: FWHM is about 8-10°, and the orientation factor $f_c = (3<\cos^2\phi>-1)/2 = 0.995$-0.998, where ϕ is the angle between stretching axis and the c-axis. The orientation distribution of the a-axis or [110] axis in the plane perpendicular to the stretching axis is somewhat broader than c-axis orientation [Fig. 4 (b)]: FWHM is 10-15° depending on film thickness, and $<\cos^2\chi> = 0.90$-0.98, where χ is lateral deviation angle from

(a)
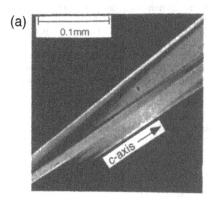

(b)
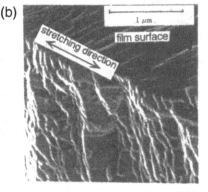

Fig. 1. (a) An optical micrograph, taken with crossed polarizers, of a SC film of P(VDF/TrFE) (75/25). (A fragment cleaved from a SC film in liquid nitrogen)

Fig. 1. (b) SEM micrograph of a uniaxially stretched lamellar crystalline film of P(VDF/TrFE) (74/26). Surface and fractured cross section are observed.

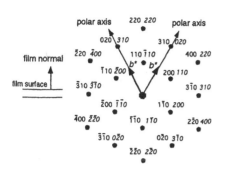

Fig. 2. Reciprocal lattice of a poled SC film (twin crystal) of P(VDF/TrFE). The lattice having different orientation is presented by roman or *italic* characters.

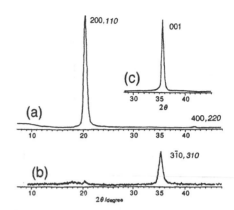

Fig. 3. X-ray diffraction profiles (θ–2θ scan) of a SC film of P(VDF/TrFE) (75/25). Scattering vector is scanned on three different directions on the reciprocal lattice.

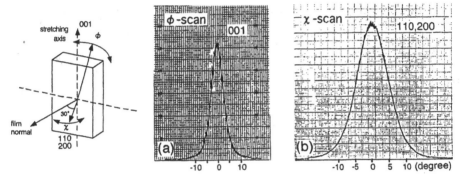

Fig. 4. Locking curves of X-ray diffraction at constant diffraction angle for (a) *orth* 110/200 and (b) *orth* 001. Scanning vectors are illustrated on left side. The distribution of *c*-axis is axially symmetrical wit h respect to the stretching axis.

the 200/110 reciprocal lattice point in a perfectly oriented film. Comparing to SC films, LC films prepared from uniaxially stretched films have lower orientation factor: $f_c = 0.8$-0.9. Inferior *c*-axis orientation in the LC film is brought about during growth of lamellar crystals in the paraelectric phase.

The SC film displays a sharp D-E hysteresis loop as shown in Fig. 5: the remanent polarization is 0.11 C/m², and the coercive field is 33 MV/m. This polarization corresponds to the *b*-axis spontaneous polarization of 0.127 C/m² in a P(VDF/TrFE) (75/25) single crystal. As shown in Fig. 6, the diffraction intensity distribution from {200}/{110} planes (locking curve at $2\theta = 20.1°$) has a 6-folded symmetry for an unpoled SC film, but it becomes orthorhombic symmetry after poling. This indicates that, in the unpoled film, the *b*-axis of crystallites or crystal domains orients to the six-fold directions ($\pm\pi/6$, $\pm\pi/2$ and $\pm 5\pi/6$ from the film normal), while in the poled film, the *b*-axis orients only to the directions $\pm\pi/6$ with equal probability. This also implies that the polar axis (*b*-axis) can be rotated under an external field by 60°-increments around the chain axis, as has been observed in PVDF [10]. Therefore, the poled SC film in the ferroelectric phase has the same symmetry property as a twin crystal with the (310) plane being one of possible twin (domain) boundaries. The SC film in the paraelectric phase is regarded as a single crystal, because the *orth* (200)/(110) planes are reduced to the *hex* (100) plane.

A qualitative explanation for formation mechanisms of SC film during crystallization in the paraelectric

25

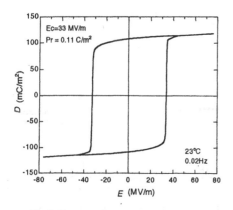

Fig. 5. D-E hysteresis loop of a SC film of P(VDF/TrFE) (75/25). (film thickness; 30μm)

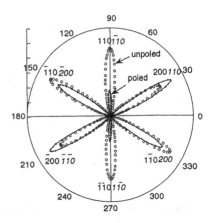

Fig. 6 Polar plots of X-ray diffraction intensity of P(VDF/TrFE) SC films before and after poling. The intensity is normalized for each curve.

phase is as follows. In a uniaxially drawn film, growth of a large extended chain crystal is more probable than the growth of a lamellar crystal, because the latter has higher free energy due to chain folding at lamellar surfaces. The chain molecules in the hexagonal lattice are very mobile along the chain axis like in a one dimensional liquid as will be described later. Therefore, the extended chain crystals are grown infinitely along the chain axis in the hexagonal phase. We found that the six-fold symmetry before poling (Fig. 6) is developed in the hexagonal phase. In this phase, the film surface is the *hex* (100) plane. This preferential orientation most probably stems from the fact that the *hex* (100) plane is the crystal surface having least surface energy, because molecular chains are packed most closely in this surface [9]. When the hexagonal film is cooled down into the ferroelectric phase, the *hex* (100) is transformed into the *orth* (110) or (200) plane.

MECHANICAL, DIELECTRIC AND PIEZOELECTRIC PROPERTIES

Pizoelectric Resonators

The SC film exhibits mechanical, dielectric and piezoelectric properties considerably different from

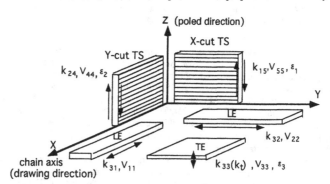

Fig. 7. Piezoelectric resonators of P(VDF/TrFE) SC films used for mesurements of electromechanical properties. Arrows indicate the directions of vibration displacement in the resonators.

those of LC films [11]. For measurements of these properties, we prepared five kinds of piezoelectric resonators as illustrated in Fig. 7: two length extensional (LE) resonators, one thickness extensional (TE) resonator, and two thickness shear (TS) resonators. To prepare the TS resonators, we stacked 100-200 sheets of poled SC film with epoxy resin. The stacked block was then cut, by a rotating diamond disc blade, into thin plates whose cut-planes are perpendicular to the chain axis (X-cut TS resonator), or parallel to the chain axis (Y-cut TS resonator). The plate surfaces were further finished by grinding and polishing with fine particles, and were electroded with Al or Au. The X-cut and Y-cut plates were also serv٤d for measurements of dielectric constants ε_1 and ε_2, respectively. The mechanical properties in the paraelectric phase cannot be measured by the piezoelectric resonance method. Instead, they were measured by a composite resonator method [12]: a SC film (15-17μm thick) was bonded onto an AT-cut quartz TS plate-resonator (resonance frequency 3579.5 kHz) with stretching axis of the SC film parallel or perpendicular to the direction of vibration displacement of the resonator. Analysis of resonance behavior in electric admittance of the composite resonator provides information of mechanical properties of the SC film.

Mechanical Properties

The longitudinal sound velocities v_{33} and v_{22}, and their temperature dependence of the SC films (75/25) are very similar to those of LC films. In contrast, reflecting the fact that the SC film is composed of infinitely extending chain crystals, the velocity v_{11} is very large as shown in Fig. 8: v_{11} is 8.0 km/s at 10K, which corresponds to the Young's modulus $1/s_{11}$ of 120 GPa, being much larger than that in LC films, where loose interaction between lamellae limits the sound velocity (v_{11}=2.8 km/s, $1/s_{11}$= 13.9 GPa). The value of $1/s_{11}$ of the SC film is not largely different from a calculated value (240 GPa) for a PVDF crystal [13]. At temperatures around -20°C (T_β), $1/s_{11}$ decreses steeply. However, it remains still much larger as compared to $1/s_{11}$ of LC films.

The velocities v_{55} and v_{44} of shear waves propagating in the direction normal to the plate surface with shear displacement vector parallel and perpendicular to the chain axis, respectively, exhibit very interesting temperature dependence in the temperature range above T_β, as shown in Fig. 9. The velocity v_{44} decreases monotonically with increasing temperature up to T_c, while v_{55} decreases more rapidly than v_{44}. Both of v_{55} and v_{44} decrease steeply at T_c. The resonance frequency response of the composite resonator composed of an AT-cut quartz plate and a SC film showed that v_{55} becomes zero at T_c. On the other hand, although v_{44} decreases stepwise at T_c, v_{44} retains a finite value at Tc and decreases gradually (400-200 m/s) with increasing temperature in the paraelctric phase, and then suddenly becomes zero at the melting point [Fig. 9]. This fact implies that chain molecules can slide easily in the chain direction like in a liquid (shear

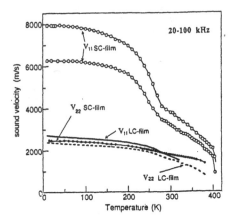

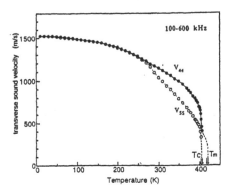

Fig. 8 Temperature dependence of velocities of longitudinal acoustic waves in SC film of P(VDF/TrFE) (75/25). Velocities in LC films are also displayed. v_{11} of SC film varies with crystalliation condition.

Fig. 9. Temperature dependence of velocities of shear waves in SC films of P(VDF/TrFE) (75/25). Data obtained above T_c with an AT-cut quartz resonator are indicated by dotted lines.

waves cannot propagate in a liquid), but they are confined in a solid lattice in the directions across the chains. Therefore, the SC film in the paraelectric phase behaves as a liquid in the direction parallel to the chain axis, while it behaves as a solid in the direction perpendicular to the chain: P(VDF/TrFE) in the hexagonal phase is a liquid crysal of 1D-liquid and 2D-solid. This property is very important both in the formation process of SC films and in dielectric properties in the paraelectric phase.

Dielectric Properties in the Ferroelectric Phase

Three principal components of the dielectric constant tensor, ε_a, ε_b, and ε_c, for a P(VDF/TrFE) single crystal along the a, b and c axes, respectively, can be obtained from ε_1, ε_2, and ε_3 measured for a X-cut plate, a Y-cut plate, and a TE film (Fig. 7), respectively. The temperature dependence of dielectric constants for the single crystal in the heating process is shown in Fig. 10 [11]. Three interesting phenomena are noticeable. First, ε_b is considerably larger than ε_a and ε_c in the ferroelectric phase, and increases steeply above T_β. This shows that the chain librational motion, which induces fluctuation of CF2 and CHF

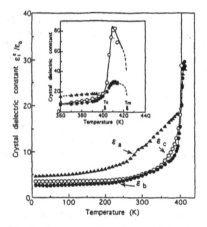

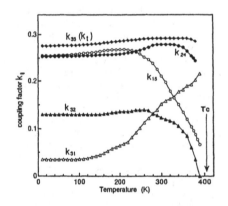

Fig. 10. Temperature dependence of dielectric constants of a single crystal of P(VDF/TrFE) (75/25) deduced from ε_1, ε_2 and ε_3 of SC film.

Fig. 11. Temperature dependence of electromechanical coupling factors for various vibration modes of SC film of P(VDF/TrFE) (75/25).

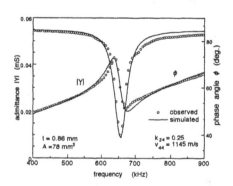

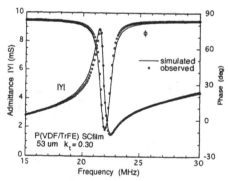

Fig. 12. Admittance-frequency curves of piezoelectric resonance of P(VDF/TrFE) SC films. Left: thickness shear mode (k_{24}), Right: thickness extensional mode (k_t). (at 25 °C)

28

dipoles mainly in the direction parallel to the a-axis and accordingly contributes mainly to ε_a, becomes more active above T_β. Second, ε_a becomes equal to ε_b in the paraelectric phase, as is expected in hexagonal crystals. Third, ε_c becomes much larger than ε_a and ε_b in the temperature range higher than T_C, as shown in the inset of Fig. 13. This is attributable to the flip-flop motion of $TGTG'$ chains in the paraelectric phase, which is described in detail in the next section. In the literature, it is often reported that the relaxation phenomena in dielectric and mechanical properties around -20°C are attributed to a glass transition in the amorphous phase. However, we attribute these phenomena to a second order phase transition associated with such molecular motion in the crysal as mentioned above: T_β is not a glass trasition temperature but a phase transtion temperature of the crystal.

Piezoelectric Properties

Figure 11 shows the temperature dependence of five electromechanical coupling factors k_{ij} [13], which were determind by analyzing freuency dependence of electric admittance of the resonators (Fig. 7) with the modified Mason's equivalent circuit in which dielectric and mechanical loss factors of resonator material are taken into account [14]. Figures 12 (a) and (b) show the resonance curves of a TE mode resonator and a Y-cut TS mode resonator, respectively. In general, the SC film has larger piezoelectric effects than those of conventional LC film in a wide temperature range. In particular, coupling factors for LE modes (k_{31} and k_{32}) and TS modes (k_{24} and k_{15}) are considerably larger than those of LC films.

These piezoelectric data suggest that the SC film is a promising sensor/transducer material not only for longitudinal acoustic waves but also for shear acoustic waves. Owing to its high Young's modulus in the direction along the chain axis ($1/s_{11}$) and to rather large coupling factor k_{31}, the SC film is also useful for bending mode sensors and actuators.

PROPERTIES OF SC FILMS IN THE PARAELECTRIC PHASE—CHAIN MOTIONS

The dielectric constant of P(VDF/TrFE) increases anomalously at temperatures around T_c. This interesting phenomenon was first found in LC films of P(VDF/TrFE) by Furukawa and his colleague, and has been explained in terms of a large collective fluctuation of polarization perpendicular to the chain, assuming that the interaction between CF_2 dipoles on the neighboring chains plays an important role even in the paraelectric phase [15-18]. This interpretation is not consistent with our experimental results that the dielectric constant along the chain axis (ε_c) becomes much larger than ε_a ($= \varepsilon_b$) in the paraelectric phase (Fig. 10), and that the SC film is a 1D-liquid along the chain axis. Our precise studies on the temperature dependence of anisotropic dielectric relaxation strength and relaxation time, and on the diffuse X-ray scattering lead to a new model for chain motions to explain the anomalous phenomena in the paraelectric phase [12]. The new model is as follows. (1) Each chain molecule of $TGTG'$ conformation undergoes a rotational motion independently of neighboring chains, and, in each chain, the $TGTG'$ sequences with up- or down-oriented dipoles take a flip-flop motion, exchanging their dipole orientation.

correlated tgtg' sequence
undergoing flip-flop motion

polymer chains in
hexagonal crystal

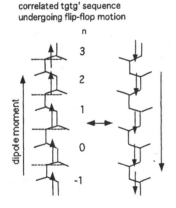

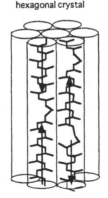

Fig. 13. A model for molecular chains undergoing flip-flop motion in the hexagonal crystal lattice of paraelectric phase.

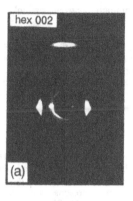
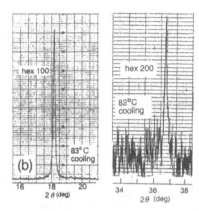

Fig. 14. X-ray diffractions of P(VDF/TrFE) (75/25) SC film in the paraelectric phase. (a) Diffraction photograph taken at 88°C with the incident beam tilted 9° from the film normal to the c-axis. (b) Line profiles of *hex* 100 and *hex* 200 diffractions detected by the symmetric reflection method in the cooling process.

(2) The length of *TGTG'* sequences increases around T_c, which results in anomalous increase in the dielectric relaxation strength and relaxation time along the chain axis. (3) The anomalous increase in dielectric constant ε_a (= ε_b) around T_c is not from the fluctuation of collective polarization in the polar axis (b-axis), but essentially from the consequence of axial rotational motion of these TGTG' chains. A model of the flip-flop motion in the paraelectric phase is shown in Fig. 13. This model is deduced from the experimental facts described bellow.

As shown in Fig. 14, the *hex* 002 diffraction is a diffuse scattering widely extending along the direction perpendicular to the chain axis, while the *hex* 100 and *hex* 200 diffractions are extremely narrow ($\Delta 2\theta \sim 0.1°$). The intensity ratio of the *hex* 100 and *hex* 200 diffractions is about 1:1/120. The analysis of these diffraction profiles shows that each chain is packed in a hexagonal lattice highly regular in the plane perpendicular to the chain axis (the coherent length is 1000-2000Å), and the chain has no positional correlation with neighboring chains in the direction parallel to the chain axis. Weak diffraction intensity indicates that, although the hexagonal lattice is very regular on the average, the chains thermally fluctuate around the lattice points with a large amplitude in the direction perpendicular to the chain axis. The *hex*002 profile scanned along the chain axis is of a Lorentzian shape, whose linewidth decreases with decreasing temperature, indicating that the length of *TGTG'* sequences, dynamically fluctuating within a chain due to the flip-flop motion, increases with decreasing temperature.

The above mentioned chain motions are more clearly reflected on the anisotropic dielectric relaxation phenomena in the hexagonal phase. The dielectric relaxation strength for parallel (//) direction $\Delta\varepsilon_{//}$ is much stronger than that for the perpendicular ($\perp$) direction $\Delta\varepsilon_{\perp}$ (Fig. 15). Furthermore, the parallel dielectric relaxation time $\tau_{//}$ is much longer than the perpendicular relaxation time $\tau_{\perp}$ (Fig. 16). The temperature dependence of $\tau_{\perp}$ is of a simple thermal excitation process for restricted rotation: $\tau_{\perp} = \tau_0 \exp(-G/kT)$. On the other hand, $\tau_{//}$ is essentially proportional to a product of $\Delta\varepsilon_{//}$, $\tau_{\perp}$ and an additional excitation factor: $\tau_{//} = A\tau_{\perp}\Delta\varepsilon_{//}T\exp(-\Delta G/kT)$. These characteristic features in the parallel dielectric relaxation phenomena are well explained by the flip-flop motion of *TGTG'* sequences in a single chain (Fig. 13). Based on the Fröhlich's dipole fluctuation theory, which relates the relaxation strength $\Delta\varepsilon_{//}$ and the pair correlation function Δ_n in a linear chain, we can estimate the correlation length ξ (or correlation number ν) of a *TGTG'* sequence undergoing the flip-flop motion [19]:

$$\Delta\varepsilon_{//} = \frac{1}{k_B T \varepsilon_0} 4\mu_{//}^2 N\Phi_n \sum_{n=-\infty}^{\infty} \Delta_n , \qquad (1)$$

where N is the number of *TGTG'* units per unit volume, Φ is the ratio of the internal field to the external field, and $\Delta_n = g_n - 1/4$, where g_n is the probability that the 0-th and n-th dipoles (*TGTG'* units) in a chain orient to the positive direction. Assuming that all the chains are composed of *TGTG'* units ($N = 7.8 \times$

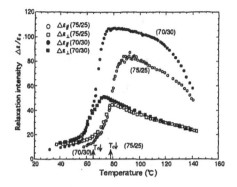

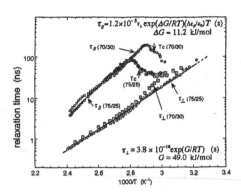

Fig. 15. Temperature dependence of dielectric relaxation strength for the directions parallel and perpendicular to the chains axis ($\Delta\varepsilon_{//}$ and $\Delta\varepsilon_{\perp}$) observed during cooling process for P(VDF/TrFE) (75/25) and (70/30) SC films.

Fig. 16. Arrhenius plots of anisotropic dielectric relaxation time ($\tau_{//}$ and $\tau_{\perp}$) in P(VDF/TrFE) (75/25) and (70/30) SC films (cooling process). Fitted curves for $\tau_{//}$ and $\tau_{\perp}$ calculated with the illustrated equations for (75/25) film are shown by broken lines.

$10^{27}/m^3$), and that $\Phi = 2.2$, which is a reasonable value as the Onsager's local field, we obtain that $\Sigma\Delta_n = 1.75$ at $88°C$ in the cooling process for P(VDF/TrFE) (75/25). The correlation number $v\ (=\xi/c)$ of $TGTG'$ units ($c = 4.55\text{Å}$, the length of a $TGTG'$ unit) can be estimated from observed value of $\Delta\varepsilon_{//}$ and an approximate expression

$$\sum_n \Delta_n = \sum_n (1/4)\,\exp(-|n|/v) \simeq v/2, \qquad (2)$$

which yields $v=3.5$ at $88°C$: the correlation length ξ is about 16Å, or each $TGTG'$ sequence is composed of, on the average, 7 $TGTG'$ units at $88°C$. The temperature dependence of ξ is shown in Fig. 17: ξ increases with decreasing temperature in proportion to $\Delta\varepsilon_{//}T$. The correlation length ξ estimated from the linewidth of hex 002 scattering profile along the c^* axis is also in good agreement with that estimated from $\Delta\varepsilon_{//}$ [12]. The diffuse X-ray scattering peak-intensity I_D of hex 002 must be proportional to $\Sigma\Delta_n$, or to $\Delta\varepsilon_{//}T$. This is also supported by the observed intensity (Fig. 17).

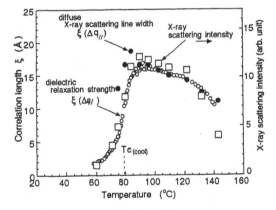

Fig. 17 Temperature dependence of correlation length ξ of $TGTG'$ sequence estimated from $\Delta\varepsilon_{//}$ and line width of diffuse X-ray scattering of hex 002 in the direction along the c^* axis for (75/25) SC film. The diffuse X-ray scattering intensity of hex 002 are also plotted.

As shown in Fig.16, the flip-flop motion requires an additional activation energy ΔG (=11.2 kJ/mol) to the energy E required for the restricted rotation. The observed value of ΔG is reasonable as a value required for the conformational trans-formation of $trans \rightleftarrows gauche$. The temperature dependence of $\tau_{//}$ indicates that the flip-flop motion is mediated by the chain rotation, that is, the transformation from $TGTG'$ to $GTG'T$ must be accompanied by the chain rotational motion. This is a very reasonable process for the flip-flop motion. Both of the relaxation strength and relaxation time depend on the composition ratio of VDF/TrFE. As seen in Fig. 16, the relax-ation time $\tau_{//}$ at Curie temperature T_c is much longer for the (70/30) film than for the (75/25) film, even though both the films have comparable correlation length ξ at respective T_c's. Since the former has lower T_c, and consequently much longer relaxation time $\tau_{\perp}$ at T_c, much longer relaxation time $\tau_{//}$ is required for the conformational transformation.

The relaxation strength $\Delta\varepsilon_{\perp}$ exhibits similar temperature dependence to that of $\Delta\varepsilon_{//}$, despite difference in their magnitude. Since the chain rotation is much faster than the flip-flop motion,

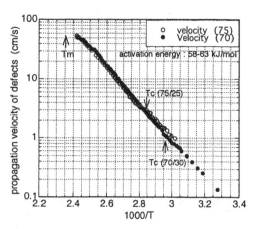

Fig. 18. Temperature dependence of velocity of defects propagating on $TGTG'$ chains in P(VDF/TrFE) (75/25) and (70/30) crystals of the paraelectric phase. The velocity is estimated from relaxation time $\tau_{//}$ and correlation length ξ of TGTG' sequences.

$\Delta\varepsilon_{\perp}$ is attributable to a system of rigid rotors, each having a dipole moment of $2v\mu_{\perp}$ in the direction perpendicular to the chain, where $\mu_{\perp}$ is the dipole moment of a $TGTG'$ unit in the direction perpendicular to the chain. (Average number of $TGTG'$ units in a $TGTG'$ sequence is $2v$). Thus, we may predict that $\Delta\varepsilon_{\perp} /\Delta\varepsilon_{//} = (1/2) (\mu_{\perp}/\mu_{//})^2 = 0.64$ for a (75/25) SC film. (The factor 1/2 is from rotational average around the c-axis. $\mu_{\perp}$ and $\mu_{//}$ for a $TGTG'$ unit are 6.2×10^{-30} and 5.5×10^{-30} Cm, respectively.) The observed values of $\Delta\varepsilon_{\perp}$ are essentially consistent with this prediction.

In a chain molecule undergoing flip-flip motion, there are boundaries or defects between up- and down-oriented $TGTG'$ sequences. The dynamical flip-flop motion of a $TGTG'$ sequence is regarded as a diffusive motion of the defect in the chain. The velocity of elementary motion of the defect propagating in a $TGTG'$ sequence, $v_d (=2\xi/\tau_{//})$, is plotted in Fig. 18. T velocity v_d varies from 80 cm/s at temperatures near T_m to about 1cm/s at T_c with activation energy of $G + \Delta G$.

All the phenomena observed in the paraelectric phase, including the shear sound velocities, are the results arising from the characteristic nature of long chain molecules in a crystal of hexagonal phase, where each rotating chain undergoes a flip-flop motion independently of neighboring chains. The correlation length of $TGTG'$ sequences increases with decreasing temperature, which seems to induce the structural phase transition from the hexagonal to orthorhombic phases. The results of X-ray scattering and dielectric studies are also consistent with shear sound velocities in the paraelectric phase: a P(VDF/TrFE) crystal in the paraelectric phase is a liquid crystal of 1D-liquid and 2D-solid.

APPLICATION OF SC FILMS TO SHEAR ULTRASONIC TRANSDUCERS

P(VDF/TrFE) is the most effective piezoelectric polymer. Although various ultrasonic transducers have been developed with P(VDF/TrFE) films [19], all of them are the transducers for longitudinal waves utilizing the TE mode of piezoelectric films. As described previously, the SC film of P(VDF/TrFE) is also the most effective transducer material for shear ultrasonic waves. We have developed transverse ultrasonic transducers using LC films [20] and SC films of P(VDF/TrFE) [21]. We describe here the performance of transducers developed utilizing the shear mode (k_{24}) of P(VDF/TrFE) SC films [21].

The structure of the transducer for transverse ultrasonic waves is essentially the same as adopted for longitudinal ultrasonic waves [19]. A Y-cut multi-layered resonator of SC films operating in TS mode is backed with a Cu plate-electrode, which is supported on a polymer block. The front surface of the

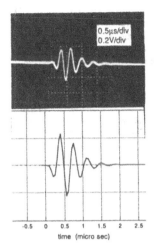

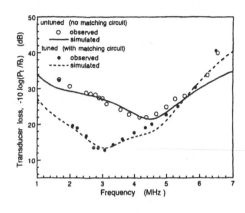

Fig. 19. Observed (top) and calculated (bottom) impulse response waveform of thickness-shear (24-mode) transducer composed of P(VDF/TrFE) SC films (tuned at 3.0 MHz). (resonator thickness: 75μm, area: 70mm².)

Fig. 20. Observed and calcuated frequency dependence of transducer loss for P(VDF/TrFE) SC film shear mode transducer with matching circuit (tuned at 3.0 MHz) and without matching circuit. The transducer is the same as used in Fig. 19.

resonator is electroded with a thin metal foil and further covered with a polymer film. For electrical impedance matching, a transformer and an inductor are inserted in series between the resonator and an electric source.

Figure 19 (a) shows an impulse response waveform of a TS transducer (tuned at 3.0 MHz), when the pulse waves were transmitted in an Al-block as a transmission and reflection medium. The pulse response waveform calculated with the modified Mason's equivalent circuit [22] well reproduces the observed waveform, as shown in Fig. 19 (b). The observed frequency dependence of the transducer loss is also shown in Fig. 20 for two cases that this transducer was operated with and without a matching circuit. The observed loss is compared with the loss calculated with the modified Mason's equivalent circuit. (The transducer loss is defined by $10\log(P_0/P_t)$, where P_0 is the maximum input power available from the electric source, and P_t is the acoustic power delivered to the transmission medium.) The observed transducer loss is in good agreement with the calculated one.

In order to clarify their feasibility for nondestructive evaluation (NDE) of industrial materials, we measured the reflection signals from test blocks, and compared these signals with the signals detected with conventional TS-mode PZT transducers for transverse ultrasonic waves. It was shown that, as compared to the PZT transducers, P(VDF/TrFE) transducers have comparable or higher sensitivity, and operate in pure TS mode. (Output signals of longitudinal waves were observed with thickness shear PZT transducers, probably due to shear-longitudinal coupling modes of the PZT resonators.) These results indicate that P(VDF/TrFE) shear mode transducers operate with effective sensitivity and high S/N ratios, and are feasible for use in the fields of NDE and physical acoustics.

CONCLUSIONS

The SC film of P(VDF/TrFE) is a fully crystallized ferroelectric polymer film, which has the same symmetry properties as that of an orthorhombic twin crystal in the ferroelectric phase and of a hexagonal single crystal in the paraelectric phase. The SC film has high Young's modulus along the stretching axis. All of five electromechanical coupling factors are largest among those of known piezoelectric polymers. In the paraelectric phase, the SC film is a liquid crystal of 1D-liquid and 2D-solid, and $TGTG'$ chains undergo rotational and flip-flop motions independently of neighboring chains, which brings about anomalous dielectric properties. Since the SC film is in the nearest distance to a bulky single crystal of P(VDF/TrFE), this material is very useful for studies of various problems in polymer crystal physics, such as chain motions, crystal growth, crystal defects, phase transition, origin of functionalities, etc. Promising potential applications are expected in this film for its high mechanical modulus, strong piezoelectricity, high optical transparency, etc. In this paper we showed that it is usable for transverse ultrasonic transducers for NDT.

ACKNOWLEDGMENTS

We thank K. Omote, A. Yamamoto, H. Abe, T. Hisaminato and S. Iwai of Yamagata University, and T. Miya of Toray Techno Co. and Y. Yokono of Non-Destructive Inspection Co. for their collaboration. We also thank Daikin Industries, Ltd. for providing P(VDF/TrFE). This work was supported in part by Grants-in-Aid for Scientific Research from Ministry of Education, Science, Culture and Sports, Japan, and by Japan Science and Technology Corporation.

REFERENCES

1. H. Ohigashi and K. Koga, Jpn. J. Appl. Phys. **21**, 475 (1982).

2. K. Koga and H. Ohigashi, J. Appl. Phys. **59**, 2142 (1986).

3. H. Ohigashi, Jpn. J. Appl. Phys. **24**, Suppl. 24-2 , pp. 23-27 (1985).

4. K. Koga, N. Nakano, T. Hattori, and H. Ohigashi, J. Appl. Phys. **67**, 965 (1990).

5. H. Ohigashi, S. Akama, and K. Koga, Jpn. J. Appl. Phys. **27**, 2144 (1988).

6. K. Tashiro, in *Ferroelectric Polymers*, edited by H. S. Nalwa (Marcel Dekker, New York, 1995), pp. 63-181.

7. T. Yamamoto, J. Macromol. Sci.-Phys. B **16**, 487 (1979).

8. M. Hikosaka, Polymer **28**, 1257 (1987), **31**, 458 (1990).

9.. H. Ohigashi, K. Omote, and T. Gomyo, Appl. Phys. Lett. **66**, 3281 (1995).

10. A. J. Bur, J. D. Barnes, and K. J. Wahlstrand, J. Appl. Phys. **59**, 2345 (1986).

11. K. Omote and H. Ohigashi, J. Appl. Phys. **81**, 2760 (1997).

12. H. Ohigashi, K. Omote, H. Abe, and K. Koga, J. Phys. Soc. Jpn. **68**, 1824 (1999).

13. K. Tashiro, M. Kobayashi, H. Tadokoro, and E. Fukada, Macromolecules, **13**, 691 (1980).

14. H. Ohigashi, T. Itoh, K. Kimura, T. Nakanishi, and M. Suzuki, Jpn. J. Appl. Phys. **27**, 354 (1988).

15. T. Furukawa, Phase Transition **18**, 143 (1989).

16. Y. Tajitsu, A. Chiba, T. Furukawa, M. Date, and E. Fukada, Appl. Phys. Lett. **36**, 286 (1980).

17. T. Furukawa, G. R. Johnson, H. E. Bair, Y. Tajitsu, A. Chiba, and E. Fukada, *Ferroelectrics* **32**, 61 (1981).

18. T. Furukawa, Y. Tajitsu, X. Zhang, and G. E. Johnson, Ferroelectrics **135**, 401 (1992).

19. T. Mitsui, T. Tatsuzaki, and K. Nakamura, *Kyoyudentai* (Maki Shoten, Tokyo, 1969) pp. 157-258 [*An Introduction to the Physics of Ferroelectrics* (Gordon and Beach, Princeton, NJ, 1976)].

20. H. Ohigashi, in *Applications of Ferroelectric Polymers*, edited by T. T. Wang, J. M. Herbert, and A. M. Glass (Blackie and Son, Glasgow, 1988) pp. 236-273.

21. K. Omote and H. Ohigashi, IEEE Trans. Ultrason. Ferroelect. Freq. Control **43**, 321 (1996).

22. H. Ohigashi, S. Iwai, T. Miya, T. Yamamoto, Y.Yokono, and T. Imanaka, *1998 IEEE Ultrasonics Symposium Proc.* (IEEE, Piscataway, NJ, 1998) pp. 1047-1050.

23. K. Kimura and H. Ohigashi, Jpn. J. Appl. Phys. **27**, 540 (1988).

STRUCTURAL CHANGE IN FERROELECTRIC PHASE TRANSITION OF VINYLIDENE FLUORIDE COPOLYMERS AS STUDIED BY WAXS, SAXS, IR, RAMAN, AND COMPUTER SIMULATION TECHNIQUES

K. TASHIRO
Department of Macromolecular Science, Graduate School of Science, Osaka University, Toyonaka, Osaka 560-0043, Japan, ktashiro@chem.sci.osaka-u.ac.jp

ABSTRACT

Structure changes in the ferroelectric phase transitions of vinylidene fluoride-trifluoroethylene (VDF-TrFE) copolymers were investigated on the basis of a set of experimental data of WAXS, SAXS, IR and Raman measurements. An intimate relation has been clarified between the structural changes in the crystal lattice and the morphological changes. These structural changes were found to originate ultimately from the remarkable change in the molecular conformation between trans and gauche forms. In order to extract the essentially important factors governing this trans-gauche conformational change in the phase transition, the molecular dynamics calculation was carried out, giving relatively good reproduction of the observed structural changes.

INTRODUCTION

Vinylidene fluoride-trifluoroethylene (VDF-TrFE) copolymers have attracted much interest since their ferroelectric phase transition was discovered for the first time as the synthetic polymers [1-6]. This ferroelectric phase transition is unique in the point that the chain conformation changes drastically between the trans and gauche forms at the so-called Curie transition temperature (Tc) [7-14]. This structure change is intimately related with the change in the physical properties such as dielectric constants, piezoelectric constants, pyroelectric constants, elastic constants, etc. In a series of papers we have discussed the structural changes occurring in the molecular chain, crystal lattice, and morphology of the bulk samples by using many kinds of experimental tools such as X-ray diffraction, infrared, Raman, etc [4, 7, 12 - 17]. The structural changes viewed from different scales were found to be intimately related with each other, as will be described in this paper. At the same time the results of the molecular dynamics calculation is also reviewed in order to clarify the important factors which govern this unique ferroelectric phase transition [18].

CHARACTERISTIC FEATURES OF PHASE TRANSITIONS

Structural Changes in Crystal Lattice

As shown in Figure 1 (a) and (b), the two types of the ferroelectric phases, the low-temperature (LT) and the cooled (CL) phases, exist at room temperature depending upon the VDF content of the copolymers as well as the sample preparation conditions [4]. The LT phase consists of the parallel arrangement of the CF_2 dipoles of the planar-zigzag all-trans chains as in the case of PVDF form I, while in the CL phase the long trans segments are connected along the chain axis by irregular trans-gauche linkages to form a kind of superstructure. The CL phase is transferred to the LT phase by an application of tensile force along the chain axis or electric field along the polar axis. Above Tc, these phases transform into the high temperature (HT) phase with the contracted chains consisted of the statistical array of TT, TG, and TḠ rotational isomers. Through the trans-gauche conformational exchange, the molecular chains rotate violently in the paraelectric HT phase [19-25], resulting in the non-polar unit cell structure of the hexagonal packing [Figure 1 (c)]. Such a large conformational change is characteristic of polymers in which the monomer units or the polar CH_2CF_2 dipoles are connected strongly by the covalent bondings, and quite different from the structural change observed in the ferroelectric phase transition of the low-molecular-weight compounds in which the small ionic groups rotate or translate slightly.

Ferroelectric Phase   Paraelectric Phase

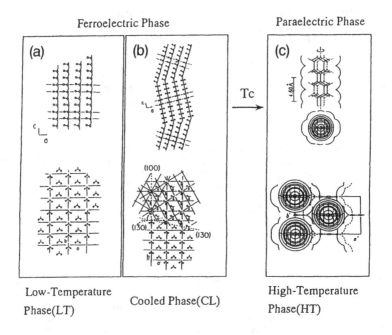

Tc

Low-Temperature
Phase(LT)   Cooled Phase(CL)   High-Temperature
Phase(HT)

Figure 1. Crystal structures of VDF-TrFE copolymers. (a) Low-temperature (LT) phase, (b)Cooled (CL) phase, and (c) High-temperature (HT) phase.

Change of Domain Size in the Crystallite through Phase Transition

As the temperature increases and the LT phase transfers into the HT phase, the X-ray profile changes remarkably, as shown in Figure 2 [15]. The reflection of the LT phase is in general broad and that of the HT phase generated at the early stage of transition is also relatively broad and rather close to that of the LT phase. The reflection width of the HT phase becomes remarkably sharper as the transition proceeds furthermore at higher temperature, while the half width of the reflection of the LT phase remains broad and almost constant during the transition. These observations suggest an existence of the domain structure and its growth during the phase transition. The crystallite of the LT phase is assumed to have a multi-domain structure, where the domains of approximately 100 Å size in the lateral direction, as evaluated from the X-ray reflection width, are gathered with their dipoles oriented into different directions to form a large crystallite as illustrated in Figure 2. As the temperature increases and approaches the Tc, the HT phase begins to appear randomly in some domains and the domain size of the HT phase reflects directly that of the LT phase, giving the reflection width close to that of the LT phase. As the temperature increases furthermore, the number of domains of the HT phase increases and these domains are fused together into a large single domain, because the large-amplitude rotation of the chains makes each domain nonpolar and so the boundary between the domains becomes unambiguous and these domains fuse into one large domain. The large size of the thus created domain reflects on the sharpness of the X-ray reflection.

VDF Content Dependence of the Transition Behavior

The transition behavior is affected by many factors. For example, the transition behavior changes depending upon the VDF content of the copolymers [4,10-13]. For the copolymers with

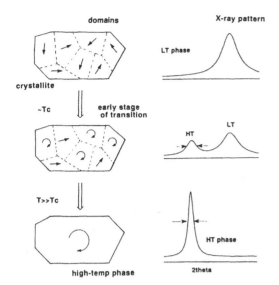

Figure 2. Temperature dependence of X-ray diffraction profile taken for VDF 73 % copolymer and the corresponding change in domain size of the crystallite during the phase transition of VDF copolymer.

the VDF content higher than 70 molar %, the transition occurs almost reversibly and discontinuously in the thermodynamically first ordered fashion between the regular LT phase and the HT phase. For the samples with VDF 70 - 60 %, the crystalline phase, obtained after the heat treatment above Tc, is a mixture of the LT phase and the CL phase at room temperature, and the complicated transitions are observed. The LT phase of VDF 50 - 60 % sample was found to show a clear transition to the CL phase with large endothermic enthalpy change, which was followed by the apparently continuous transition from the CL phase to the HT phase with smaller enthalpy change [26] For VDF 0 - 50 % copolymers, the transition occurs apparently continuously over a wide temperature range between the CL phase and the HT phase.

Heat Treatment Effects on the Phase Transition Behavior

The heat treatment affects also remarkably the Tc as well as the melting point (mp) [4, 15, 27-30]. We measured the DSC thermograms for a series of VDF 73, 65, and 55 % samples, which were prepared by annealing the cold-drawn samples at the various temperatures, and showed that the Tc and mp changed systematically depending on the annealing temperature. For example, Figure 3 shows the DSC thermograms in the transition temperature region (heating process) measured for cold-drawn VDF 73% and 65% samples annealed at the various temperatures. As the annealing temperature increased, a sharp peak could be observed and shifted to lower temperature side. In relation with these thermal data, the sizes of the crystalline domain in the lateral direction as well as along the chain direction were evaluated by measuring the half-width of the X-ray reflections. Figure 4 (a) shows the annealing temperature dependence of the crystalline domain sizes of VDF 73% sample. The domain sizes in the both directions increased when the samples were annealed at higher temperature, in particular above the trans-gauche conformational transition (region A of the DSC thermogram in Figure 4 (a)). The Raman spectra were measured for the samples annealed at the various temperatures and the intensities of the bands characteristic of the long trans sequence (> T3), short trans sequence (T3), and gauche bond were evaluated as shown in Figure 4 (b) [31]. In the region A, where the phase transition began to start, the intensity of the trans-zigzag conformation increased remarkably, corresponding well to the increase of the crystallite size in the chain direction. When the sample was annealed in the temperature region close to the mp, the trans bands decreased intensity and the gauche bands increased the intensity. This indicates a generation of structural defects in the chains, resulting in the shift of the

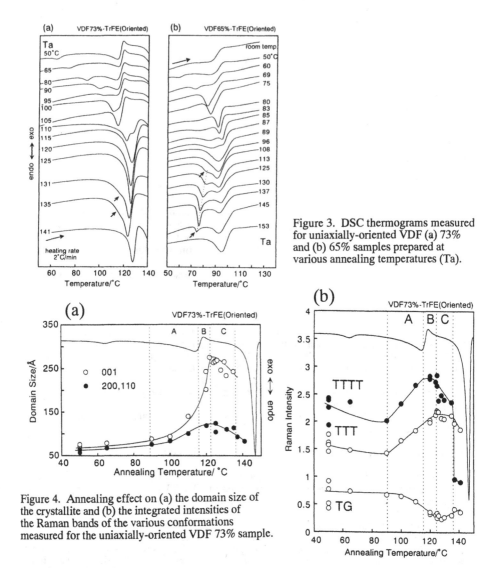

Figure 3. DSC thermograms measured for uniaxially-oriented VDF (a) 73% and (b) 65% samples prepared at various annealing temperatures (Ta).

Figure 4. Annealing effect on (a) the domain size of the crystallite and (b) the integrated intensities of the Raman bands of the various conformations measured for the uniaxially-oriented VDF 73% sample.

DSC peak toward lower temperature side. In case of VDF 65%, the behavior was similar to VDF 73%. However, in addition to such a change in the crystallite size, the tilting phenomenon of the chain from the draw axis was also observed when the LT phase transferred to the CL phase.

Morphological Changes in the Phase Transitions

In order to investigate the effect of heat treatment on the morphology as well as on the inner structure of the lamella, the 2-dimensional WAXS and SAXS patterns were measured simultaneously in the heating process of the as-drawn unannealed samples, where both the patterns of WAXS and SAXS were recorded on one imaging plate as shown in Figure 5 (for VDF 65% sample) [16, 17]. The X-ray reflections in the WAXS pattern splitted into two (for the equatorial

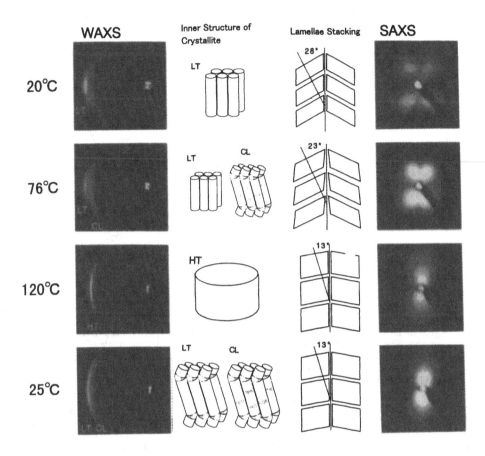

Figure 5. Temperature dependence of WAXS and SAXS patterns measured for uniaxially-oriented and unannealed VDF 65% copolymer sample and the corresponding structures of the crystallite and the stacked lamellae. The WAXS pattern (left) shows the behavior of 110 + 200 reflection. When the LT phase changed to the CL phase, new reflection appeared, which splitted into two along the vertical axis, in addition to the spot-like reflection of the original LT phase. They changed finally into the sharp reflection of the HT phase. On the other hand, the SAXS pattern (right) changed also drastically from the four-point pattern to the meridional scattering pattern in the phase transition process. In the middle part of this figure, the change in the aggregation structure of the domains in the crystallite (left side) and the change of the lamellar stacking structure are illustrated. The multi-domain structure of the LT phase transferred finally to the single domain of the HT phase. In parallel to it, the long spacing and the tilting angle of the lamellae changed. The tilting of the chains within the crystallite is related intimately with the tilting of the lamellae.

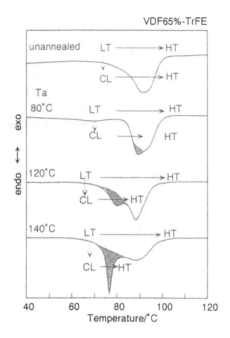

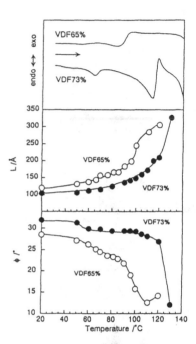

Figure 6. DSC thermograms and the corresponding phase transitions observed for the uniaxially-oriented VDF 65% copolymer samples prepared at the various annealing temperatures.

Figure 7. Temperature dependence of the long period and tilting angle of the stacked lamellae estimated for the uniaxially-oriented VDF 73% and 65% copolymer samples.

reflections as well as the layer reflections) above 89°C, indicating an occurrence of the tilting phenomenon in the CL phase. The quantitative analysis of the WAXS data allowed us to assign the DSC peaks of Figure 3 to the transitions from LT to CL, CL to HT, or LT to HT, as indicated in Figure 6. The SAXS profile changed also remarkably. The 4-points scattering pattern of the unannealed sample, which indicates the tilting of the lamellar planes from the draw axis, changed to the pattern extending along the meridional direction. The tilting angle ϕ and the long spacing L of the lamellae were evaluated from these patterns as shown in Figure 7 (VDF 65 and 73% samples). In the ferroelectric transition temperature region, the long spacing increased drastically and the tilting angle of the lamellae decreased also remarkably. In parallel to these X-ray measurements, the temperature dependence of Raman and infrared spectra was also measured to detect the conformational changes, giving a good correspondence between the molecular chain conformational change and the morphological change.

By combining all the data mentioned above, the structural change occurring in the phase transition temperature region can be described as illustrated in the middle part of Figure 5, where the structural change within the crystallite [15] and the change in the lamellar stacking structure [16, 17] are shown for the case of VDF 65% sample. At room temperature the uniaxially-oriented and unannealed VDF 65% sample contains the crystalline lamellae, where the normals of the lamellar planes tilt from the draw axis and the lamellar thickness is relatively thin. Each lamella is consisted of the multiple domains. In each domain the planar-zigzag chains are packed with the dipoles parallel to the b axis of the crystal lattice corresponding to the LT phase. By heating the sample, the trans-to-gauche conformational change begins to start and the tilting of the "molecular chains" is observed in the lamellae. At the same time the long period and the thickness of the lamellae increase and the lamellae direct their normals toward the draw axis. In higher temperature region, the molecular chains of the gauche form rotate vigorously around the chain axis and the lamellae transform as a whole to the HT phase, where the domains of small size are fused into one large domain. That is to say, the trans-gauche conformational change and the associated molecular motion are found to relate quite intimately but in a complicated manner with the change in the internal structure of an individual lamella (the tilting angle of the chain axis, the domain size, etc.) and the change in the aggregation structure of the lamellae (the long period of the stacked lamellae, the tilting angle of the lamellar plane, the growth of the lamellar thickness, etc.).

COMPUTER SIMULATION OF PHASE TRANSITION

As discussed in the preceding sections, the phase transitional behavior of the copolymers is quite complicated. But it must be emphasized again that the most important structural feature in the phase transitional phenomena of the copolymers can be seen in such a point that the chain conformation changes drastically between the all-trans-zigzag form and the disordered form constructed by statistically distributed trans and gauche bonds. This conformational transformation becomes a trigger for all the structural changes observed from the different scales. In order to understand the essential feature of this conformational transition, we need to carry out theoretical or computational research based on the microscopic information on structure. So far the several papers were reported for this purpose [32 - 35], but they were more or less phenomenological and could not describe enough well the details of the three-dimensional trans-gauche conformational change coupled with the molecular motion. In order to trace the structural change in the ferroelectric phase transition more quantitatively from the molecular level, the computer simulation may be one of the useful methods to attain the above-mentioned purpose. Especially, the molecular dynamics (MD) simulation seems one of the best ways to study the complicated structural change of the polymer chain system.

In the molecular dynamics simulation, good choice of the potential function parameters is essentially important in order to reproduce the experimental data as reasonably as possible. In the calculations on PVDF and VDF-TrFE copolymers, we refined the potential functions reported by Karasawa and Goddard [36] so that the observed X-ray and infrared/Raman spectral data could be reproduced as well as possible [31, 37, 38]. Then we carried out the MD simulation of the phase transition of the VDF-TrFE copolymers by using the thus-established force field parameters at an aim to clarify the factors governing the phase transitional behaviors [18].

Construction of Models/Energy Minimization/Molecular Dynamics Calculation

In the MD calculation the three-dimensional periodic boundary condition was introduced and the MD unit cell was constructed by setting 16 or 32 chains, each of which was composed of random sequences of 10 - 20 VDF and TrFE monomeric units.   The monomeric units included in the adjacent unit cells along the c axis were connected by covalent bondings to form the infinitely long chains. In the MD cell, some chains were packed upwards and the other chains downwards along the c axis, where the upward direction was defined in such a way that the direction vector of CH_2CF_2 monomeric units was pointed into the positive direction of the chain axis. For example, the 8u/16 model means that the 8 chains were packed upwards and the other 8 chains downwards at random positions.

The thus constructed crystal structures were energetically minimized at first. Molecular dynamics (MD) simulations were performed based on the Parrinello-Rahman method of the constant pressure and temperature with a Nose-Hoover thermostat at the various temperatures between 250 and 500 K.

Simulation of Conformational Transition

Figures 8 and 9 show, respectively, the snapshots of crystal structure and molecular chain conformation obtained in the course of MD calculation for VDF 50% copolymer (8u/16 model) at 300 K and 350 K. When starting from the initial model obtained by the minimization method the

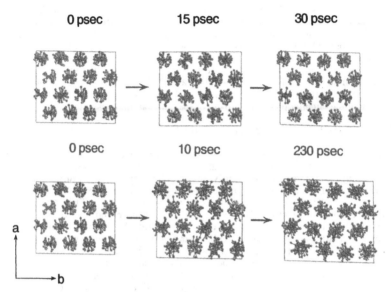

0 psec   **15 psec**   **30 psec**

0 psec   **10 psec**   **230 psec**

a

b

Figure 8. Time dependence of the crystal structure calculated for VDF 50% copolymer model at (a) 300 K and (b) 350 K.

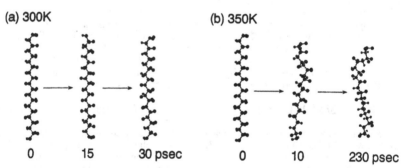

(a) 300K   **(b) 350K**

0   15   30 psec   0   10   230 psec

Figure 9. Time dependence of the molecular conformation calculated for VDF 50% copolymer model at (a) 300 K and (b) 350 K.

conformational change from the trans to gauche forms could not be seen at such a low temperature as 300 K. The molecular chains took the deflected trans-zigzag conformation and the CF_2 groups and/or the whole chains were found to librate with large amplitudes around the chain axes. When the temperature was set to 350 K, remarkable structural changes were found to occur. The molecular chain changed the conformation from the extended trans-zigzag form to the twisted and kinked form containing both the trans and gauche bonds. At the same time the rotational motion of the chains were also detected. These changes resulted in the orientational change of CF_2 dipoles or the change of the electric dipoles. The population of trans and gauche bonds of the skeletal chains is plotted against time as shown in Figure 10, where the notations of trans and gauche were used for the torsional angles in the ranges of 150 ~ 210° (T), 30 ~ 90° (G+), and 270 ~ 330° (G-), respectively. The fraction of T was 1.0 at the starting point and decreased steeply in a time region

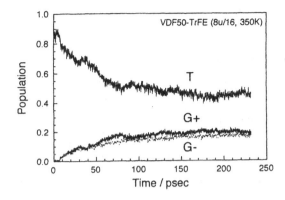

Figure 10. Time dependence of trans (T) and gauche (G+, G-) isomers of VDF 50% copolymer model calculated at 350 K.

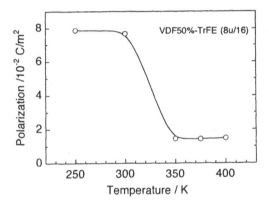

Figure 11. Temperature dependence of the electric polarization calculated for VDF 50% copolymer model.

of about 70 psec. In parallel, the populations of G+ and G- increased in the same manner. At the equilibrated stage of ca. 200 psec, the distribution to the T, G+, and G- became almost constant; T: G+: G- = 2: 1: 1. The distribution of longer conformational sequences can be estimated similarly by focusing into the sequences of TG+, TG-, T3G+, and T3G-. The results are as follows.

TG+/TG-	T3G+/T3G-	others	TG+TG-	T3G+T3G-	others
60.6 %	14.8 %	24.6 %	44.0 %	0.0 %	56.0 %

where "others" include the conformational sequences such as TG+S+G+, TTS+TG+, etc. By measuring the infrared and Raman spectra at the various temperatures, the chain conformation of the copolymer with ca. 50 molar % VDF content was reported. For example, Tashiro et al. [4, 12] made the detailed description of the molecular chains in the HT phase: the chains were constructed by a random combination of TG+, TG-, T3G+, and T3G- isomers. Their observation was reproduced well by the present MD simulation.

The averaged lattice constants of the basic unit cell in the HT phase at 350 K were a = 10.67 Å, b = 6.16 Å, c = 4.58 Å and a/b = 1.73, which were found to correspond fairly well to the data observed for VDF 55% sample [4, 13]: a = 9.75 Å, b = 5.63 Å, c = 4.60 Å and a/b = 1.73. The a/b ratio = 1.73 indicates the haxagonal packing of the chains in the HT phase as clarified by the X-ray studies.

Temperature Dependence of Electric Polarization

The electric polarization of the unit cell was calculated on the basis of the atomic coordinates obtained by the MD calculations. Figure 11 shows the temperature dependence of the electric polarization estimated at the equilibrated stage of each MD calculation. Around 330 K the polarization decreases drastically. This reproduces quite reasonably the actually observed behavior of the polarization [3]. (Strictly speaking, the calculated polarization did not reach zero even in the HT phase. This may come from the imperfect cancellation of the dipole moments of individual chains along the chain axes because of the limited number of chains in the MD cell.)

Role of Intramolecular and Intermolecular Interactions

In order to estimate the role of various kinds of interactions in the structural transition of the VDF 50% copolymer, the MD calculations were made by neglecting particular interactions.
(1) Internal torsional energy: The energy barrier of the skeletal torsional motion was about 4 kcal/mol in the above MD calculation. When this barrier was increased by several times, no conformational change was induced even when the temperature of the system was increased to 400 K.
(2) Coulombic interactions: When the long-range Coulombic interactions were neglected in the calculation, the molecular chains did not show any conformational transition from trans to gauche but they rotated rigidly around the chain axes even when the calculation was made for a long time at 350 K where the transition was observed in the original calculation including the Coulombic interactions.
(3) Intermolecular van der Waals interactions: When these interactions were neglected, the neighboring chains were found to collide together and the system was broken because of the explosive increment of the total energy.
From (1) - (3), we can notice an importance of the suitable balance between the short-range interactions (van der Waals forces) and long-range interactions (Coulombic forces) in the realization of the conformational transition phenomenon, in addition to the significant role of the torsional barrier between the trans and gauche forms.

VDF-Content Dependence of the Phase Transition Behavior

As already mentioned, the ferroelectric phase transition of VDF-TrFE copolymers is affected remarkably by the content of the VDF monomeric units. We built the several models for VDF 70 % sample and carried out the MD calculation in the same way as mentioned for the case of VDF 50% sample. Figure 12 shows the temperature dependence of the specific volume of the cell calculated for VDF 70% copolymer (8u/16 model) at the equilibrated stage. At about 470 K the cell was found to expand drastically. In this figure the temperature change of the specific volume of VDF 50% copolymer is also shown. A tendency of VDF 70 % copolymer to give a higher transition temperature than VDF 50 % copolymer is recognized reasonably, although the calculated transition temperatures are as a whole higher than the experimental value (The transition temperature averaged for the various models is ca. 390 K for VDF 50% copolymer and ca. 470 K for VDF 70% copolymer. The latter is higher by 70 K than the experimental value).

CONCLUSIONS

In this paper we reviewed a series of our researches concerning the structural changes in the ferroelectric phase transition of the VDF-TrFE copolymers from the various dimensions of the molecular conformation, the crystal lattice, the crystallite, and the stacked lamellae. These structural changes observed in the different scales are intimately related to each other, and they can be ascribed ultimately to the trans-gauche conformation change of the molecular chains. The

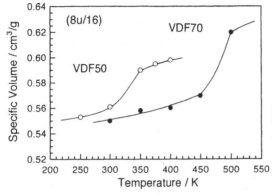

Figure 12. Temperature dependence of the specific volume calculated for VDF 50% and 70% copolymer models.

computer simulation described in the above section has clearly shown the importance of the good balance between the internal torsional potential barrier, the van der Waals interactions, and the Coulombic interactions in the crystal lattice. In this simulation, however, we have still many unsolved problems. For example, the copolymers with low VDF content exhibits the generation of the CL phase. But we have not yet succeeded to reproduce this CL phase by the molecular dynamics calculation. The stacking structure of the lamellae changes remarkably as already shown in the above sections. The relation between the tilting phenomenon of the molecular chains and the trans-gauche conformational transition also needs to be clarified concretely. The ferroelectric phase transition of the VDF copolymers is quite complicated but is really challengeable theme.

ACKNOWLEDGMENTS

The author greatly acknowledges the contribution of Mr. Kohji Takano, Mr. Yukihiro Abe, and Ms. Rieko Tanaka in a series of researches about VDF-TrFE copolymers. He wishes also to thank Dr. Hiroyuki Tadokoro, Emeritus Professor of Osaka University, the late Dr. Masamichi Kobayashi, Emeritus Professor of Osaka University, and Dr. Yozo Chatani, Emeritus Professor of Tokyo University of Agriculture and Technology for their encouragement of us in these researches.

REFERENCES

1. T. Furukawa, M. Date, E. Fukada, Y. Tajitsu, and A. Chiba, Jpn. J. Appl. Phys., **19**, p. L109 (1980).
2. E. Fukada, Phase Transitions, **18**, p. 135 (1989).
3. T. Furukawa, Phase Transitions, **18**, p. 143 (1989).
4. K. Tashiro, Chapter 2, "Ferroelectric Polymers" ed. H. S. Nalwa, Marcel Dekker, New York, 1995.
5. T. Yamada, T. Ueda, and T. Kitayama, J. Appl. Phys., **52**, p. 948 (1981).
6. Y. Higashibata, J. Sako, and T. Yagi, Ferroelectrics, **32**, p. 85 (1981).
7. K. Tashiro, K.Takano, M. Kobayashi, Y. Chatani, and H. Tadokoro, Polymer Commun., **22**, p. 1312 (1981).
8. A. J. Lovinger, G. T. Davis, T. Furukawa, and M. G. Broadhurst, Macromolecules, **15**, p. 323 (1982).
9. G. T. Davis, T. Furukawa, A. J. Lovinger, and M. G. Broadhurst, Macromolecules, **15**, p. 329 (1982).

10. A. J. Lovinger, T. Furukawa, T., G. T. Davis, and M. G. Broadhurst, Polymer, 24, p. 1225 (1983).
11. A. J. Lovinger, T. Furukawa, G. T. Davis, and M. G. Broadhurst, Polymer, 24, p. 1233 (1983).
12. K. Tashiro, K. Takano, M. Kobayashi, Y. Chatani, and H. Tadokoro, Polymer, 25, p. 195 (1984).
13. K. Tashiro, K. Takano, M. Kobayashi, Y. Chatani, and H. Tadokoro, Ferroelectrics, 57, p. 297 (1984).
14. K. Tashiro and M. Kobayashi, Polymer, 29, p. 4429 (1988).
15. K. Tashiro, R. Tanaka, K. Ushitora, and M. Kobayashi, Ferroelectrics, 171, p. 145 (1995).
16. K. Tashiro, R. Tanaka, and M. Kobayashi, Polym. Prepr. Jpn., 45, p. 3119 (1996); 46, p. 3585 (1997).
17. R. Tanaka, K. Tashiro, and M. Kobayashi, Polymer, 40, p. 3855 (1999).
18. Y. Abe, K. Tashiro, and M. Kobayashi, Comp. Theor. Polym. Sci., in press.
19. F. Ishii, A. Odajima, and H. Ohigashi, Polym. J., 15, p. 875 (1983).
20. V. J. McBrierty, D. C. Douglass, and T. Furukawa, Macromolecules, 15, p. 1063 (1982).
21. V. J. McBrierty, D. C. Douglass, and T. Furukawa, Macromolecules, 17, p. 1136 (1984).
22. F. Ishii and A. Odajima, Polym. J., 18, p. 539 (1986).
23. F. Ishii and A. Odajima, Polym. J., 18, p. 547 (1986).
24. J. F. Legrand, Ferroelectrics, 91, p. 303 (1989).
25. J. F. Legrand, B. Frick, B. Meurer, V. H. Schmidt, M. Bee, and J. Lajzerowicz, Ferroelectrics, 109, p. 321 (1990).
26. K. Tashiro, R. Tanaka, and M. Kobayashi, Macromolecules, 32, p. 514 (1999).
27. G. R. Li, N. Kagami, and H. Ohigashi, J. Appl. Phys., 72, p. 1056 (1992).
28. G. M. Stack and R. Y. Ting, J. Polym. Sci.: Part B: Polym. Phys., 26, p. 55 (1988).
29. K. J. Kim, G. B. Kim, C. L. Vanlencia, and J. F. Rabolt, J. Polym. Sci.: Part B: Polym. Phys., 32, p. 2435 (1994).
30. R. Gregorio Jr. and M. Botta, J. Polym. Sci: B, Polym. Phys. 36, p. 403 (1998).
31. K. Tashiro, M. Kobayashi, and H. Tadokoro, Macromolecules, 14, p. 1757 (1981).
32. A. Odajima, Ferroelectrics, 57, p. 159 (1984).
33. N. C. Banic, F. P. Boyle, T. J. Sluckin, P. L. Taylor, S. K. Tripathy, and A. J. Hopfinger, J. Chem. Phys., 72, p. 3191 (1980).
34. R. Zhang and P. L. Taylor, J. Appl. Phys., 73, p. 1395 (1993).
35. S. Ikeda and H. Suda, Phys. Rev. E, 56, p. 3231 (1997).
36. N. Karasawa and W. A. Goddard III, Macromolecules, 25, p. 7268 (1992).
37. K. Tashiro, Y. Abe, and M. Kobayashi, Ferroelectrics, 171, p. 281 (1995).
38. M. Kobayashi, K. Tashiro, H. Tadokoro, Macromolecules, 8, p. 158 (1975).

EFFECTS OF SAMPLE PROCESSING AND HIGH-ENERGY ELECTRON IRRADIATION CONDITIONS ON THE STRUCTURAL AND TRANSITIONAL PROPERTIES OF P(VDF-TrFE) COPOLYMER FILMS

Vivek Bharti, Z.-Y Cheng, H. S. Xu, G. Shanthi, T.-B. Xu, T. Mai and Q. M. Zhang
Materials Research Laboratory, The Pennsylvania State University
University Park, PA 16802

ABSTRACT

Recently it has been demonstrated that using a high electron energy irradiation, the electro-mechanical properties of polyvinylidene fluoride – trifluoroethylene, P(VDF-TrFE) copolymers can be improved significantly for example; high electric field induced strain (~ 4.5%) with high elastic energy density, high dielectric constant (~ 65 with loss less than at 1kHz), high piezoelectric coefficient (~ 350 pm/V) and high electro-mechanical coupling coefficient (~0.45). It was found, depending on sample processing and electron irradiation conditions these properties can be controlled. In this talk we will present the experimental data on the effect of processing and irradiation conditions on structural and transitional behavior of copolymers. These structural information coupled with the electro-mechanical properties will be presented in order to show the possibilities for further improvement.

INTRODUCTION

We have shown recently that after the proper electron irradiation treatment, P(VDF-TrFE) copolymers could exhibit high electrostrictive strain with high elastic energy density [1-4]. It is also observed that under the bias electric field, copolymer possses a high electromechanical coupling coefficient and high piezoelectric state [2]. In addition, we have shown that in contrast to other electroactive polymers, the electrostrictive copolymer can maintain their strain level even under the high tensile stress and hydrostatic pressure [4]. Also, we found by mechanical stretching the electro-mechanical properties of irradiated copolymers can be tuned. For example for stretched films, the electric field induced strain and electromechanical coupling coefficient are higher along the stretching direction in comparison to the thickness or applied electric field direction whereas for unstretched films the strain and piezoelectric coefficient along the field is always higher than the perpendicular direction [2].

Although irradiated P(VDF-TrFE) copolymer films possess an excellent combination of dielectric and electro-mechanical properties, there are still several issues that need to be addressed for further improvement. For example; the loss in crystallinity due to irradiation as in the copolymer system, the ferroelectric behavior is mainly coming from the crystalline regions. Also as the electro-mechanical properties in copolymer were found to be very sensitive to the irradiation conditions therefore, the understanding of structure-property relationship with the irradiation conditions is essential. Furthermore, the challenge for us is to chemically synthesize the same defective structure without electron irradiation.

In order to address these issues it becomes very important to investigate the possible causes which transform these ferroelectric copolymers into electristrictive copolymers with exceptional dielectric and electro-mechanical properties. A systematic structural investigation was carried out using X-ray, DSC and FTIR techniques to observe the effect of irradiation dose on the structural and transitional behavior of stretched and unstretched films.

EXPERIMENT

The PVDF (x) - TrFE (1-x), copolymers powder having x = 50 mol % was supplied by Solvay and Cie, Bruxelles. The unstretched films were prepared either by hot pressing followed by slow cooling. The stretched films were prepared by uniaxially stretching the quenched hot pressed or solution cast unstretched films up to 5 times to their initial length at 25°C. In order to enhance the crystallinity, the stretched films were annealed under the clamped condition at 140°C for 16 hours. The electron irradiation was carried out in a nitrogen atmosphere with 2.55 MeV electrons at 95°C for several irradiation doses ranges from 40 to 100 Mrad.

The polarization hysteresis loops were measured using a Sawyer-Tower technique. An external electric field was applied in the form of triangular waveform with a frequency 10 Hz and amplitude 150 MV/m. The DSC measurements were taken at a scanning rate 10°C/min under a nitrogen atmosphere using differential scanning calorimeter (TA instrument, model no. 2010). The X-ray pattern were taken at room temperature (20°C) using Scintag diffractometer (model PAD-V) with Ni filtered CuKα radiation. The IR measurements were carried out in the range of 4000 to 400 cm^{-1} spectral region using Bio-Rad (model no. FTS 175) FTIR spectrophotometer. The measurement on the crosslinking factor was performed by measuring the gel content of irradiated films using the American standard test method (D2765-95) [4].

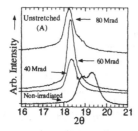 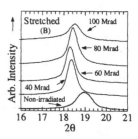

Figure 1 : X-ray diffraction scans of (A) unstretched (B) stretched, P(VDF-TrFE) 50/50 mol % copolymer films irradiated with different doses at 95°C, using 2.55 MeV electrons.

RESULTS AND DISCUSSIONS

X-ray Results

Figures 1 (A) and (B) show X-ray patterns obtained from unstretched and stretched films irradiated for different doses, respectively. Consistent with the earlier studies [5,6], before irradiation, the unstretched films exhibit two peaks at 2θ =18.79° (4.72 Å) and 19.28° (4.59 Å). The peak at 4.72 Å is due to the hexagonal packing of 3/1-helical chains generated due to the presence of TG and TG′ defects and second is due to similarly packed trans-planar chains. However, the stretched films show only one peak at 4.68 Å spacing and therefore show that the stretching not only transform the chain segments containing 3/1 helical conformation but also pack more closely and more regularly the chain segments that already are in the trans-planar conformation [6].

After 40 Mrad of irradiation, only one peak is observed at a lower angle for both the unstretched and stretched films and thus clearly indicates the expansion of the lattice due to the introduction of defects in the crystalline phase during the irradiation. This is responsible for the observed change in the polarization hysteresis from a typical normal ferroelectric hysteresis loop to a slim polarization loop (figure not shown). Upon 60 Mrad irradiation, the peak appears at 4.84 Å for both unstretched and stretched irradiated films. The corresponding lattice spacing is close to the paraelectric phase of the non-irradiated copolymer films determined from X-ray above the Curie temperature and therefore indicates the conversion of ferroelectric phase of the copolymer to the paraelectric phase [6]. The non-polar phase induce by irradiation is permanent and cannot be switched back even after poling the film using a high electric field and also heating the films above 220°C temperature (figure is not shown). For both unstretched and stretched films, the peaks become sharper and more intense upon 60 Mrad irradiation in comparison to non-irradiated and 40 Mrad irradiated films. The increase in the coherence length of X-ray diffraction is due to the disappearance of the ferroelectric ordering and hence, ferroelectric domains. As the irradiation increases further, the shape of the X-ray peak becomes Lorentz type suggesting that the crystallite-amorphous interface is also quite diffused and crystallite size becomes very small.

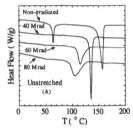 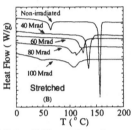

Figure 2 : DSC thermogram (A) unstretched, and (B) stretched, P(VDF-TrFE) 50/50 mol % copolymer non-irradiated, and irradiated with different doses at 95°C.

DSC Results
The thermal properties of the copolymer also undergo a great change upon irradiation. Figures 2 (A) and (B) summarize the DSC results obtained from the unstretched and stretched copolymer films, respectively. Being ferroelectric in nature, the peak appeared at 65°C is due to ferroelectric to paraelectric (F-P) transition while the peak at 160 °C is due to the melting of crystallites [7]. Consistent with X-ray results the crystal melting peak for stretched films is sharper and more intense than unstretched films and thus, reflects the higher crystalline ordering for stretched films. After the irradiation, the temperature and the enthalpy of the melting peak decreases continuously with increasing the irradiation dose for both unstretched and stretched films. The shift in temperature and broadening of the melting peak is higher for the unstretched films in comparison to the stretched films. The apparent shift and broadening of the melting peak indicates the presence of broad distribution in crystallite sizes and reduction in crystal ordering in irradiated films which is due to the lattice defects and crosslinking within the crystallites or at crystallites-amorphous boundaries. Assuming that the enthalpy of the melting is directly proportional to the crystallinity in the sample, the change of crystallinity with dose can be deduced and is presented in figure 3(A). Clearly, the change of crystallinity with the dose acquired from the X-ray data, is consistent with that from the DSC data, although the DSC data yields a slightly lower crystallinity.

Interestingly, a broad DSC peak reappears at temperatures near the original F-P transition peak position of non-irradiated films when the dose is increased to 80 Mrad and beyond. This is consistent with the X-ray data where at high doses the X-ray peak moves back to higher angle.

The finding here indicates that the structural defects introduced by irradiation in the crystalline region depends on the crystallite size and also the boundary conditions at the crystalline-amorphous interface.

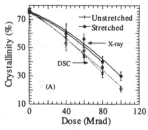 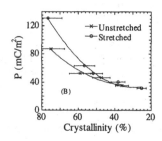

Figure 3 : Change (A) in the crystallinity with irradiation dose, calculated from X-ray data (figure 1) and DSC data (figure2) and, (B) in induced polarization (P) at 160 MV/m field with the averaged value of the crystalinity obtained from figure 3A.

It is well known that the morphology of P(VDF-TrFE) copolymers is that of crystallites embedded in an amorphous matrix, which is analogous to a composite structure. It is interesting to compare the results obtained here on the polarization and crystallinity (approximately the volume fraction of the crystallites in the polymer). Figure 3(B) presents the polarization P induced by 160 MV/m field as a function of crystallinity, which is taken from the averaged value of the DSC and X-ray data. Apparently, the initial drops of the crystallinity at low dose range (40 Mrad) causes a large reduction in the polarization for both stretched and unstretched films. As has been pointed out, the initial precipitant decrease of the polarization in the stretched films is also partly caused by the reduction in the crystallite orientation in the irradiation process. Therefore, if the crystallinity of the polymer under 40 Mrad dose can be raised, due to the high ratio of the polarization/crystallinity for the copolymer studied here at this particular dose the polarization level can be improved markedly, especially in stretched films.

IR Results

Since IR is very sensitive to the molecular conformation, the FTIR measurements were carried out in order to further quantify the changes in the copolymer due to irradiation. Although there are several peaks for each type of conformation, the peaks at 1288, 614 and 510 cm^{-1} which correspond to the presence of all trans ($T_{m \geq 4}$), TG and T_3G chains sequences respectively, were selected to analyze the irradiation effect. The Lorentizian function was fitted to separate the component from each individual bands. The samples used for investigation have slight variation in thickness and also the aborbity for the same conformation might be different for each sample due to different crosslinking densities. In order to eliminate the effects due to these factors, the band at 3022 cm^{-1} which is due to the C-H band viberation and is almost propotional to the thickness of the sample is selected as an internal standard [8]. The change in the absorbance of each vibration with respect of internal standerd were calculated. Furthermore, in order to quatify the cahnges due to irradiation, the change in molecular fraction (F_i) corresponding to $T_{m \geq 4}$, T_3G and TG chain sequences were calculated using the following expression [9]:

$$F_i = \frac{A_i}{A_I + A_{II} + A_{III}} \qquad (1)$$

Where I= I, II and III and A_I, A_{II}, and A_{III} are the absorbance of all trans ($T_{m \geq 4}$), T_3G and TG sequences respectively. As can be observed in figure 4(A) after 40 Mrad irrdaiation molecular fraction of all trans sequences decreases drastically. While the molecular fraction for T_3G and TG chains sequences increases. These results support the fact that at first irradiation breaks all trans chains ($T_{m \geq 4}$) and generate T_3G and TG chains sequences by introducing the gauche bonds. Consistent with the polarization data the rate of decrease in fractional absorbance corresponding to all trans conformation is higher for stretched films than unstretched films.

In addition, after irradiation, a peak appeared at 1717 cm^{-1}, indicating the presence of the double bond structure of CH=CF as well as carboxyl group (-C=O) and also a broad peak of hydroxyl (-OH) group between 3200 to 3600 cm^{-1}[10]. It is believed that there is small chain scissioning as well as formation of carboxylic acid due to presence of oxygen during the irradiation. Even though irradiation was carried out under a continuous flow of nitrogen gas, a small amount of oxygen may still exist in the irradiation chamber.

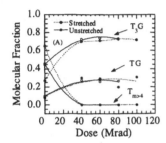 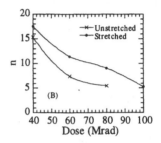

Figure 4 : Change in (A) molecular fraction corresponding to all trans ($T_{m \geq 4}$), T_3G and TG chain sequences and, (B) number of repeating units (n) between two crosslinks; for unstretched and stretchd, P(VDF-TrFE) 50/50 mol % copolymer non-irradiated, and irradiated with different doses at 95°C.

Crosslinking Results

Crosslinking density was measured to provide information on the effect of crosslinking between the chains on the properties investigated here. Figure 4(B) presents the change in the number of repeating units (n) between two crosslink points along the chain in unit of $-CH_2-CF_2-CHF-CF_2-$, as a function of dose for both stretched and unstretched films. The lower the value of n the greater the crosslinking density. Clearly, for stretched films at doses below 80 Mrad, the crosslinking increases with dose at a much lower rate compared with unstretched films. This indicates that the chain orientation introduced by stretching reduces the rate of crosslinking in the irradiation process. Also for stretched films at doses from 80 to 100 Mrads, the rate of the crosslinking with dose becomes much higher than that for unstretched films. As has been shown, irradiation has the effect of randomizing the orientation (crystallites and polymer chains) induced by stretching. At higher dosages the local chain orientation in stretched films may not be very much different than in unstretched films.

It was observed that for both unstretched and stretched films, the crosslinking increases in proportion to the decrease in the crystallinity. Furthermore, even with the same crystallinity, the crosslinking of unstretched films is higher than that in stretched films. In other words, the reduction of the crystallinity in the copolymer under irradiation is not directly controlled by the crosslinking density, but it is expected that the crosslinking process has a significant role here in the conversion from the crystalline to amorphous phase in the copolymer.

The influence of crosslinking on the ferroelectric behavior and polar ordering in the crystalline region is not clear. From the data presented, it seems that as far as the field induced polarization is concerned, the effect of crosslinking density is not significant and direct. However, the crosslinking density have a direct effect on the crystallite size, which may affect the polar response in the copolymer when the size of crystallites becomes small.

CONCLUSIONS

The effect of irradiation on the transitional behavior, crystalline and molecular structure were investigated using X-ray, DSC, FTIR crosslinking and polarization measurements. It was observed that the irradiation induced structural and conformational changes are not a monotonic process. The effect of irradiation is more pronounced for stretched films in comparison to unstretched films. It was found, initially up to 40 Mrad dose, irradiation breaks ferroelectric domains into micro polar regions and thus cause dramatic drop in the polarization level. Structurally, this process takes place due the break–up of all-trans ($T_{m \geq 4}$), chains into T_3G and TG chains by inducing the gauche bonds. However, at higher doses, due to heavy crosslinking, the chains tend to come closer and thus favor trans bonds instead of gauche bonds and therefore regains their micro-polar region slightly.

ACKNOLEDGEMENTS

This work was supported by the Office of Naval Research through Grant No N00014-97-1-0900 the National Science Foundation through Grant No ECS-9710459 and DARPA through Grant No. N00173-99-C-2003.

REFERENCES

1. Q. M. Zhang, V. Bharti and X. Zhao, Sicence **280**, p. 2,101 (1998).
2. Z. Y. Cheng, T. B. Xu, V. Bharti, S. Wang and Q. M. Zhang, Appl. Phys. Lett. **74**, 1901 (1999).
3. V. Bharti, Z.Y. Cheng, S. Gross, T. B. Xu and Q. M. Zhang, Appl. Phys. Lett. **75**, 2653 (1999).
4. V. Bharti, H. S. Xu, G. Shanthi, Q. M. Zhang and K. Liang, J. Appl. Phys. **87**, (2000).
5. K. Tashiro, K. Takano, M. Kobayashi, Y. Chantani and H. Tadokoro, Ferroelectrics **57**, 297 (1984).
6. A. J. Lovinger, T. Furukawa, G. T. Davis and M. G. Broadhurst, Polymer **24**, 1233 (1983).
7. T. Yamada, T. Ueda and T. Kitayama, J. Appl. Phys. **52**, 948 (1981).
8. K. Kobayashi, K. Tashiro and H. Tadokoro, Macromolecules **8**, 158 (1975).
9. S. Osaki and Y. Ishida, J. Polym. Sci. : Polym. Phys. **13**, 1071 (1975).
10. K. J. Kuhn, B. Hahn, V. Percec, M. W. Urban, Appl. Spect. **41**, 843 (1987).

SYNTHESIS AND ELECTRIC PROPERTY OF
VDF/TrFE/HFP TERPOLYMERS

A. Petchsuk, T. C. Chung
Department of Material Science and Engineering, The Pennsylvania State University, University Park, PA 16802

ABSTRACT

This paper focuses on the molecular structure-electric property relationships of VDF/TrFE/HFP terpolymers, containing vinylidene difluoride (VDF), trifluoroethylene (TrFE) and hexafluoropropene (HFP) units. Several terpolymers were synthesized and evaluated with the corresponding VDF/TrFE copolymers. In general, a small amount of bulky HFP units in the polymer chain prevents the long sequence of crystallization and results in smaller ferroelectric (polar) micro-domains, which show the improved electric properties. The resulting terpolymer possesses a interesting combined properties with good processibility, low Curie transition temperature and high dielectric constant, narrow polarization hysteresis loop, and high strain response (2.5%) at relatively low electric field (50 MV/m).

INTRODUCTION

In the past decade, many research activities in ferroelectric fluorocarbon polymer have been focussing on VDF/TrFE copolymer with a general goal of reducing the energy barrier for ferroelectric-paraelectric (Curie) phase transition and generating large and fast electric-induced mechanical response at ambient temperature. Although VDF/TrFE copolymer exhibits a high piezoelectric and high pyroelectric constant [1], the response of the dipoles to the electric field is very slow at the ambient temperature, the polarization hysteresis loop of the copolymer is very large. The close connection between the crystalline structure and electric properties led to many attempts to alter copolymer morphology by mechanical deformation [2], electron-radiation [3], crystallization under high pressure [4], crystallization under high electric field [5], etc.

Zhang et al. [6] at Penn State University recently perfect the electron-radiation process with systematical study of the radiation conditions, such as dosage, temperature, inert atmosphere, stretching sample, etc. They revealed an exceptionally high electrostrictive response (> 4%) of the irradiated copolymer that behaves like a relaxor ferroelectric. The polarization hysteresis loop became very slim at room temperature, comparing with that of the sample before irradiation. However, the polarization was also significantly reduced and the sample became intractable because of the crosslinking. The increase of hardness of the copolymer sample was also revealed in its electric response, very high electric field was required (150 MV/m) for the irradiated sample to get a high strain response. It appears that the radiation process not only reduces the polar crystalline domain size but also involves some undesirable side reactions that increases the amorphous phase and diminishes the processibility of sample.

Another method of altering the morphology of VDF/TrFE copolymer is to introduce third (bulky) comonomer, such as hexafluoropropene (HFP), into the polymer chain. Freimuth et al. [7] studied VDF/TFE/HFP terpolymer. They concluded that the incorporation of HFP did not affect the crystalline structure, but strongly reduced the degree of crystallinity. The terpolymer exhibited thermal behavior similar to those observed in copolymer. Tajitsu et al. [8] reported the switching phenomena in VDF/TrFE/HFP terpolymers having a low HFP content (< 2.5 mol%).

Mat. Res. Soc. Symp. Proc. Vol. 600 © 2000 Materials Research Society

With the increase of HFP content, both polarization and dielectric constant decreased at the Curie temperature, and the switching time was very much independent on the HFP content.

Although there are experimental results on switching time, little information is available on the structure and the piezoelectric and pyroelectric properties of the VDF/TrFE/HFP terpolymers. It is very interesting to understand their structure-property relationship, especially the sample containing only a very small content of HFP termonomer (enough to reduce the size of polar crystalline domains, but not the overall crystallinity). The interested terpolymers have the VDF/TrFE mole ratio between 50/50 and 70/30, that provide high polarization at low Curie temperature (~ 60 °C). In this study, we will investigate their thermal and electrical properties, such as dielectric constant, polarization hysteresis loop, and electrostrictive response.

RESULTS AND DISCUSSION

Table 1 summarizes several VDF/TrFE/HFP terpolymers that were prepared by borane/O_2 or trichloroacetyl peroxide initiators at low temperature. Two commercial copolymers with VDF/TrFE mol% of 50/50, and 62/38 were also investigated for comparison.

Table 1 A Summary of VDF/TrFE/HFP Co- and Ter-polymers.

Run No.	Polymer Comp.(%)			Tm (°C)	ΔHm (J/g)	Tc (°C)	ΔHc (J/g)	Mw (x10^{-4})
	VDF	TrFE	HFP					
50/50	50	50	-	158.2	28.1	63.8	6.3	19.67
62/38	62	38	-	152.0	32.7	103.1	22.7	34.30
103	54.3	43.8	1.83	128.3	18.0	40.6	4.1	3.40
104	69.2	30.0	0.84	137.9	25.1	44.8	7.4	3.35
106	65.9	33.4	0.71	139.6	28.8	61.2	15.3	3.43
108	59.7	38.2	2.1	138.4	19.6	40.9	4.7	3.36
112	73.3	26.5	0.14	143.0	27.5	55.6	13.3	11.3
113	57.7	41.96	0.3	142.0	25.7	45.5	5.7	8.89
114	52.59	46.63	0.8	139.3	23.2	49.5	9.3	8.61
115	63.4	35.8	0.8	139.7	22.7	42.4	4.6	18.20
119	60.49	37.2	2.3	131.5	20.9	42.4	4.8	5.19
121	55.17	42.35	2.46	129.1	15.6	35.6	2.6	17.20

Thermal Properties

DSC measurement was used to investigate the thermal behaviors of the terpolymers. Figure 1 compares three terpolymers having VDF/(TrFE+HFP) mole ratio in the vicinity of 55/45 (run no. 114,103, 121) with the copolymer having VDF/TrFE 50/50 mole ratio. It is very clear that as the HFP molar content increased the Curie and melting temperatures of the terpolymer decreased. The Curie temperature decreased from 63 °C (commercial VDF/TrFE 50/50 copolymer) to 49.5, 40.6, 35.6°C (our VDF/TrFE/HFP terpolymers with 0.18, 1.8, 2.46 mol%HFP, respectively). The bulky HFP units clearly effect the crystalline domains. The smaller polar domains may result in the lower energy for the conformation transition, hence the Curie transition temperature decreases. In addition, the bulky groups may also reduce the intermolecular interaction between the polymer chains.

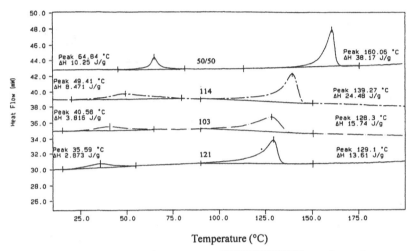

Figure 1. Comparison of DSC thermograms of VDF/TrFE/HFP terpolymer (run no. 114,103, 121) with the copolymer having VDF/TrFE 50/50 mole ratio.

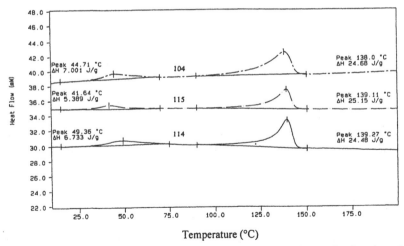

Figure 2. Comparison of DSC thermograms of VDF/TrFE/HFP terpolymers having about 0.8 mol% HFP and the TrFE content, 30 (104), 35.8(115), and 46.6 (114) mol%, respectively.

Figure 2 compares DSC curves of the terpolymers having about 0.8 mol% HFP and the TrFE content, 30, 35.8, 46.6 mol% respectively. Both Tm and Tc are not very sensitive to the TrFE contents. VDF and TrFE units in the terpolymer must co-crystallize well in the lattice, and the crystal structure and polar domain size maintains invariable despite the change of the VDF/TrFE mole ratio. Comparing with the corresponding VDF/TrFE copolymers, it appears that a small amount (< 1 mol%) of bulky HFP in VDF/TrFE/HFP terpolymers may only reduce the polar domain size, but not overall crystallinity and the magnitude of the dipole.

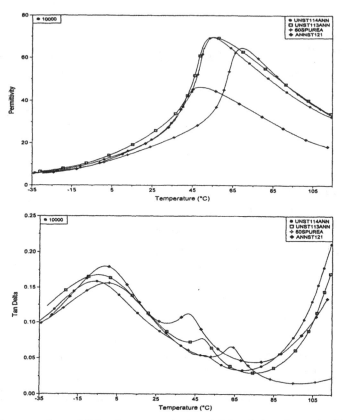

Figure 3. The comparison of dielectric constant (top) and dielectric loss (bottom) of three terpolymers with HFP content of, 0.3, 0.8, 2.46 mol%, designated as □, ●, ♦, respectively, and the commercial copolymer, P(VDF/TrFE) 55/45, ✧

Dielectric Properties

Figures 3 compares permittivity and dielectric loss at 10,000 Hz, respectively, of three unstretched and annealed VDF/TrFE/HFP terpolymers containing 0.3, 0.8, 2.46 mol% HFP (run no. 113, 114, 121)and a VDF/TrFE copolymer. All polymers have a similar VDF/TrFE mol ratio 55/45. The results clearly show a significant permittivity dependence on the HFP content. The correlated Tg (at γ relaxation) of the polymer, associated with the amorphous phase, almost no change in dielectric loss with small amount of HFP units. However, small quantity of HFP units in the crystalline domains has a large impact to the crystallization process of the polar domains. The Curie transition temperature shifts to the lower temperature as the HFP content increases. It is worth noting that the dielectric constant values of the terpolymers are about the same as comparing with copolymer. This result indicates the fact that the introduction of HFP units into polymer do not alter the conformation of the polymer chain (polymer chains still possess an all-trans ferroelectric conformation at the room temperature). Indeed, it facilitates the rotation of the dipoles upon heating by presumably breaking up the large polar domain size into microdomains. The small ferroelectric domain only need a small amount of energy to change the polymer conformation from polar to non-polar.

Ferroelectricity

Figure 4 compares the electric displacement vs. electric field of three terpolymers (run no. 115, 108, 119) with similar VDF/TrFE (~ 60/38) and different HFP content, 0.8, 2.1, 2.3 mol % HFP, respectively. As HFP content increases, the coercive field decreases as well as the polarization. For the terpolymer, the coercive field (Ec) required to change the direction of dipoles is much lower comparing with those of the commercial copolymer. Coercive field of the terpolymer is in the range of 20-30 MV/m while the coercive field of the commercial copolymer is in the range of 50-60 MV/m. This is presumably due to HFP termonomer breaks large polar domain size into smaller domain size, and allows reversal of the dipoles at lower field.

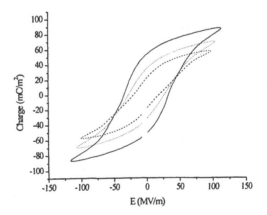

Figure 4. Comparison of electric displacement vs. electric field at 22°C of unstretched VDF/HFP/TrFE terpolymers with various HFP contents, 0.8, 2.1, 2.3 mol % HFP, designated as —, ..., ----, respectively.

It is interesting to note that polarization of terpolymer is also dependent on the relative amounts of α- and β-phase. [12] For some terpolymers with relatively low TrFE contents, the

57

remnant polarization of the stretched polymer film increases dramatically compared to that of unstretched film. Mechanical drawing clearly increases the relative amount of β-phase by increasing the amount of oriented dipoles resulting in an increase in the remnant polarization.

Figure 5 compares the coercive field and the polarization of VDF/TrFE/HFP (55.17/42.35/2.46) terpolymer 121 under various temperatures. As the temperature increased from 21°C to 42°C, the coercive filed reduce since the higher temperature facilitate the change of the direction of the dipoles. As the temperature increased to 50°C, the coercive field slightly increased arising from space charge trapped in samples. As the temperature increases further beyond the Curie transition temperature, the hysteresis loop obtained was very large with "rounding" at the ends indicative of conduction losses. The ability of switching the direction of the dipoles is reduced as the temperature goes beyond the phase transition temperature.

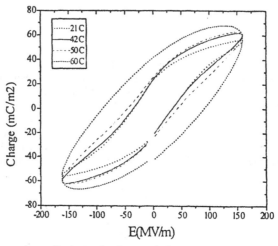

Figure 5. Comparison of polarization hysteresis loop of stretched VDF/TrFE/HFP (55.17/42.35/2.46) terpolymer 121 under various temperatures.

Piezoelectricity

One of the most desirable properties of the eletroactive polymer is to possess large mechanical deformation at low external field. Figure 6 shows the deformation of the VDF/TrFE/HFP (55.17/42.35/2.46) terpolymer under electric field at various temperatures. The deformation of the terpolymer has a highest value at 50°C, near the Curie temperature (40°C). As the temperature increases further, the deformation drops at low field. After the temperature increases beyond the Curie transition temperature, the polymer chains lose the piezoelectric properties. Consequently, the deformation drops.

It is very interesting to quantify the effect of HFP content to the strain. Figure 7 compares a deformation under electric field of stretched VDF/TrFE/HFP (55.17/42.35/2.46) terpolymer with that of the irradiated VDF/TrFE copolymer at Curie temperature. The deformation of the terpolymer indeed is higher than that of the irradiated sample at low field. Never before has the strain of almost 2.5% been achieved at 50 MV/m. It should be noted that the high strain response of the terpolymer was obtained at 50°C because the Curie temperature of the terpolymer is still above room temperature.

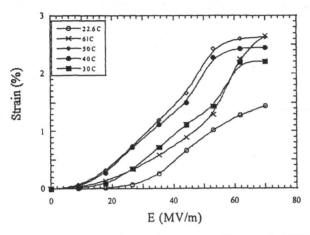

Figure 6. The dependence of the strain response of the stretched VDF/TrFE/HFP (55.17/42.35/2.46) terpolymer on the temperature.

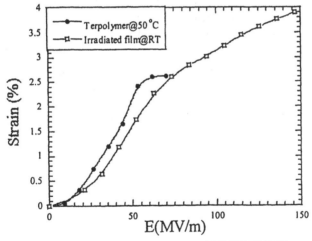

Figure 7. Comparison of the strains of stretched VDF/TrFE/HFP terpolymers measured at the Curie temperature with that of irradiated VDF/TrFE copolymer at room temperature* (*data are reproduced from ref. 6).

CONCLUSIONS

All the experimental results of VDF/TrFE/HFP terpolymers consistently show that a small amount of HFP units have a significant effect to the polymer morphology and electric properties. Few bulky groups effectively reduce the size of polar crystalline domains, but not the overall crystallinity and conformation of the polymer. The flexible dipoles in the smaller polar domains reflect in the piezoelectric property. A highest value of strain (about 2.5%) at very low field (50 MV/m) was observed in the terpolymer, which is significantly higher than that of the irradiated copolymer sample under the same conditions.

ACKNOWLEDGMENTS

The authors would like to thank Professor Q. M. Zhang for his technical assistance in the electric property studies. We gratefully acknowledge the financial supports of DARPA and Office of Naval Research.

REFERENCES

1. K. Koga, and H. Ohigashi, J. Appl. Phys., **59**, 2142 (1986).
2. K. Tashiro, S. Nishimura, and M. Kobayashi, Macromolecules, **21**, 2463 (1988), and **23**, 2802 (1990).
3. B. Daudin, and M. Dubus, J. Appl. Phys., **62**, 994 (1987).
4. T. Yuki, S. Ito, T. Koda, and S. Ikeda, Jpn. J. Appl. Phys., **37**, 5372 (1998).
5. S. Ikeda, H. Suzaki, and S. Nakami, Jpn. J. Appl. Phys., **31**, 1112 (1992).
6. Q. M. Zhang, V. Bharti, and X. Zhao, Science, **280**, 2102 (1998).
7. H. Freimuth, C. Sinn, and M. Dettenmaier, Polymer, **37**, 832 (1996).
8. Y. Tajitsu, A. Hirooka, A. Yamagish, and M. Date, Jpn. J. Appl. Phys., **36**, 6114 (1997).
9. T. C. Chung, and W. Janvikul, J. Organomet. Chem., **581**, 176 (1999).
10. J. E. Leffler, and H. H. Gibson, jr., J. Amer. Chem. Soc., **90**, 4117 (1968).
11. P. K. Isbester, J. L. Brandt, T. A. Kestner, and E. J. Munson, Macromolecules, 31, 8192 (1998).
12. B. A. Newman, C. H. Yoon, K. D. Pae, And J. I. Scheinbeim, J. Appl. Phys, **50**, 6095 (1979).

Giant Electrostrictive Response in Poly(vinylidene-fluoride hexafluoropropylene) copolymer

Xiaoyan Lu, Adriana Schirokauer and Jerry Scheinbeim
Dept. of Chemical and Biochemical Engineering,
Rutgers University, Piscataway, NJ 08855

ABSTRACT

In the present study, the strain response of a new class of copolymers of PVF_2 is investigated. Electrostrictive strains were measured in poly(vinylidene-fluoride hexafluoropropylene), $P(VF_2-HFP)$, using a capacitance method (air-gap capacitor), as a function of electric field. Three different thermal treatments (ice water quenched, air quenched and slow cooled) were given to samples of composition 5% and 15% HFP. Strains greater than 4 % were observed in the 5% HFP ice water quenched $P(VF_2-HFP)$ copolymer. Values of elastic modulus were lower for the quenched 5 % films than for the slow cooled ones, and in both cases they were higher than previously studied polyurethane elastomers and poly(vinylidene-fluoride trifluoroethylene) copolymers.
Key words: Electrostrictive Polymer, Poly(vinylidene-fluoride hexafluoropropylene) copolymer, Ferroelectic, X-ray, DSC.

INTRODUCTION

Field-induced electrostrictive strains can be observed in a material upon application of an electric field. These strains are known to be proportional to the square of the applied field. If the strains are high enough, materials having this property offer great promise in applications such as sensors, actuators, artificial muscle, robotics and MEMS.
Giant electrostrictive strains were first observed in a polyurethane elastomer [1]. The polyurethane exhibited strains greater than 3% under electric fields of up to 40 MV/m and an elastic modulus on the order of 0.01 GPa. Recently, strains up to 4% were observed in a copolymer of PVF_2, Poly(vinylidene fluoride/ trifluoroethylene), $[P(VF_2/TrFE)]$ upon application of electric fileds up to 150 MV/m. These films were subjected to a two step process in which the materials were first melt-pressed and slow cooled and then irradiated with a high energy electron beam [2]. The irradiated $P(VF_2/TrFE)$ films were observed to have an elastic modulus of approximately 0.4 GPa.
In the present study, the electrostrictive strains were measured in poly(vinylidenefluoride/hexafluoropropylene), $P(VF_2/HFP)$, random copolymers under electric fields up to 60 MV/m. The strain response of ice water quenched, air quenched and slow cooled samples was compared for 5% and 15% HFP copolymer compositions. The thickness strain constant, d_t, was then calculated from the strain vs. electric field curve. Dielectric constant of the materials, elastic modulus, d_{31} and e_{31} were studied, as well as the material's melting behavior and crystallinity observed from DSC and X-Ray Diffraction data.

Mat. Res. Soc. Symp. Proc. Vol. 600 © 2000 Materials Research Society

EXPERIMENTAL

Films of the 5% and 15% mol % HFP copolymers of PVF_2/HFP were prepared in a Carver Laboratory Press®. The copolymers were obtained in pellets from Soltex, and converted then into powder in a Freezer Mill. The powder was melted at 190 °C and pressed at 4,000 psi. For each composition, three different thermal treatments were used: ice water quench, air quench and slow cool. All samples were 50 to 60 µm in thickness and were cut into strips of 3 x 2 cm. Two strips of the same film were cut for each kind of sample. Gold electrodes 30 nm in thickness were deposited on the two sides of the films using a Sputter Coater EMS 650. The electrode area was 2.5 x 1.5 cm. The two films were placed on the sides of an air-gap capacitor, sandwiched between the capacitor plates (Fig.1). One electrode of each film was connected to the high voltage supply and the other electrode was grounded. Electrostrictive strains were measured as a function of dc electric field up to 60 MV/m using the air-gap capacitor.

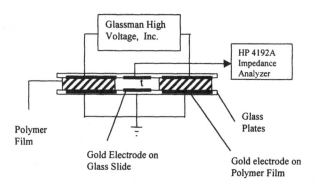

Fig. 1- Schematic of the air-gap capacitance system. Air-gap electrode area is 4.0 cm^2.

The capacitance of the air-gap, measured using an HP 4192A Impedance Analyzer, was then related to the strain change in the film. The measured capacitance is related to the air-gap thickness through equation 1.

$$t = \varepsilon_o \cdot A/C \qquad [1]$$

where ε_o is the permittivity of air, A is the air-gap electrode area and C is the measured capacitance. Given the arrangement of the polymer films sandwiched between the capacitor plates, the air-gap thickness is approximated to the film thickness, t. The percent strain of the film is then approximated to the percent change of the air-gap. Calibration of the system was first performed using a 1.00 pF standard capacitor. The precision of the measurement was 2%. The accuracy of the capacitance system was also confirmed using the previously tested polyurethane films. The results obtained with this system matched the ones previously obtained [1]. For a given composition, the response of ice water quenched, air quenched and slow cooled samples were compared. Elastic modulus and relative dielectric constant were measured at 100 Hz and room temperature, as well as a function of frequency (ranging from 0.01 to 10,000 Hz) and temperature (ranging from -80°C to 100°C) using a Rheolograph Solid® Toyoseiki, Japan.

X-Ray Diffraction patterns were obtained using a Diffraktometer Kristalloflex® Siemens. The melting behavior of the material was studied using a Differential Scanning Calorimeter (DSC), TA Instruments®. Values of remanent polarization and coercive field were obtained by poling the films using a high voltage source Trek® Model 610C and Keithley® 617 electrometer and 195A digital multimeter, and the aid of computer software.

RESULTS AND DISCUSSION

Both P(VF$_2$-HFP) copolymers of 5 % and 15 % mol % HFP content exhibited the same behavior with respect to thermal treatment; slow cooled and air quenched samples showed the smallest response (strains $\approx$ 1 %), whereas the ice water quenched samples exhibited the largest response in both, the 5 % and the 15% HFP compositions (Fig. 2). Strains greater than 4 % were observed in the 5% HFP ice water quenched P(VF$_2$-HFP) copolymer, the largest response observed among all of the samples. Following the same behavior, 15 % HFP ice water quenched films showed the largest strains ($\approx$ 3%) among the samples of the same composition. All curves were proportional to the square of the electric field until saturation. The ice water quenched 5% HFP copolymer also shows the highest thickness strain coefficient, d_t, among all of the samples (Fig. 3). The thickness strain coefficient increased linearly with increasing electric field up to the beginning of saturation indicating electrostrictive response. The d_t value for the 5% HFP ice water quenched copolymer was approximately 16.5 Å/V. The value of the electrostrictive constant, M_{33}, was 5.5×10^{-17} m^2/V^2. Both values are larger than the ones observed in previously studied polyurethane elastomers and P(VF$_2$-TrFE) copolymer.

To gain an understanding of the structure of these copolymers and the differences between them that can be accounted for in terms of their strain responses, change in dielectric constant with temperature and frequency, DSC scans and X-Ray diffraction were performed. Figure 4 shows the change in dielectric constant with temperature. Both 5% and 15% HFP ice water quenched copolymers showed an increase in dielectric constant with increasing temperature with a pattern similar to that of PVF$_2$ homopolymer. The dielectric loss curves showed the glass transition temperature peak at around -40 °C for the 5% HFP ice water quenched copolymer. This peak shifts to higher temperatures as the HFP content is increased as shown in the curve for the 15% HFP ice water quenched copolymer with glass transition temperature at around -20°C. The dielectric constant frequency scans showed a decrease in dielectric constant with frequency. The change in dielectric constant with frequency is also similar to that of PVF$_2$ homopolymer. No Curie transition temperature is observed on the dielectric constant-temperature scan.

The X-Ray diffraction patterns of the 5% and 15% ice water quenched samples did not lead to a characterization of the crystal structure of these copolymers. It was known

63

from previous work [3] that PVF$_2$ homopolymers exhibit two phases, phase I, a non-polar phase and phase II which is a polar ferroelectric phase. To gain more information on

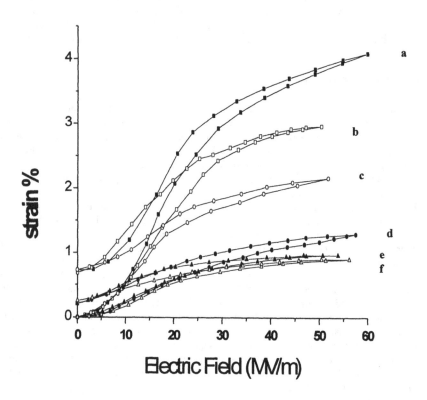

Fig. 2- Strain response of PVF2/HFP copolymers with applied electric field. a) ice water quenched PVF$_2$/HFP 5%, b) ice water quenched PVF$_2$/HFP15%, c) air quenched PVF$_2$/HFP 15%, d) air quenched PVF$_2$/HFP 5%, e) slow cooled PVF$_2$/HFP 5% and f) slow cooled PVF$_2$/HFP15%.

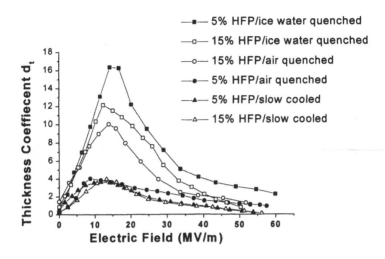

Fig. 3- Thickness strain constant, d_t, as a function of electric field.

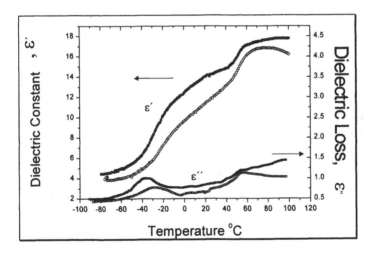

Fig. 4- Dielectric constant vs. temperature for ice water quenched samples, 5% HFP (●) and 15% HFP (o). The frequency of measurement is 104 Hz.

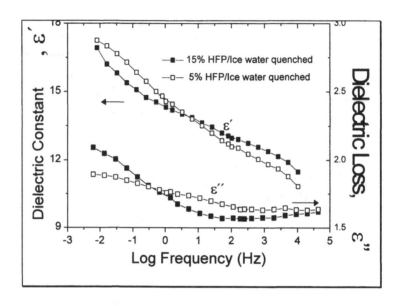

Fig. 5- Dielectric constant vs. frequency for ice water quenched samples
measured at room temperature.

ferroelectric behavior of the 5% and 15% HFP ice water quenched copolymer a poling experiment was performed in the range of -150 MV/m to 150 MV/m. At the higher electric fields, the 5% and 15% HFP copolymers showed ferroelectric behavior. The ferroelectric switching occurred at electric fields that are higher than the range used for strain measurements (electrostrictive strains were measured up to 60 MV/m). Figure 6 shows the electric displacement, D, vs. electric field curves for the ice water quenched 5% and 15% HFP copolymers. The remanent polarization, Pr, decreased about 20% from the 5% HFP copolymer ($\approx$ 60 mC/m^2) to the 15% HFP copolymer ($\approx$ 48 mC/m^2) due to the decrease in crystallinity when adding more HFP. For all other samples the remanent polarization was negligible.The previous result suggests that the material exhibits ferroelectric behavior at higher electric fields, however no conclusion can be drawn as to weather or not this property plays a role in the mechanism of electrostriction.

The crystallinity of the material was studied using Differential Scanning Calorimeter (DSC) to analyze the melting behavior and X-Ray Diffraction (WAXD) to find out changes in crystal structure. Fig 7 shows the increase in crystallinity as well as the crystal form from before and after the application of high field ($\approx$ 50 MV/m). Also as suggested by the enthalpy values, crystallinty decreases with increasing HFP content in the copolymer, which is due to the bulky size of the CH_3 groups.

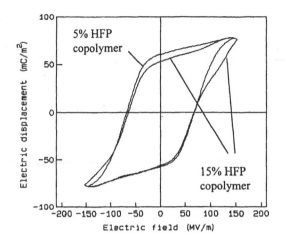

Fig. 6- Electric Displacement vs. Electric Field for ice water quenched 5% and 15% mol % HFP copolymers.

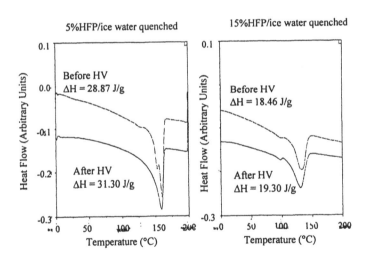

Fig. 7- DSC Thermograms of ice water quenched 5% and 15% HFP, P(VDF-HFP) samples before and after application of high voltage (HV).

This observation was confirmed by the X-Ray diffraction patterns (WAXD) of the samples. (Figure 8). For 5% HFP/ice water quenched sample, both curves before and after application of field can be matched to the pattern of PVF$_2$ polar phase II [3], showing the (010) peak at around $2\theta = 19$ and (110) peak at around $2\theta = 20$. For 15% HFP/ice water quenched sample, the diffraction pattern is unsolved. Further characterization of these crystal structures will be performed using IR spectroscopy.

Values of elastic modulus and dielectric constants for both samples before and after application of field were measured and tabulated in tables 1 and 2. Values of elastic modulus for both 5% and 15% HFP ice water quenched samples were higher than the previously studied polyurethane elastomers (elastic modulus = 0.068 GPa), [1], and poly(vinylidene-fluoride trifluoroethylene) copolymers (0.38 GPa), [2]. The energy denisty of 5% HFP ice water quenched sample is also much higher compared to previously studied polyurethane elastomers and high energy electron irradiated electrostrictive poly(vinylidene-fluoride trifluoroethylene) copolymers .

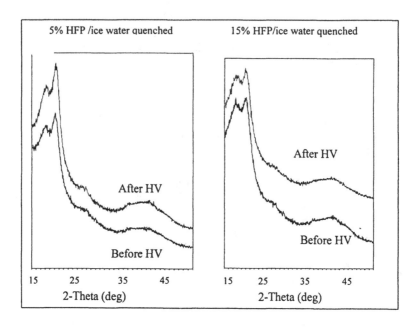

Fig. 8- X- Ray diffraction patterns of ice water quenched samples before and after application of high voltage.

Material P(VDF-HFP)	Y (GPa)	ε_r	$YS^2/2$ (J/cm^3)
5% HFP/Ice Water Quenched	0.88	13.5	0.7
15% HFP/Ice Water Quenched	0.55	12.7	0.25

Table 1. Elastic modulus ,dielectric constant and energy density for ice water quenched P(VDF-HFP) copolymer with 5 and 15% HFP before poling .

Material P(VDF-HFP)	Y (GPa)	ε_r	$YS^2/2$ (J/cm^3)
5% HFP/Ice Water Quenched	1.10	13.3	0.88
15% HFP/Ice Water Quenched	0.50	13.3	0.23

Table 2. Elastic modulus , dielectric constant and energy density for ice water quenched P(VDF-HFP) copolymer with 5 and 15% HFP after poling .

Further research will investigate other structure property relationships in these copolymers and will be directed towards gaining an understanding of the molecular mechanisms responsible for the observed electrostrictive response.

REFERENCES

[1] Ma, Z, J. Scheinbeim, J. Lee and B. Newman. High Field Electrostrictive Response of Polymers, Journal of Polymer Science: Part B: Polymer Physics, **32**, 2721 (1994).

[2] Zhang, Q. M, V. Bharti and X. Zhao. Giant Electrostriction and Relaxor Ferroelectric Behavior in Electron-Irradiated Poly(vinylidene fluoride-trifluoroethylene) Copolymer, Science, **280**, 2101 (1998).

[3] B. A. Newman and J. Scheinbeim. Polarization Mechanism in phase II Poly(vinylidene fluoride) films, Macromolecules, **16**, 60 (1983).

ELECTRO-MECHANICAL PROPERTIES OF ELECTRON IRRADIATED P(VDF-TrFE) COPOLYMERS UNDER DIFFERENT MECHANICAL STRESSES

Z.-Y. Cheng,* S.J. Gross, V. Bharti, T.-B. Xu, T. Mai and Q. M. Zhang
Materials Research Laboratory, The Pennsylvania State University
University Park, PA 16802 (*E-mail: zxc7@psu.edu)

ABSTRACT

The electro-mechanical properties of high energy electron irradiated poly(vineylidene fluoride-trifluorethylene) (P(VF-TrFE)) copolymers under different mechanical stress conditions are reported. In stress free condition, the electric field induced longitudinal and transverse strains of the irradiated P(VDF-TrFE) copolymer films at room temperature (RT) can reach about 5% and more than 3% respectively. The longitudinal strain response of the material under hydrostatic a pressure up to 83 atmospheses was studied at RT. The transverse strain response of the material at RT was studied under uniaxial tensile stress. It was found that the material has a high load capability and for stretched films along the stretching direction the induced strain remains high, even at about 45 MPa. The temperature dependence of the strain re onse is also characterized. Both the temperature and stress dependence of the strain response indicate that the electric field induced strain response originates from the electric field induced local phase transition.

INTRODUCTION

Electroactive materials have been used in numerous applications, such as robotics, active damping, vibration control and isolation, manipulation, ultrasonic transducers for medical diagnosis and sonar. However, most of the activities in the area of electroactive materials was focused on inorganic ceramics/crystals [1]. Not much attention was paid to electroactive polymer (EAP) due to the small number of available materials, as well as their limited actuation capability.

For electro-mechanical applications, a material exhibiting a high electric field induced strain response is always highly desirable. Recently, a very high electric field induced strain response was observed in many EAP systems [2], such as heavily plasticized poly(vineylidene fluoride-trifluorethylene [P(VDF-TrFE)] [3], polyurethane and silicone [3], [4]. Very recently, we reported that high energy electron irradiated P(VDF-TrFE) copolymer film exhibits a very large electric field induced strain response [5], [6]. The irradiated P(VF-TrFE) copolymers possesses a very high elastic energy density besides the high strain response since the material is of relative high elastic modulus compared to other EAP [5], [6]. In addition, these polymers have many advantages over ceramics/crystals, such as low cost, flexible, easy to process and shape. All these make the electroactive polymer (EAP) very attractive for a broad range of electromechanical applications [2].

In most electromechanical applications, the material is under some mechanical load condition. Thus, for a soft polymeric material, the electromechanical properties of the material under different stresses conditions or the load capability of the material is always an important concern. In this paper, the studies of the electric field induced strain response of irradiated P(VDF-TrFE) copolymer under different mechanical boundary conditions are reported. The results show that the material has high load capability.

Mat. Res. Soc. Symp. Proc. Vol. 600 © 2000 Materials Research Society

EXPERIMENT

P(VDF-TrFE) copolymer powders were purchased from Solvay and Cie, Bruxelles, Belgium. The compositions used here are 50/50, 65/35, and 68/32 mol%. The copolymer films studied here were prepared using two approaches: melt press and solution cast. In the melt press process, the copolymer powders were pressed between two pieces of aluminum foil at 215 to 240 °C. In the solution cast method, the copolymer powders were dissolved in dimethyl formanide (DMF) and then the solution was cast on a flat glass plate and dried in an oven at 70 °C. Two types of films, the stretched and unstretched films, were studied in this work. For the unstretched films, the films prepared above were annealed at 140 °C for a time period between 12 to 14 hrs to improve the crystallinity. In order to prepare the stretched films, the films made from solution cast or quenched from melt press were uniaxially stretched up to 5 times at a temperature between 25 to 50 °C. The stretched films were then annealed at 140 °C for 12 to 14 hrs to increase the crystallinity. During the annealing process, the two ends of the stretched films were mechanically fixed. The crystallinity of the final films is about 75%. The thickness of the films was in the range from 15 to 30 μm. The electron irradiation was carried out in nitrogen atmosphere at different temperatures with an electron energy of 2.55 MeV, 1.2 MeV, and 1.0 Mev.

In order to characterize the electric filed induced strain response, gold electrodes of a thickness of about 40 nm were sputtered on both surfaces. The longitudinal strain response of the film under a stress free condition was characterized using a specially designed strain sensor based on the piezoelectric bimorph [7]. The longitudinal strain response of the films under hydrostatic pressure was characterized using a specially designed set-up [8]. The transverse strain of the films under different tensile conditions was characterized using a cantilever-based dilatometer that was newly developed for characterizing transverse strain response of polymeric film [9].

RESULTS AND DISCUSSION

Strain Response of Films under Stress Free Condition

The electric field induced strain responses of some irradiated copolymer films at RT are shown in Fig. 1. A sinusoidal electric field with a frequency of 1 Hz was applied and the strain was measured at 2 Hz using a lock-in amplifier since the material is electrostrictive. The longitudinal strain is a result of the thickness change of the film, while the transverse strain is related to the length change of the film perpendicular to the applied field direction. The transverse data reported here was measured along the stretch direction of the stretched samples. Clearly, the material exhibits a very high strain response. It should be noticed that the longitudinal strain is always negative. For unstretched films, the transverse strain, which is about 1/3 of the absolute value of the longitudinal strain, is always positive. However, for the stretched sample, it is found that the transverse strain along the stretch direction, which is comparable with the strain level of the longitudinal strain, is positive, while the transverse strain along the direction perpendicular to the stretch direction, which is about or less than 1/3 of the longitudinal strain and is negative. That is, the volume strain of the irradiated copolymer is also very high [6]. It should also be mentioned that in many electroactive materials, such as polyurethane, silicone, and piezoelectric ceramics, the volume strain is very small [1], [4]. The large volume strain generated by the irradiated copolymer indicates that the material also has a potential for hydrostatic applications.

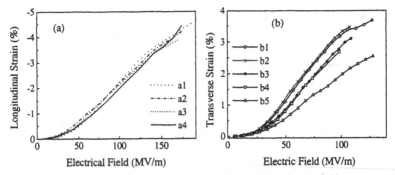

Fig. 1. Amplitude of the strain response vs. amplitude of the electric field.
(a). Longitudinal strain of samples irradiated with 2.55 MeV electrons. Curves a1, a2, a3 and a4 are the stretched 65/35 irradiated at 120 °C with 80 Mrad dose, unstretched 65/35 irradiated at RT with 100 Mrad dose, stretched 50/50 irradiated at 77 °C with 60Mrad dose, and unstretched 50/50 irradiated at RT with 100 Mrad dose.
(b). Transverse strain of stretched samples. Curves b1 and b2, are 68/32 film irradiated at 100 °C with 65 Mrad and 70 Mrad doses respectively using 1.2 MeV electrons. Curve b3 is 65/35 irradiated at 105 °C with 70 Mrad dose using 1.0 MeV electrons. Curves b4 and b5 are 65/35 irradiated at 95 °C with 60 Mrad dose and at 77 °C with 80 Mrad dose respectively using 2.55 MeV electrons.

Longitudinal Strain of Material under Hydrostatic Pressure

The hydrostatic pressure dependence of the longitudinal strain of an unstretched 65/35 copolymer film, irradiated at 95 °C with 60 Mrad dose using 2.55 MeV electrons, is shown in Fig. 2. The pressure range applied here is limited by the experimental set-up. Even at high pressure, the material can still produce the high strain response. With regard to the relation between the strain response and pressure, it is found that the strain does not change much with the pressure at low driving electric field and that the strain response shows an increase with the pressure at high electric field.

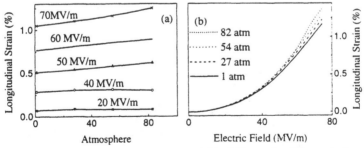

Fig. 2. (a). Longitudinal strain vs. electric field of film under different hydrostatic pressures. (b). Longitudinal strain vs. hydrostatic pressure for the material under different electric fields. The electric field was 1 Hz sine wave.

Transverse Strain of Stretched Film under Uniaxial Tensile

The transverse strain along stretch direction of a stretched 65/35 copolymer film, irradiated at 95 °C with 60Mrad dose using 2.55 MeV electrons, under different tensile loads along the stretch direction is shown in Fig. 3. It was found that the film broke down after working at the tensile stress of 45 MPa for a while. That is, the elastic strength of the film is about 45 MPa which is comparable to that of commercial nylon. Even at the tensile stress of 45 MPa, the film still exhibits a very large strain response. These results indicate that the materials studied here have a high load capability. This is a great advantage of the irradiated copolymer over the other EAP. The data in Fig. 3 clearly shows that the transverse strain response increases with tensile stress at the beginning and then decreases with the stress. There is a tensile stress (S_M) at which the strain exhibits a maximum for the film under a constant electric field. This tensile stress, S_M, is strongly dependent on the electric field. The higher the electric field, the lower S_M.

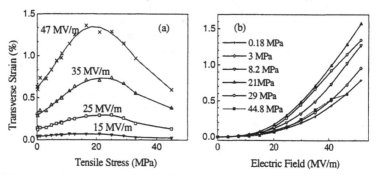

Fig. 3. (a). Transverse strain amplitude vs. tensile for the film under different electric fields. (b). Transverse strain amplitude vs. electric field amplitude for the film under different tensile. The electric field is 1 Hz since wave.

Temperature Dependence of Strain Response

In order to understand the stress dependence of the electric field induced strain response and to determine the working temperature range for the material, the temperature dependence of the strain response was measured. Two typical results are shown in Fig. 4. In both longitudinal and transverse strain measurements, the temperature range used in this investigation was limited by the set-up.

From the data in Fig. 4(b) one can see that the strain response decreases with increasing temperature beginning at RT. When decreasing the temperature, the strain response increases at the beginning and then decreases. There is a temperature (T_m) at which the strain response exhibits a maximum for the film under a constant electric field. It seems that the T_m increases slightly with the electric field. The data in Fig. 4(a) show a similar behavior. However, since the strain at temperatures lower than RT can't be measured, the Tm can't be determined here. However, the trend is similar to that in Fig. 4(b).

The temperature dependence of the electric field induced strain reported here can be easily understood by considering the fact that the irradiated copolymer is relaxor ferroelectrics [5], [9], [10]. For relaxor ferroelectrics, the electric field induced strain is related to the electric field induced breathing of the polar regions [11], [12]. At high temperatures, the concentration of the polar region increases with decreasing temperature. The electric field induced strain therefore increases with decreasing temperature. At low temperatures, a freezing of the polar region happens. With decreasing temperature, an increased amount of the frozen polar regions results in the decrease of the electric field induced strain response. Therefore, the T_m is related to the freezing process of the polar regions. However, the electric field has some influence on the freezing process of the polar region. That is why the strain maximum temperature changes slightly with the driving electric field.

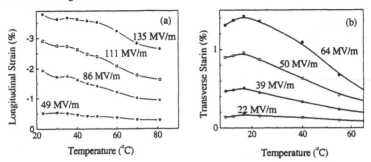

Fig. 4. Temperature dependence of the strain response for the films under a constant electric field. (a). Longitudinal strain of the film characterized in Fig. 2. (b) Transverse strain of the film studied in Fig. 3.

Relationship Between Stress Effect and Temperature Effect

For unirradiated copolymers, there is a Curie temperature, at temperatures below which the material is the ferroelectric phase. After irradiation, the material is transferred from a normal ferroelectric to a relaxor ferroelectric. Although the relaxor ferroelectric does not have Curie temperature, there are polar regions whose behavior is very similar to ferroelectrics. Thus, we can discuss approximately the relaxor ferroelectric using the Devonshire phenomenological theory of the ferroelectric. For a normal ferroelectric, the Curie temperature changes with external stress as following [9], [13]:

$$\Delta T = 2\varepsilon_0 C Q_{i3} X_i \qquad (1)$$

Where ΔT is the change of Curie temperature corresponding to the stress (X_i). Both C and Q_{i3} are the material properties, the Curie-Weiss constant and the electrostrictive coefficients. ε_0=is the vacuum permittivity. For the material under a hydrostatic pressure (negative stress), the stress and Q in Eq. (1) are the pressure and $Q_h(=Q_{33}+Q_{13}+Q_{23})$ respectively. For the irradiated copolymer studied here, it has been reported that $Q_{13}>0$, $Q_{33}<0$, and $Q_h=<0$ [6]. Thus, in both the hydrostatic pressure and the uniaxial tensile stress (positive stress), the Curie temperature increase with increase pressure or tensile.

Now we discuss the stress effect on the strain for the material shown in Fig. 3 at a constant electric field. With an increasing tensile stress, the T_m shifts toward RT, at first due to T_m being lower than RT. This will result in the strain response of the material at RT increasing with tensile stress at the beginning. If the tensile stress increases further, the T_m will be higher than RT. In this case, the increase of tensile stress will result in the T_m deviating away from RT. Thus, the electric field induced strain response of the film at RT will decrease with increasing tensile stress. These observations are consistent with the experimental data shown in Fig. 3(a).

When the electric field changes, as shown in Fig. 4, the T_m increases with increasing tensile stress. In a stress free condition, the higher the electric filed, the closer the T_m is to RT. Therefore, the higher the electric field, the lower the stress needed to shift the T_m to RT. That is, the higher the electric field, the smaller the S_M. This is consistent with the data shown in Fig. 3(a).

All the above discussions indicate that the Devonshire theory can be used to explain the stress effect on the strain response of the irradiated copolymer films studied here. The consistence of these results confirms again that: 1) the ferroelectric nature of the irradiated copolymer studied here; 2) the strain response is mainly from the electric field induced phase transition of the local polar region.

SUMMARY

In summary, high energy (MeV) electron irradiated P(VDF-TrFE) copolymer films have very large electric field induced strain response (about 5% longitudinal strain and more than 3% transverse strain). Compared to the other electroactive materials, a great advantage of the irradiated copolymers is that the material has a very high volume strain, which makes the material very well suitable for hydrostatic applications. The strain response of the material under different stress conditions shows that the irradiated copolymer films have very high load capability. The temperature dependence of the strain response was presented. The relationship between the stress dependence and temperature dependence of the strain responses is discussed with the Devonshire phenomenological theory. The consistence between the Devonshire theory and experimental data indicates that the strain response obtained in the irradiated copolymer originates from the electric field induced phase transition of the local polar regions.

ACKNOWLEDGEMENTS

The work was supported by ONR through Grant No. N00014-97-1-0900, NSF through Grant No. ECS-9710459, and DARPA through Grant No. N00173-99-C-2003.

REFERENCES

1. L. E. Cross, Ceramic Trans. **68**, 15(1996).
2. Y. Bar-Cohen, *Elelctroactive Polymer Actuators and Devices*, Proc. of SPIE, Vol.3669(1999).
3. Z. Ma, J. I. Scheinbein and B. A. Newman, J. Polym. Sci. B, Polym. Phys. **32**, 2721 (1994).
4. R. E. Pelrine, R. D. Kornbluh and J. P. Joseph, Sens. Actuators A, **64**, 77(1998).
5. Q. M. Zhang, V. Bharti and X. Zhao, Science **280**, 2101 (1998).
6. Z. -Y. Cheng, T. -B. Xu, V. Bharti, S. Wang and Q. M. Zhang, Appl. Phys. Lett. **74**, 1901(1999).
7. J. Su, P. Mouse and Q.M. Zhang, Rev. Sci. Instrum. **69**, 2480 (1998).
8. S. Gross, M.S. thesis, The Pennsylvania State University, 1999.
9. Z.-Y. Cheng, V. Bharti, T.-B. Xu, S. Wang and Q.M. Zhang, J. Appl. Phys. **86**, 2208 (1999).
10. V. Bharti, X. Zhao, Q.M. Zhang, T. Ramotowski, F. Tito and R. Ting, Mater. Res. Innovat. **2**, 57 (1998).
11. Q.M. Zhang and J. Zhao, Appl. Phys. Lett. **71**, 1649 (1997).
12. Z.-Y. Cheng, R.S. Katiyar, X. Yao and A.S. Bhalla, Phys. Rev. **B 57**, 8166 (1998).
13. M.E. Line and A. M. Glass, *Principles and Applications of Ferroelectrics and Related Materials* (Oxford University Press, New York, 1977).

ELECTRON IRRADIATED p(VDF-TrFE) COPOLYMERS FOR USE IN NAVAL TRANSDUCER APPLICATIONS

T. RAMOTOWSKI*, K. HAMILTON*, G. KAVARNOS**, Q. ZHANG***, V. BHARTI***
*Transduction Materials Branch, Naval Undersea Warfare Center, Newport, RI 02841
**EG&G, Inc., Groton, CT 06340
***Materials Research Laboratory, Pennsylvania State University, University Park, PA 16802

ABSTRACT

Recently, it was discovered that the transduction / strain capability of p(VDF-TrFE) copolymers can be enhanced by more than an order of magnitude by irradiating the copolymer with large doses (40-300 Mrads) of β-particles (1-3 MeV electrons). The goal of this project is to understand how irradiation improves the electromechanical properties of p(VDF-TrFE) copolymers while simultaneously identifying the secondary effects of the irradiation process on the material (not critical to the electromechanical properties) and attempting to separate the two contributions. It has been found that β-irradiation affects the material profoundly in several different manners. Reduction in the melt temperature, degree of crystallinity, and the resulting crystal quality have been observed for increasing doses of β-particles. Similar results have been observed for the Curie transition, especially in the energy associated with, and the breadth of, the transition. In addition, thermogravimetric analysis indicates that irradiation causes both chain scission and network polymer formation. Solid-state NMR results are discussed in reference to postulated dehalogenation, dehydrohalogenation, and olefinic bond formation activities.

INTRODUCTION

Ferroelectric polyvinylidene fluoride p(VDF) and vinylidene fluoride-trifluoroethylene copolymers p(VDF-TrFE) have earned a niche as choice transducer materials for undersea sensors and projectors[1]. This interest is in part due to the lighter weight, greater conformability and flexibility, and easier fabrication of these electroactive polymers as compared to lead-based ceramics that have been the traditional materials of choice in Navy sonar applications. Still piezoelectric polymers such as p(VDF) and p(VDF-TrFE) suffer from relatively low energy densities and low coupling. Nonetheless, recent research at Penn State University has provided compelling evidence that p(VDF-TrFE) copolymer films can be converted into very high strain and high energy density materials by electron irradiation treatment[2]. The electron irradiation is thought to convert the normal ferroelectric copolymer into a relaxor ferroelectric exhibiting electrostrictive strains exceeding 4%. The discovery of so-called "giant" electrostrictive strains in p(VDF-TrFE) copolymers has renewed interest for the use of these materials in a variety of applications such as transducers, actuators, and sensors.

To fully achieve the potential of electrostrictive P(VDF-TrFE) for these applications, it is desirable to understand the effects of irradiation on the copolymer structure and crystallinity and the consequences of these changes on the electromechanical properties. With a knowledge of these structural and morphological changes, the mechanisms by which electron irradiation transforms the copolymer into a high-strain relaxor can be identified and used to establish structure-property relationships as well as an enhancement in electromechanical properties.

EXPERIMENT

P(VDF-TrFE) copolymer powder, 65 mol% VDF and 35 mol% TrFE was obtained from Solvay and Cie (Brussels, Belgium). The mean molecular weight was 200,000. Films were fabricated by solvent-casting from methyl ethyl ketone, followed by annealing for one hour at 140° C and cooling to room temperature. The film thicknesses ranged from about 5 μm to over 200 μm. The p(VDF-TrFE) films were irradiated with 1.2 and 2.55-MeV electrons in a nitrogen atmosphere at selected temperatures. The total absorbed doses were from 20 to 200 Mrads. The electron irradiation experiments were performed at the Massachusetts Institute of Technology's (MIT's) High Voltage Research Laboratory in Cambridge, MA.

Differential scanning calorimetry (DSC) measurements were performed on the films before and after irradiation using a TA Instrument Model 2910 DSC. Thermogravimetric analysis (TGA) data were collected using a TA Instruments model 2950 TGA. Fourier transform infrared (FTIR) spectra were recorded at room temperature using a Nicolet model 510 Fourier transform infrared spectrometer. Ultraviolet (UV) absorption spectra of the films were obtained with a Perkin-Elmer Lambda 14 ultraviolet-visible spectrometer. Nuclear magnetic resonance (NMR) experiments were performed under the direction of Professor Karen Gleason of MIT's Department of

77

Chemical Engineering using a homebuilt spectrometer tuned to the ¹⁹F Larmor frequency of 254.0 MHz. A Chemagnetics 3.2 mm variable temperature solids probe capable of a maximum spinning speed of 25 kHz was used to achieve high resolution magic angle spinning (MAS).

RESULTS

The DSC thermogram (heating and cooling cycles) of a p(VDF-TrFE) film previously annealed for 1 hour at 140° C before irradiation is shown in figure 1.

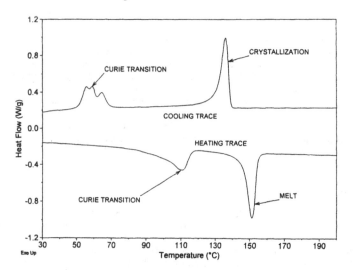

The thermogram exhibits a Curie transition at about 100° C that represents the complex solid-solid transformation from an ordered ferroelectric β-phase into a more disordered, paraelectric α-phase[3]. The ferroelectric β-phase consists of ordered, all-*trans* monomer units where the fluorine atoms are arranged on one side of the packed polymer chains, giving rise to a large permanent dipole[4]. In the paraelectric α-phase, the introduction of *gauche* bonds forces the carbon-fluorine dipoles to rotate so that the net polarization of the packed crystal cells is practically zero. Figure 2 shows typical DSC traces of p(VDF-TrFE) when subjected to electron irradiation with 1.2 MeV electron energy.

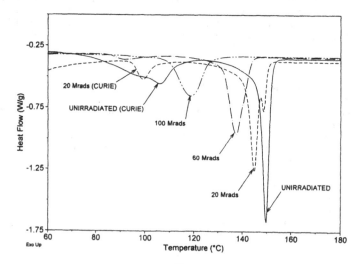

What is particularly striking in figure 2 is the disappearance of the Curie transition and the broadening and shifting to lower temperature of the melt endotherm. The disappearance of the Curie transition on irradiation is consistent with the loss of ferroelectricity in the irradiated films. The loss of ferroelectricity suggests also the loss of long-range ordered crystal domains due to the formation of structural defects. Interestingly, the sample irradiated with the 20 Mrad dose still displays a vestige of the original melt endotherm.

Figure 3 shows TGA weight loss curves for an unirradiated and several irradiated films.

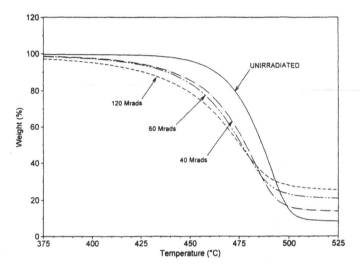

Initially, the irradiated samples are less thermally stable, suggesting chain scission into lower molecular weight (and hence, more volatile) polymer chain fragments. At higher temperatures, the irradiated samples degraded to a lesser extent than the unirradiated sample. This observation suggests the presence of a higher crosslink density in the irradiated samples. Thus, on the basis of the TGA evidence it appears that chain scission is accompanied by formation of crosslinks during irradiation.

Ultraviolet spectra of an unirradiated and several irradiated are shown in figure 4. The unirradiated film shows a slight rise in absorption in the region from 200 nm to 360 nm. Irradiation, however, produces a broad, but distinct absorption in this region. As the irradiation dose increases, this absorption becomes more intense. λ_{max} occurs below 200 nm and may represent a band due to an isolated carbon-carbon double bond. The shoulder at ~230 nm can be tentatively assigned to absorption by conjugated double bonds or allyl radicals. The formation of unsaturated bonds may be due to irradiation-induced dehydrohalogenation.

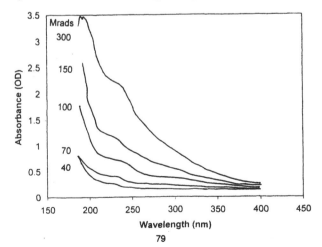

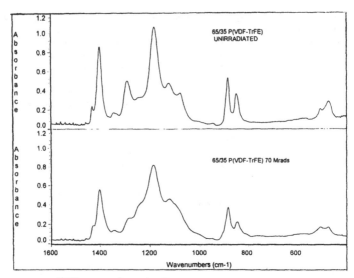

The room-temperature FTIR spectra of an unirradiated and an irradiated film are shown in figure 5. The peaks at 1420, 1290, 884, and 848 cm^{-1} are assigned to symmetric C-F vibrations in *trans* sequences[5]. The shoulders at 1250 and 811 cm^{-1} result from C-F stretching vibrations in a *gauche* sequence. The absorbance ratio A_{811}/A_{848} displays a discernible increase, and the *trans* peaks at 1290 and 884 cm^{-1} undergo a reduction in intensity upon electron irradiation. In the irradiated films, the reduction of the peaks at 1420, 1290, 1180, 884, and 848 cm^{-1} suggests the irreversible formation of *gauche* bonding. Experiments were performed to establish that the increased *gauche* bonding was not caused by heating the films to the temperature at which the electron-irradiations were performed. Given the absence of clear evidence of double bond absorption in the FTIR spectra, the support for formation of double bonds during electron irradiation (as suggested by the UV spectra) can be regarded at best as only tentative.

^{19}F-NMR was chosen to evaluate the p(VDF-TrFE) films in view of the 100% isotopic abundance and high gyromagnetic ratio of the ^{19}F nucleus.[6] That the isotropic chemical shift range for ^{19}F is about 20 times broader than ^{1}H enhances the ability to resolve differences in chemical bonding configurations. Magic angle sample spinning reduces homonuclear dipolar coupling and chemical shift anisotropy that broaden solid-state spectra. High resolution ^{19}F spectra were obtained by using spinning speeds exceeding 25 kHz. In addition, the spectra were collected at elevated temperatures to achieve better resolution than is possible at ambient temperatures. At elevated temperatures, increased molecular mobility results in additional motional averaging.

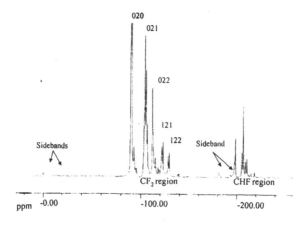

Figure 6 shows the ^{19}F-MAS NMR spectra of unirradiated p(VDF-TrFE) obtained at a spinning speed of 23 kHz and 170° C. There are two main regions: peaks due to fluorine in CF_2 groups (region 1) and peaks due to fluorine in CHF groups (region 2). Since the peaks in region 2 can only result from TrFE units (-CHF-CF_2-) and the peaks in region 1 can be attributed to both VDF (-CF_2-CH_2-)and TrFE, it is possible to eluciate up to five carbon bond lengths.[7] By analyzing these spectral intensities, much can be learned concerning the stereochemistry of the copolymer. Figure 7 shows the changes in the ^{19}F-NMR spectra that 2.55 MeV electron-irradiation generates.

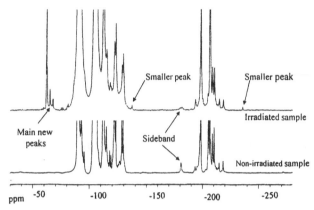

New peaks are observed, most notable of which are a quartet at about -65 ppm, representing CF_3 endgroups to -CHF-CF_3 and -CF_2-CF_3. The intensity of these peaks increases with irradiation dose. Also present in figure 7 are two less intense peaks at -138 ppm and -237 ppm caused by terminal -CHF_2 and -CH_2F endgroups, respectively. The presence of these endgroups is consistent with chain scissioning processes. A point of interest is that the resolution in regions 1 and 2 is observed to be less in the irradiated sample (figure 7). This loss of resolution may be due to decreased molecular mobility caused by crosslinking. No absorption was noted at 15-30 ppm due to carbonyl fluoride (COF). The formation of oxygenated polymer fragments then is unlikely due to the lack of evidence for COF, which is commonly formed when fluoropolymers are subjected to irradiation.

CONCLUSIONS

From the observations accumulated to date in this study, a picture is emerging of the effects of electron-irradiation on p(VDF-TrFE) and the consequences of these changes on the ferroelectric and relaxor properties of this material.

On the basis of the experiments in this study, electron irradiation of P(VDF-TrFE) films appears to breakup of the large crystal domains into smaller microregions by the production of structural defects. The irradiated copolymer thus can be described as a distribution of polar crystalline microregions consisting of largely *trans* bonding separated by amorphous regions of smaller polymer chain fragments, linked chains, and perhaps some unsaturated linkages. Analogous to relaxor lead-magnesium-niobate (PMN) ceramic where polar microregions separated by nonpolar disordered regions can impart a large electrostrictive strain, irradiated p(VDF-TrFE) behaves as a "relaxor" capable of exhibiting high strain when stressed by large electric fields. Furthermore, considering that the structural transformations can "freeze" in defects, it is of great interest to probe the relative effects of chain scission and cross-linking on the polymer's electromechanical properties. Experiments are now underway to determine the effects of electron energy and absorbed doses on the properties of the copolymer, specifically whether the relative degree of chain scissioning and crosslinking can be controlled. In addition, the NMR studies are being expanded to include ^{13}C-NMR, which can yield direct evidence of cross-linking and other carbon-backbone-affecting structural defects. Finally, both ^{19}F- and ^{13}C-NMR experiments will be designed and performed to determine the relative percentage of *trans* and *gauche* bonding in the crystalline and non-crystalline fractions of irradiated copolymer. These experiments will provide an improved understanding on the effects of irradiation on the polymer microstructure as well as better ways to control the electromechanical responses of the copolymer.

ACKNOWLEDGEMENTS

We gratefully acknowledge Professor Karen Gleason and Dr. Pierre Mabboux of MIT's Department of Chemical Engineering for the performance of the NMR experiments as well as for fruitful discussions, and Kenneth Wright of MIT's High Voltage Research Laboratory for assistance with the irradiation experiments. This work was supported by the Office of Naval Research.

REFERENCES

1. J. Lindberg, Mat. Res. Soc. Symp. Proc. **459**, 509 (1997).

2. Q. Zhang, V. Bharti, and X. Zhao, Science **280**, 2, 101 (1998).

3. J. Kim and G. Kim, Polymer **38**, 4, 881 (1997).

4. A. Lovenger, in *Developments in Crystalline Polymers – 1*, edited by D. C. Bassett (Applied Science, London, 1982), chapter 5.

5. K. Tashiro, R. Tanaka, K. Ushitora, and M. Kobayashi, Ferroelectrics **171**, 145 (1995).

6. F. Bovey and P. Mirau, *NMR of Polymers* (Academic Press, San Diego, 1996), p. 4.

7. K. Lau and K. Gleason, J. Phys. Chem. B **102**, 5, 977 (1997).

DYNAMICS OF PYROELECTRICITY OF A COPOLYMER OF VINYLIDENE FLUORIDE WITH TRIFLUOROETHYLENE

Y. TAKAHASHI, J. TANIWAKI and T. FURUKAWA
Department of Chemistry, Faculty of Science, Science University of Tokyo,
1-3 Kagurazaka, Shinjuku, Tokyo, 162-8601 Japan, ytakahas@ch.kagu.sut.ac.jp

ABSTRACT

The pyroelectric response of a copolymer of vinylidene fluoride with trifluoroethylene was measured as a function of time by applying a laser pulse to give an abrupt increase in temperature. The response curve shows several processes: the thickness extensional mode TEM, the length extensional mode LEM and the heat transfer to the environment. By fitting a theoretical curve for the TEM mode, the pyroelectric primary and secondary effects are separated. The primary effect is found to be small at room temperature. Additional gradual response shows that there exists a surface layer which has smaller polarization than in the bulk.

The temperature dependence of the pyroelectric response curve shows that the primary effect increases with temperature. It dominates the response at just below the ferroelectric-paraelectric transition temperature.

INTRODUCTION

Ferroelectric polymers have been studied very widely and their properties are now well known.[1, 2] Nevertheless, there remain a lot to be discussed. The mechanism of pyroelectricity is one of such problems.[3] The pyroelectricity reflects the temperature dependence of the remanent polarization. Therefore the knowledge of the mechanism of pyroelectricity is important for understanding of the ferroelectricity.

One difficulty in studying the pyroelectricity is that it is composed of two components: the primary effect and the secondary effect. The temperature derivative of the electric displacement at no stress boundary condition can be written as

$$\left(\frac{\partial D_i}{\partial T}\right)_X = \left(\frac{\partial D_i}{\partial T}\right)_x + \left(\frac{\partial D_i}{\partial x_j}\right)\left(\frac{\partial x_j}{\partial T}\right)$$
$$= \left(\frac{\partial D_i}{\partial T}\right)_x + e_{ij}\alpha_j$$

where D_i, the dielectric displacement, T, the temperature, x, the strain, X, the stress, e_{ij}, the piezoelectric stress constant and α_j is the thermal expansion constant. The first term is the primary effect and the second term is the secondary effect that is a piezoelectric response coupled with the thermal expansion.

Kepler et al.[4, 5] studied thoroughly the pyroelectricity in polyvinylidene fluoride. They used a laser pulse to heat the sample instantaneously and observed the time development of the pyroelectric response. They discussed the separation of the primary and secondary effect. We also developed a system to measure the pyroelectric response as a function of time by irradiating a laser pulse.[6] In this paper, we report our recent results on a copolymer of vinylidene fluoride with trifluoroethylene by using a laser generated by a mode-locked laser system. We also measured the pyroelectric response curve at various temperatures and discuss the temperature dependence of the primary and secondary effects.

EXPERIMENT

The sample was the copolymer of vinylidene fluoride VDF and trifluoroethylene TrFE of 75 mol% VDF content which was supplied by Kureha Chemical Industry Co., Ltd. The polymer was dissolved in 2-butanone together with crystal violet which well absorbs light of wavelength used in this study. The concentration of the dye was 0.001 mol/dm^{-3} in the film. After cast at room temperature, the dyed polymer film was annealed at 145 °C for 1 hour and, after

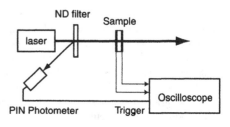

Fig. 1. Experimental System

coated with gold on both surfaces, subjected to an ac field of 0.1 Hz at room temperature. The amplitude of the field was 100 MV/m. The electric field was removed at the maximum. The poled film was cut to 4 mm square. Typical thickness was 16 μm.

A laser pulse of wavelength of 532 nm, the pulse width of 7 ns and the pulse energy of 200 mJ was generated by a Q-switched frequency-doubled yttrium-aluminum-garnet, YAG, laser (Surelite-I, HOYA Continuum Co.). A neutral density, ND, filter, was used to attenuate the energy to 4 mJ. Also used was a laser pulse of 532 nm, 20 ps, 12 mJ of a mode-locked Nd:YAG laser generator(PL2143, EKSPLA) which was attenuated to 1.2 mJ by an ND filter.

Figure 1 illustrates the experimental system.[6] A laser pulse was applied to the sample through a gold electrode which was semitransparent to the light used. Thus, the absorption of the laser

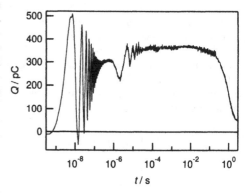

Fig. 2. Pyroelectric response curve at room temperature. The pulse width is 20 ps.

pulse energy occurs at the electrodes and inside the polymer film. With using a PIN photometer signal as a trigger, the electric response was measured by the digitizing oscilloscope TDS620 (Tektronix) at the rate of 2×10^9 samplings per second at the fastest. By repeating the application of a laser pulse at an interval of several seconds, the electric response was accumulated with decreasing the sampling rate. Obtained electric response was converted to the charge response by computation. Thus we obtained the time development of the pyroelectric response from 1 ns up to 10 s.

RESULTS AND DISCUSSION

Figure 2 shows the semi-logarithmic plot of the pyroelectric charge response curve on an application of a 20 ps laser pulse at room temperature. It is composed of several processes. The oscillatory mode in 10 ns region is caused by the proper oscillation of thickness extensional mode TEM of the film that is excited by an abrupt thermal expansion. This mechanical oscillation causes electric response because of the piezoelectricity, that is, the pyroelectric secondary effect. The oscillation in 10 μs region is the length extensional mode LEM. Frequencies of these oscillations agree exactly to the piezoelectric resonance frequencies which are observed in the dielectric permittivity measure-

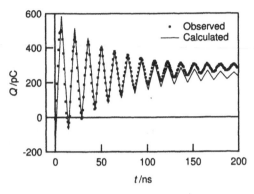

Fig. 3. Pyroelectric response curve. Dots are the observed data and the solid line is the calculated curve.

ment. The decrease after 100 ms is the cooling process, *i.e.*, the heat transfer to the environment.

The pyroelectric response in the TEM region is plotted linearly against time in Fig. 3. The waveform of TEM is a damping triangular wave. In order to analyze the wave form, the equation of motion, or, the one-dimensional wave equation of the film along the thickness direction is solved analytically when a change of temperature $\Delta T(t)$ is applied. The basic relation is

$$X = cx - A\Delta T$$

where $X = X_3$, the stress, $x = x_3$, the strain, $c = c_{33}$, the elastic stiffness and $A = A_3$ is the thermal stress coefficient. The wave equation is

$$\rho\frac{\partial^2 u}{\partial t^2} + \gamma\frac{\partial u}{\partial t} = c\frac{\partial^2 u}{\partial \xi^2} \tag{1}$$

with a boundary condition

$$X = 0 \qquad \text{at} \qquad \xi = \pm\frac{l}{2}$$

where ξ, the coordinate, u, the displacement, t, the time, ρ, the density, γ, the damping factor and l is the thickness of the film. By solving Eq. (1), we obtain the displacement as a function of time $u(t)$. The electric displacement is expressed as

$$D_{\text{TEM}}(t) = \frac{e_{33}}{l} \int_{-l/2}^{l/2} x d\xi = \frac{e_{33}}{l} \left[u\left(\frac{l}{2}\right) - u\left(-\frac{l}{2}\right) \right].$$

Using a solution of $u(t)$ we obtain

$$D_{\text{TEM}}(t) = \frac{e_{33}A_3}{c_{33}} \left[\Delta T(t) - \sum_{\substack{n=1 \\ \text{odd}}}^{\infty} \frac{8}{\pi^2 n^2 \omega_n} F_n(t) \right] \tag{2}$$

where

$$F_n(t) = \int_0^t \left(\frac{d^2 \Delta T}{dt^2} + \frac{2}{\tau} \frac{d\Delta T}{dt} \right) e^{t'/\tau} \sin\{\omega_n(t - t')\} dt' e^{-t/\tau},$$

$$\omega_n = \sqrt{n^2 \omega_0^2 - 1/\tau^2}, \qquad \omega_0 = \frac{\pi}{l} \sqrt{\frac{c_{33}}{\rho}},$$

$$\tau = \frac{2\rho}{\gamma}.$$

For heating by a short laser pulse, this solution becomes a damping triangular wave.

With fitting parameters of the amplitude ΔQ_{TEM}, the frequency f_{TEM}, the damping time constant τ_{TEM} and an offset constant Q_0, we fitted the function of Eq. (2) to the experimental data as a solid line shown in Fig 3. Here, $Q_0 = -150\text{pC}$, $\Delta Q_{\text{TEM}} = 390\text{pC}$ and $f_{\text{TEM}} = 69\text{MHz}$.

ΔQ_{TEM} corresponds to the secondary effect. On the other hand, the offset constant Q_0 means a response which occurs faster than the thermal expansion. Therefore Q_0 corresponds to primary effect. The fitting result shows that Q_0 is rather small, or slightly negative. Thus, it is shown that the pyroelectric primary effect is small at room temperature.

There remains a gradual increase as a discrepancy in the longer time region in Fig. 3. This process can be attributed to the heat transfer from the gold electrodes into the polymer. The pyroelectric response curve of a sample that contains no dye shows such gradual response more prominently. The heat absorbed at the gold electrodes transfers normally to the film surface. If the polarization is uniform in the film, the pyroelectric response does not change in the course of the heat diffusion in the sample: the further the heat transfer, the smaller ΔT becomes, the average increase in temperature being constant. Thus, the gradual increase indicates the existence of a surface layer that

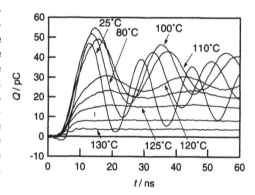

Fig. 4. Pyroelectric response curves at various temperatures. The laser pulse width is 7 ns.

has smaller polarization than in the bulk. The thickness of this layer is estimated to 0.1 μm.

The pyroelectric response curve changes with the ambient temperature. The oscillation frequency decreases corresponding to the decrease in elastic stiffness. As the temperature becomes near the transition point, the shape of the pyroelectric response curve changes drastically. Figure 4 shows the response curves at various temperatures. In this figure, the data were taken using Surelite-I whose pulse width was 7 ns. The amplitude of TEM oscillation decreases abruptly above 110 °C. We roughly estimated Q_0 and ΔQ_{TEM}. In Fig. 5, Q_0 and ΔQ_{TEM} are plotted against temperature. Also shown are the remanent polarization P_r, the dielectric permittivity ε' and the endothermic chart of differential scanning calorimetry all on heating. DSC shows that the ferroelectric-paraelectric transition point is ca. 120 °C. This figure clearly shows that Q_0 gradually increases while ΔQ_{TEM} remains almost constant up to 100 °C. The increase in Q_0 well corresponds to the decrease in P_r. Q_0 finally becomes dominant at just below the transition temperature where P_r strongly depends on temperature. Thus, it is concluded that Q_0 reflects the temperature dependence of the remanent polarization. This may correspond to the excitation of defects in the chain as the temperature becomes near the transition temperature.

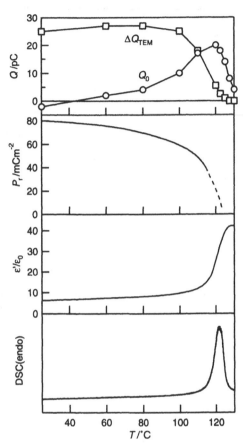

Fig. 5. The primary and secondary effects as functions of temperature, compared with the remanent polarization, dielectric permittivity and DSC chart on heating.

CONCLUSIONS

We measured the pyroelectric response curve on an application of a laser pulse. Several processes are observed: the thickness extensional mode, the length extensional mode and the heat transfer to the environment By fitting a TEM damping oscillation, the primary effect is shown to be small at room temperature. A gradually increasing response is also shown to exist, which is attributed to the heat transfer from the surface. It shows that there is a surface layer which has smaller polarization than in the bulk.

The temperature dependence of the pyroelectric response curve shows that the pyroelectric primary effect gradually increases with temperature corresponding to the decrease in remanent polarization. At high temperature, the contribution of the primary effect becomes dominant.

ACKNOWLEDGEMENT

We are grateful to Mr. N. Moriyama of Kureha Chemical Industry Co., Ltd. for kindly supplying samples. We also thank to Mr. K. Tsunoda of Department of Applied Chemistry, Faculty of Science, Science University of Tokyo for helping us use a mode-locked Nd:YAG laser. This work was partly supported by a Grant-in-Aid from the Ministry of Education, Science and Culture, Japan.

REFERENCES

1. T. Furukawa, Phase Transitions, **18**, 143 (1989)

2. R. G. Kepler and R. A. Anderson, Adv. Phys., **41**, 1 (1992)

3. T. Furukawa, IEEE Trans. Electr. Insul., **24**, 375 (1989)

4. R. G. Kepler and R. A. Anderson, J. Appl. Phys., **49**, 4918 (1978)

5. R. G. Kepler, R. A. Anderson and R. R. Lagasse, Ferroelectrics, **57**, 151 (1984)

6. Y. Takahashi, K. Hiraoka and T. Furukawa, IEEE Trans. Dielectr. Electr. Insul., **5**, 957 (1998)

ANALYSIS OF FERROELECTRIC SWITCHING PROCESS IN VDF/TrFE COPOLYMERS

Kenji Kano, Hidekazu Kodama, Yoshiyuki Takahashi, Takeo Furukawa
Department of Chemistry, Faculty of Science, Science University of Tokyo,
1-3 Kagurazaka, Shinjuku, Tokyo, 162-8601 Japan, kc@furukawa.ch.kagu.sut.ac.jp

ABSTRACT

The ferroelectric switching curve of vinylidene fluoride / trifluoroethylene copolymer exhibits a characteristic time evolution consisting of two processes; an initial gradual increase in proportion to $t^{0.5}$ followed by a rapid increase according to an exponential function with particularly large exponent 6. Such a switching curve was analyzed by means of computer simulation based on a modified nucleation-growth mechanism. It was found that the initial gradual increase is attributed to generation of considerably large nuclei that grow according to a random-walk scheme. Once such nuclei gain a critical size, they start to grow automatically either one or two dimensionally. The time required to generate critical nuclei serves as an incubation time for the later growth process to result in the large exponent.

INTRODUCTION

Copolymers of vinylidene fluoride (VDF) and trifluoroethylene (TrFE) containing 50-80 mol% former are ferroelectric polymers undergoing fast polarization reversal. In the ferroelectric phase, molecules assume an all-trans conformation and are packed in a parallel manner to form a spontaneous polarization. Polarization reversal occurs as a result of 180° rotation of individual molecules about their chain axes.

In general, polarization reversal in ferroelectrics progresses via. a nucleation-growth mechanism[1]. On an application of an electric field, nuclei of reversed domains are generated at a certain probability. They grow with growth dimension m until they coalesce with each other to reverse entire polarization. Such a process can be expressed by an exponential function, $1 - \exp(-(t/\tau)^n)$. Exponent n called Avrami index is related to growth dimension via $m < n < m + 1$. It was shown that the switching curve of polyvinylidene fluoride(PVDF) is consistent with one-dimensional growth of reversed domain[2].

Copolymers of VDF and TrFE have received much attention because they exhibit Curie transition which is absent in PVDF. More interestingly, they are imparted extremely high crystallinity by annealing in the paraelectric phase. If the uniaxially oriented films are annealed at a temperature just below the melting point, we obtain considerably large lamellae regularly stacked along the film thickness. We have previously reported a finding of very sharp switching curve in that n value reaches 6[3, 4]. A simple nucleation growth mechanism fails to explain such a large n value. In this paper, we analyzed the shape of the switching curve characteristic for uniaxially oriented and annealed VDF/TrFE copolymer in detail on the basis of modified nucleation-growth mechanism by means of computer simulation.

EXPERIMENTAL SWITCHING CURVE

The samples used in this investigation were a VDF/TrFE copolymer with composition 65/35 mol% supplied by Daikin Kogyo Co. Ltd., Japan. Film samples were obtained by melt-extrusion and annealed at 147°C for one hour. Scanning electron microscope images showed that such samples consisted of regularly stacked large lamellae 60nm in thickness and $600 \times 600 \text{nm}^2$ in area. Gold was evaporated on both surfaces to serve for electrode.

The time domain measurements of polarization reversal consist of an application of a stepwise electric field in the direction opposite to the existing remanent polarization and a detection of charge response as a function of time. Details of the experimental setup have been described previously [5]. As a result, we obtain a switching curve which is approximately expressed by

$$P_{sw} = \Delta P[1 - \exp\{-(t/\tau)^n\}] \tag{1}$$

Here ΔP is the amount of reversed polarization equaling double the remanent polarization $2P_r$ and being independent of applied field strength E. The time constant τ, called a switching time, rapidly decreases with increasing E obeying an exponential law. Exponent n is related to the shape of the switching curve. Its value is normally greater than unity.

In practice, the observed switching curve $D(t)$ contains considerable contributions from linear and nonlinear dielectric responses and is written as follows

$$D(t) = P_{sw} + \varepsilon_1 E + \varepsilon_3 E^3 + \ldots + (\varepsilon_2 E^2 + \varepsilon_4 E^4 + \ldots)\left(\frac{P_{sw}}{P_r} - 1\right) \tag{2}$$

Here, ε_1 is the linear permittivity and ε_n with $n \geq 2$ is the n-th order nonlinear permittivity. It should be noted here that the even order permittivity change sign in accord with polarization reversal. We have thus incorporated their contributions assuming that they are proportional to $P_{sw} - P_r$. In order to extract P_{sw} from the observed switching curve, we removed contributions from dielectric responses in a manner as described in ref. 4.

Figure 1 shows a double logarithmic plot of P_{sw} vs time t obtained for a uniaxially oriented and annealed 65/35 mol% copolymer at 50MV/m. It is

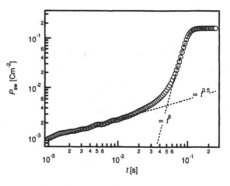

Fig. 1 Two step character of Switching Curve for VDF(65) / TrFE(35).

found that P_{sw} consists of two components, an initial gradual increase in proportion to $t^{0.5}$ and a latter rapid increase according to eq. (1) with exponent $n = 6$. Assuming the nucleation-growth mechanism, we can naturally attribute the latter to the growth process. However, such a large exponent is beyond the prediction of conventional theory. Its maximum value is 4 for the case of three dimensional growth of homogeneous nucleation. We have recently suggested that such a large exponent can be explained by introducing an incubation time [3]. In the following, we try to reproduce the observed P_{sw} by means of computer simulation.

COMPUTER SIMULATION

Computer simulation was made with respect to an array of N cells and $N \times N$ cells describing 1-D and 2-D growth processes, respectively. Each cell corresponds to a unit of polarization reversal as well as a nucleus. We assign quantity $S(x)$ or $S(x,y)$ to each cell in order to express its polarization state. Its value is 0 before switching and becomes S_c after switching. At every simulation step, the value of S is subjected to changes according to a certain scheme. Summing up S over the entire cells yields the time evolution of P_{sw}. For the case of 1-D growth,

$$P_{sw} = \frac{2P_r}{NS_c} \sum_{x=1}^{N} S(x) \tag{3}$$

In order to reproduce the observed rapid charge in polarization, we employed an assumption that a nucleus has a critical size so that it induces reversal of adjacent molecules. The simulation scheme based on this assumption is illustrated in Fig. 2. If a cell is chosen to be a nucleation site, its S value gains an increment of 1 at every step. Once it reaches S_c to become a critical nucleus, adjacent cells gain an increment of S_c at the next step.

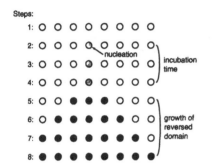

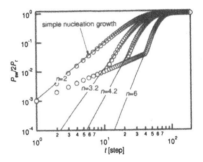

Fig. 2 Schematic representation of computer simulation based on modified nucleation-growth mechanism with $S_c = 3$.

Fig. 3 Simulated switching curve with different $S_c = 3$ value.

Figure 3 shows the result of simulation using various S_c values. It is seen that simulated curves exhibit two precesses, an initial gradual increase (nucleation) and a latter rapid increase (growth). The slope of the latter process increases with increasing S_c. When S_c is chosen to be 40, n becomes 6 being largely consistent with experimental results shown in Fig. 1. On the other hand, if S_c is chosen to be 1, the switching curve shows a continuous increase with $n = 2$. These results indicate that S_c is related to incubation time.

Although simulation qualitatively well-reproduced the experimental switching curve, quantitative agreement is not satisfactory. The initial slope associated with the nucleation process is 0.5 experimentally whereas simulation yields 1. In simulation, we assumed that the chosen nucleation site grew at a constant rate, As a result S increased in proportion to the steps which yielded a slope of 1 in $\log P_{sw}$ vs $\log t$ plots.

It is known that the mean square displacement of a random-walk process is proportional to the step number. We have, therefore, modified the nucleation scheme so that the S value of the nucleation cell is subject to a random increment of $+1$ or -1. As a result, the S value increases in proportion to the square of steps on average as shown in Fig. 4. Once S reaches S_c, it starts to grow as domain wall motion.

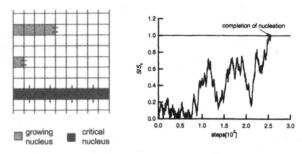

Fig. 4 Nucleation process based on random-walk scheme.

Figure 5 shows results of simulation based on modified nucleation scheme. It is found that the initial slope is strongly affected by the initial nucleation process. If the nucleation initiates randomly (a: homogeneous nucleation), the initial slope is again 1. On the other hand, if all the nucleation sites are activated instantly (b: heterogeneous nucleation), we obtain a slope of 0.5. It is seen that the simulated switching curve based on heterogeneous nucleation reasonably well-reproduce the observed over the entire time range.

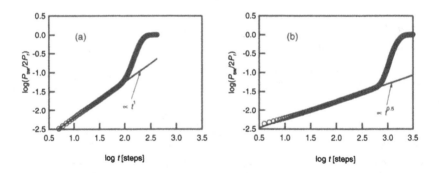

Fig. 5 Results of simulation with homogeneous nucleation(a) and heterogeneous nucleation(b).

We have shown so far the results of computer simulation for the case of one-dimensional growth. We have also examined the case of two-dimensional growth and obtained equally satisfactory results by choosing a smaller S_c value as long as modified nucleation process is introduced. In other words, the question whether the growth dimension is one or two remains uncertain.

CONCLUSIONS

We have successfully simulated the ferroelectric switching curve of VDF(65)/TrFE(35) copolymer on the basis of modified nucleation-growth mechanism. One of the most characteristic features is the nucleation process that starts with heterogeneous nucleation and completes when the nuclei gain a critical size via the random-walk scheme. Such nucleation process would be related to the long chain nature of ferroelectric polymers.

REFERENCES

1. W. J. Merz, Phys. Rev., **95**, 690 (1954)

2. M. Date and T. Furukawa, Ferroelectrics, **57**, 37 (1984)

3. H.Kodama, Y. Takahashi and T. Furukawa, Ferroelectrics, **203**, 433 (1997)

4. T. Furukawa, H.Kodama, O. Uchinokura, and Y. Takahashi, Ferroelectrics, **171**, 33 (1995)

5. T. Furukawa and G.E. Johnson, Appl. Phys. Lett., **38**, 1027 (1981)

6. K. Kano, H. Kodama, Y,Takahashi, T. Furukawa, Rep. Progr. Polym. Phys. Jpn, **40**, 389 (1997)

FERROELECTRIC AND PIEZOELECTRIC PROPERTIES OF BLENDS OF POLY(VINYLIDENE FLUORIDE-TRIFLUOROETHYLENE) AND A GRAFT ELASTOMER

J. SU*, Z. OUNAIES**, J. S. HARRISON***
*National Research Council, NASA-LaRC, Hampton, VA 23681, USA
**ICASE, NASA-Langley Research Center, Hampton, VA 23681, USA
***NASA-Langley Research Center, Hampton, VA 23681, USA

ABSTRACT

A piezoelectric polymeric blend system has been developed. The system contains two components: ferroelectric poly(vinylidene fluoride-trifluoroethylene) and graft elastomer. The remanent polarization, P_r, and the piezoelectric strain coefficient, d_{31}, of the blends have been studied as a function of relative composition of the two components, temperature and frequency. Both blended copolymer and graft unit in the elastomer contribute to the total crystallinity of the blend-system, and hence to the remanent polarization and piezoelectricity. The piezoelectric strain coefficient, d_{31}, of the blend systems shows dependence on both the remanent polarization and the mechanical stiffness, which in turn are determined by the fraction of the two components in the blends. This mechanism makes it possible for the piezoelectric strain response of the blend to be tailored by adjusting the relative composition. Although P_r of the copolymer is higher than that of the blends, the blend films containing 75 wt.% copolymer exhibit a higher d_{31} at room temperature, possibly due to their lower modulus. The blend films containing 50 wt.% copolymer exhibit a constant value of d_{31}, from room temperature to 70°C.

INTRODUCTION

Ferroelectric and piezoelectric poly(vinylidene fluoride-trifluoroethylene) copolymers have been extensively studied in the last two decades due to their electromechanical properties[1-4]. Recently, the electrostrictive properties of high energy electron irradiated poly(vinylidene fluoride -trifluoroethylene) copolymers have been reported and the results show promising properties for applications in actuation technologies[5,6]. In the present work, a ferroelectric copolymer-elastomer blend system was developed to improve the toughness of the electromechanical copolymers since the brittleness of the pure copolymers may limit their applications. Both the ferroelectricity and piezoelectricity of the blend-system were investigated. The results show that the piezoelectric strain response and the modulus of the blend films can be tailored by adjusting the ratio of both phases in the system. By careful selection of the composition, an enhanced piezoelectric strain response and a temperature-independent piezoelectric strain response can be obtained.

EXPERIMENT

Film preparation: The blend films were prepared by solution casting.   The ferroelectric poly(vinylidene fluoride-trifluoethylene) copolymer (50/50 mol.%) and graft elastomer powders were added to N,N-dimethylformamide. The mixture was heated to 60°C while stirred to make a 5 wt.% polymer solution containing the desired fraction of the two components. The solution was then cooled to room temperature, cast on glass substrates, and placed in a vacuum chamber. After drying overnight under vacuum, tack-free films were obtained. In order to increase their crystallinity, and possibly their remanent polarization, the blend films were thermally annealed at 140 °C for 10 hours. The thickness of the films was around 20μm. The films were tested using X-ray diffraction (XRD) and differential scanning calorimetry (DSC) in order to assess their composition and degree of crystallinity.

Poling and Measurements: Gold electrodes were sputtered on the opposing surfaces of the films to establish electrical contact. The films were poled using a triangular wave signal with a peak value of 100MV/m at 30 mHz. The modulus, E_{11}, and the piezoelectric strain coefficient, d_{31}, were measured using a modified Rheovibron. The capacitance of the blend films was measured using a HP Analyzer 4192A, and the dielectric constant, ε, was calculated from the value of the capacitance. These measurements were performed as a function of the relative composition of the blends (wt.% copolymer content), temperature, and frequency.

RESULTS AND DISCUSSION

Figure 1 shows the crystallinity as a function of copolymer content in the blend. The calculated crystallinity of the blend system is found from

$$X_{total} = f_{copolymer}X_{copolymer} + f_{elastomer}X_{elastomer} \qquad (1)$$

where f is the fraction of the components and X is the crystallinity. Both the measured and calculated crystallinities increase with increasing copolymer content in the blend, however, the measured crystallinity is lower than the calculated one. This indicates that the presence of both components in the blend may reduce their crystallization as compared to each individual one.

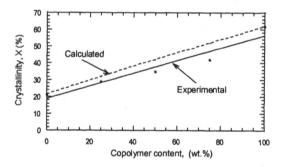

Figure 1. The relationship between the copolymer content and the
crystallinity in the blends

To determinate the remanent polarization, the measurement of the polarization, P, versus the electric field, E, was carried out. Corrections were made to eliminate the effects of conductivity on the ferroelectric hysteresis loops [7]. The measured remanent polarization, P_r, as a function of the copolymer content in the blends is shown in Figure 2 and compared with the remanent polarization calculated using the following equation:

$$P_{r(total)} = f_{copolymer} \, P_{r(copolymer)} + f_{elastomer} \, P_{r(elastomer)} \qquad (2)$$

where f is the relative fraction of the components, $P_{r(copolymer)}$ is the remanent polarization in the pure copolymer, $P_{r(elastomer)}$ is the remanent polarization in the elastomer, and $P_{r(total)}$ is the resulting remanent polarization of the blend film. As can be seen, both the measured and the calculated remanent polarization increase with increasing copolymer content in the blends. The value of the measured remanent polarization is very close to the calculated one. This is an indication of the linear relationship between P_r and the polar crystallinity in the blends.

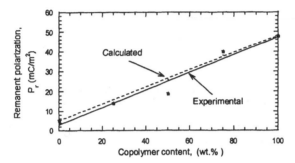

Figure 2. Relationship between the copolymer content in the blend and the remanent polarization.

Figure 3 shows the mechanical modulus, E_{11}, for all the blends as a function of temperature at 1 Hz. As expected, the mechanical modulus of the blends increases with the increase of the copolymer content and the copolymer has the highest modulus. It is also noted that due to the copolymer's brittleness, samples tended to fail at 65 °C.

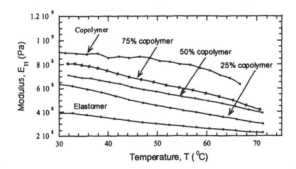

Figure 3. Comparison of the mechanical modulus E_{11} of blend films and homopolymer films.

Figure 4 shows the temperature dependence of the piezoelectric strain coefficient, d_{31}, for blend films with various compositions. The piezoelectric strain coefficient, d_{31} increases with increasing copolymer content. However, the blend film with 75 wt.%

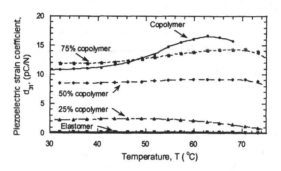

Figure 4. The temperature dependence of the piezoelectric strain coefficient, d_{31}, of the blend films (1 Hz) as a function of the various compositions.

copolymer exhibits the highest d_{31} from room temperature to about 45°C. Additionally, the blend film with 50 wt.% shows an almost constant piezoelectric response from room temperature to 70°C. These results reflect the influence of both the electrical polarization and mechanical modulus of the films on the piezoelectric strain response. As observed in the case of the 75 wt.% copolymer blend, even though it had a lower remanent polarization than the pure copolymer, it showed a higher piezoelectric strain response due to its lower modulus. Improvement in the toughness of the materials is also observed. Under the present experimental conditions, the pure copolymer film breaks at a temperature close to 65°C, while the rest of the blend films maintain their piezoelectric response up to 75°C without mechanical failure. In particular, the piezoelectric strain response of the 75 wt.% copolymer and 50 wt.% copolymer blend films is still significantly high up to 75°C.

When the dependence of the piezoelectric strain coefficient, d_{31}, on the relative composition of the two components in the blend is examined, different trends were observed at 30°C and 65°C (Figure 5). The reason for the non-linear dependence may be

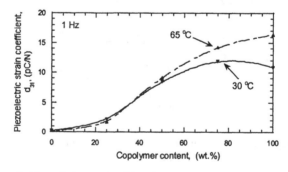

Figure 5. The relative composition dependence of the piezoelectric strain coefficient, d_{31}, at 30°C and 65°C.

attributed to the nature of the piezoelectric strain response of the material. As previously stated, the intrinsic contributions of both the mechanical properties (through the modulus) and the electrical properties (through the polarization) may yield this non-linear behavior.

Figures 6a and 6b show the temperature dependence and composition dependence of the dielectric constant at 10 Hz for the copolymer-elastomer blend films. The temperature dependence of the dielectric constant shown in Figure 6a gives a reasonable

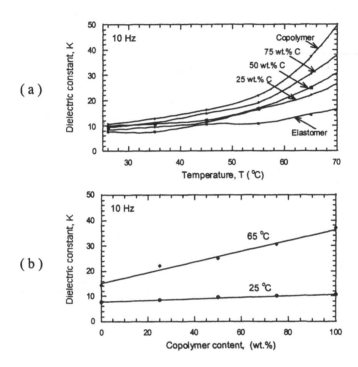

Figure 6. (a) Temperature dependence of the dielectric constant, and (b) Composition dependence of the dielectric constant.

trend for a blend system. The elastomer shows a transitional change in the temperature range from 40˚C to 50˚C and less temperature dependence than the copolymer in the measured temperature range. The transitional change is the second glass transition of the elastomer due to the molecular motion of the graft crystal cross-linking sites. It is obvious that the addition of the copolymer in the blend decreases the second glass transition of the graft elastomer significantly. This might be attributed to the molecular interaction between the added copolymer and graft unit in the elastomer. This interaction may also be the reason that the measured crystallinity of the blend is lower than the calculated one. The dielectric constant of the copolymer shows an obvious increase above 50˚C due to the ferroelectric-paraelectric phase transition [8]. For the blend system, as the copolymer content increases, the transition behavior in the dielectric constant becomes more apparent. Figure 6b shows the inter-relationship between the dielectric constant and the relative composition of the two components in the blend. Unlike the piezoelectric strain response, the dielectric constant shows a linear dependence to the relative composition at both 25°C and 65°C.

SUMMARY

The ferroelectric copolymer-graft elastomer blend system exhibited a marked improvement in toughness as compared to the pure copolymer. The blends also offer a potential of modifying the relative fraction of the two components in the blend in order to tailor the properties to desired applications. By combining the contributions of both the polarization and the mechanical stiffness, the blend system can give a higher piezoelectric strain response than the pure copolymer. As an example, the blend containing 75 wt.% copolymer exhibits a better piezoelectric strain response from ambient to 45 C than the pure copolymer. Furthermore, by adjusting the relative fraction of the two components in the blend, the piezoelectric strain response of the blend film (such as in the case of the 50 wt.% copolymer) can become temperature-independent.

REFERENCE

1. Y. Higashihata, J. Sako, and T. Yagi, *Ferroelectrics*, **32**, 85 (1981).

2. A. J. Lovinger, *Science*, **20**, 1115 (1983).

3. T. Furukawa, *Phase Transition*, **18**, 143 (1989).

4. T. T. Wang, J. M. Herbert, and A. M. Glass, *"Applications of Ferroelectric Polymers"*, Blakie, Glasgow (1988).

5. Q. M. Zhang, V. Bharti, and X. Zhao, *Science*, **280**, 2101 (1998).

6. V. Bharti, X. Zhao, Q. M. Zhang, T. Romotawski, F. Tito, and R. Ting, *Mat. Res. Innovat.*, **2**, 57 (1998).

7. B. Dickens, E. Balizer, A. S. DeReggi, and S. C. Roth, *J. Appl. Phys.*, **72**, 4258 (1992).

8. G. T. Davis, T. Furukawa, A. J. Lovinger, and M. G. Broadhurst, *Macromolecules*, **15**, 329 (1982).

STRUCTURAL AND ELECTRONIC CHARACTERIZATION OF EPITAXIALLY-GROWN FERROELECTRIC VINYLIDENE FLUORIDE OLIGOMER THIN FILMS

K. Ishida, K. Noda, A. Kubono*, T. Horiuchi, H. Yamada and K. Matsushige

Department of Electronic Science and Engineering, Graduate School of Engineering, Kyoto University, Yoshidahonmachi, Sakyo-ku, Kyoto 606-8501, Japan
*Department of Polymer Science and Engineering, Kyoto Institute of Technology, Matsugasaki, Sakyo-ku, Kyoto, 606-8585, Japan
E-mail : kishida@kuee.kyoto-u.ac.jp

ABSTRACT

The structural and electric properties of newly synthesized vinylidene fluoride (VDF) oligomer thin films evaporated on various substrate were investigated. The structural behavior of the VDF oligomer thin films strongly depended on the kinds of substrate and substrate temperature during film preparation. In particular, the VDF oligomers epitaxially grew on KBr(001), aligned their molecular chain along the <110> direction of substrate surface, and were similar to polar Form I-type crystals of poly(VDF). While, the thin films evaporated on NaCl(001), SiO_2/Si were non-polar Form II-type or mixture of both Form I- and Form II-type crystals with their molecular chain normal to the surface. These facts indicated that crystal field of substrate, based on van der Waals and electrostatic interactions, strongly influenced to ferroelectric phase transition of the VDF oligomer. In addition, we demonstrate the formation of the local polarized domains in the epitaxial crystals by applying electric pulses from a conducting AFM probe used as a positionable top-electrode, and confirm their piezoelectricity of the VDF oligomers for the first time.

1. INTRODUCTION

Ferroelectric ultra thin films have a potential in various application areas, such as sensors, memory and electroactive devices. Poly(vinylidene fluoride)(PVDF) and its copolymer with trifluoroethylene [P(VDF-TrFE)] are well known to be ferroelectic polymer with large electric dipole moment. These polymers have several crystal phases with different crystal parameter, molecular packing and conformation. Three of them, called as Form I(β), Form III(γ) and Form IV(δ) -type crystals, possess ferroelectric property, the Form I crystals give the best performance. We have studied on a local polarized domain of P(VDF-TrFE) cast films by scanning probe technique, and discuss their piezoelectric behavior at nano scale[1-3]. A. V. Bune and coworker have reported piezoelectricity and pyroelectricity in ferroelectric Langumuir-Blodgett polymer films[4-6]. However, it is difficult to characterize the dipole switching and diffusion in the polymer due to their amorphous region and low crystallinity.

In this study, we selected a newly synthesized vinylidene fluoride (VDF) oligomer as a simple ferroelectric organic molecule. The molecule with low molecular weight and its narrow distribution has the possibilities for fabricating the high crystalline thin films and for controlling their crystal phase. We attempt to fabricate the epitaxially grown ferroelectric thin films by using this sample, and also focus on ferroelectricity of VDF oligomer. In such film, it is interesting to evaluate the local piezoelectric properties and switching mechanism in microscale, in order to study more about the local ferroelectric domains in epitaxial organic crystals.

2. EXPERIMENTS

2-1. Sample and film fabrication

The VDF oligomer [CF_3-$(CH_2CF_2)_n$ -I, (n=17)] was newly synthesized by the telomerization method at Daikin Kogyo Co., Ltd. The molecule consists of seventeen units of vinylidene fluoride monomer with an electric dipole moment, and terminal groups are CF_3 and iodide. The VDF oligomer thin films were fabricated by vacuum evaporation on air cleaved KBr(001), NaCl(001) and SiO_2/ Si. The substrates were preheated at 200°C for 2 hours to obtain a clean surface, and then maintained at different substrate temperatures (Ts), room temperature(RT ; about 25°C), 50, 70°C, under a vacuum of 10^{-4} Pa during evaporation. The evaporation rate and average thickness were controlled to be 0.3-0.5nm/min and 30nm respectively, by monitoring them with a quartz oscillator. After the deposition, the substrates were slowly cooled to RT in vacuum.

2-2. Structural characterization

Surface morphology was observed by atomic force microscopy (AFM) in contact mode. IR spectra of the films were measured by fourier transform-infrared spectroscopy (FT-IR). Structural investigations were carried out by conventional θ-2θ X-ray diffraction (XRD) and energy dispersive grazing incidence X-ray diffraction (ED-GIXD) system. The ED-GIXD, developed in our laboratory, can effectively analyze the crystal structures as well as the molecular orientation in the thin film without obstructive scattering noise from the substrate, because the grazing incidence X-rays were totally reflected onto the interface of air/film or film/substrate. Moreover, the epitaxial growth can be evaluated by the variation of the relative intensities of the diffraction peaks observed at different azimuthal angles ω. Experimental details in our ED-GIXD system have been described elsewhere[7-9]. In this study, the crystal structure and the vibration spectrum are identified as those of Poly(vinylidene fluoride), because the configuration in main chain of them are the same.

2-3. Local poling and its observation

In order to form and observe the local poling areas, the AFM system was modified as shown in Fig.1[1]. We used a gold-coated Si_3N_4 cantilever as a positionable top electrode. Locally poled areas were formed by applying a DC bias between the cantilever in contact with the film and the conducting substrate. After the poling, the piezoresponse and topographic images were measured simultaneously. A small oscillating voltage was applied between the tip and conducting substrate while the tip scanned a local

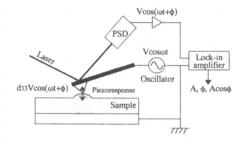

Fig. 1. The modified AFM system detecting piezo-response

polarized area in contact with the film. If the VDF oligomer films are crystallized with ferroelectric phase (Form I), the thickness will be changed with the applyed voltage, in other word the piezoresponse will appear. In this study, the oscillating voltage amplitude and frequency were typically 1V and 7kHz. The piezoelectric vibration can be detected with position sensitive detector, and analyzed the amplitude (A) and the phase difference(ϕ) by a lock-in amplifier. Then, the phase response (Acosϕ) images were used as the piezoresponse one.

3. RESULTS AND DISCUSSIONS

3-1. Surface morphology

Figure 2 shows AFM images of the VDF oligomer thin film deposited on KBr(001), NaCl(001) and SiO$_2$/Si at different Ts. The surface morphology of thin films depended strongly on Ts and kind of substrates. In the case of KBr substrate, the "rod-like" crystals, which oriented with fourfold symmetry, were observed. It was suggested that the VDF oligomer molecules epitaxially grew on KBr(001) surface. On the other hand, the molecules grew with layer-by-layer on NaCl(001) at Ts>70°C, though a grainy crystal was formed at Ts<70°C. The height of each layer was about 4.5nm, corresponding to single molecular step. Thus, the chain axes of VDF oligomer molecules were

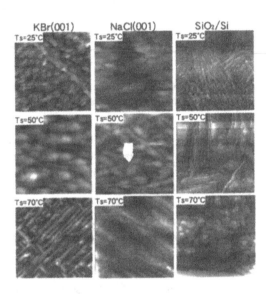

Fig. 2. AFM images(5μm*5μm) of VDF ologomer thin films evaporated at various Ts.

normal to the substrate surface. The layer-by-layer growth on SiO$_2$/Si was observed at lower Ts than that of NaCl. We could also observe the hexagonal crystal habit. One of possible packing is that of the VDF oligomer molecules which form the Form I crystal with pseudo-hexagonal symmetry. The molecular orientation is very sensitive to the total balance of inter- and intra-molecular interaction as well as molecular activities. Considering the dependence of surface morphology on Ts and variety of substrate, it was expected that the order of interaction between the VDF oligomer and each substrate is KBr(001) > NaCl(001) > SiO$_2$.

3-2. IR spectra of thin films on various substrate

Figure 3 shows the Ts dependence of FT-IR transmission spectra of VDF oligomer thin film evaporated on various substrates. The characteristic absorption peaks assigned to Form I (840 and 1270cm^{-1}) and Form II (790, 875, 1180 and 1210 cm^{-1}). The original powder of VDF oligomer consists of the mixture of both Form I and II crystals, although the Form II is dominant.

The absorption peaks of a thin film evaporated on KBr(001) were corresponding to the one of Form I. The IR spectra did not change by substrate temperature. While, the absorption peaks of

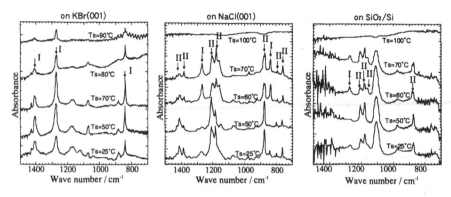

Fig. 3. Transmission IR spectra of VDF oligomer thin films evaporated on KBr(001), NaCl(001) and SiO₂/Si substrate at various Ts.

a thin films evaporated on NaCl(001) and SiO₂/Si at Ts<50°C indicated that the film consists of Form II crystals. With increasing Ts, the molecular conformation of the some crystals changed from that of Form II(TGTG') to Form I(TT). We could also observed the same results in VDF oligomer thin films evaporated on Au coated MgO.

From the above results, it is concluded that the VDF oligomer evaporated on KBr(001) formed the ferroelectric Form I crystal. In order to obtain more detailed information on the epitaxial growth of VDF oligomer, the ED-GIXD measurement was performed.

3-3. Crystal structure and epitaxial growth

The stacking structure of VDF oligomer thin films evaporated on KBr(001) at various Ts was measured by the conventional θ-2θ XRD, as shown in Fig. 4. In an X-ray diffraction profile, a peak corresponding to 200 reflection of Form I crystal of PVDF was observed. This revealed that the crystals in the VDF oligomer thin films orient their (200) plane parallel to the surface ; in other words, the chain axes of VDF oligomer molecules were parallel to the substrate. The VDF oligomer molecules on KBr substrate can be represented as Fig. 5(a) from the results of XRD and FT-IR measurement.

Next, we analyzed the epitaxial growth behavior of VDF oligomer thin films evaporated on KBr(001). The series of VDF oligomer 110 reflection were detected at different azimuthal angles by rotating the substrate to evaluate the relative orientational relationship between VDF oligomer crystal and KBr substrate. Figure 6 shows an angular variation of the relative intensities of the VDF oligomer 110 reflection. The zero degree of the azimuthal angle (ω=0°) was predetermined as the point of the maximum value of the KBr 200 reflection from the in-plane diffraction. The VDF oligomer 110 reflections showed the maximum value at ω=±45° and symmetrically and gradually decreased. This angular variation occurred repeatedly at each 90° azimuth of the KBr substrate, and have the full width at half-maximum (FWHM) of about 10°. It is suggested that the evaporated VDF oligomer molecules grow epitaxially on the KBr (001), aligning their chain axes in KBr<110> with a fluctuation of about 10°, as shown in Fig. 5(b), similarly to the thin films evaporated on KCl(001)[10].

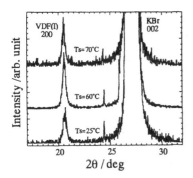

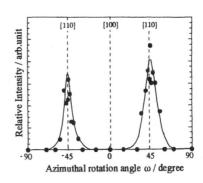

Fig. 4. The θ-2θ x-ray diffraction patterns of VDF thin films evaporated on KBr(001) at various Ts.

Fig. 6. Angular variation of the relative intensities of 110 reflection VDF oligomer thin films evaporated on KBr at Ts=50°C.

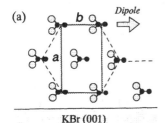

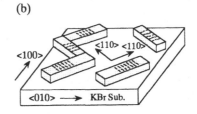

Fig. 5. Structural and orientational models for the VDF thin films evaporated on KBr(001).
(a) stacking model and (b) epitaxial growth model

3-4. Local polarized domains of VDF oligomer

The VDF oligomer thin film supported on conducting substrate was prepared by a following procedure in order to form and observe the local poled domain. The gold film of 100nm was evaporated on the VDF oligomer films, which evaporated on KBr(001) at Ts=50°C. The gold-coated side of VDF oligomer films attachs to an another conducting substrate with epoxy resin. Then, the KBr crystal was removed in pure water to design the VDF oligomer thin films on gold film used as bottom electrode.

A local poling in the epitaxial crystals were conducted by applying a DC bias between the AFM tip in contact with the film and the bottom Au films. The tip was scanned with 1Hz while the voltage of 10V was applied. Supposing that the thickness of VDF oligomer thin films was 40nm (about 46 monolayers), the magnitude of applied electric fields was 250MV/m. By poling the films using a AFM tip with positive bias, the dipole moment is normal to the substrate, because the hydrogens in VDF units orient toward substrate and the fluorines away. Figure 7 shows topographic and piezoresponse image(Acosφ) of VDF oligomer thin films, which were patterned the three stripes of alternating bias polarity. The positive and negative poled areas represented as the bright and dark, respectively, because the piezoresponse phase(φ) of negative domain was delayed 180°

 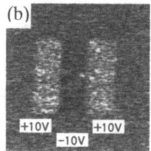

Fig. 7. Topographic and piezoresponse image (6μm*6μm) after poling the area of the epitaxially-grown VDF oligomer thin films with +10V,-10V and +10V, respectively

in comparison with that of positive. Since no topographic change was found in thin films after the poling, the poling did not affect the surface morphology. And when a reverse voltage was applied to the poled area, the direction of dipole moment changed. Thus, the local poled domain of the VDF oligomer thin films switched to the opposite polarity. From the piezoresponse and reversible switching properties, we could prove the ferroelectricity of VDF oligomer molecules for the first time.

ACKNOWLEDGMENTS

The author is grateful to Dr. M. Tatemoto of Daikin Kogyo Co., Ltd. for preparing the VDF oligomer. This work was partly supported by a Grant-in-Aid from the Ministry of Education, Science and Culture, by the Mazda Foundation's Research Grant and by KU-VBL(Kyoto University-Venture Business Laboratory) project.

REFERENCES

1. X. Q. Chen, H. Yamada, T. Horiuchi and K. Matsushige, Jpn. J. Appl. Phys., 37, 3834 (1998)

2. X. Q. Chen, H. Yamada, Y. Terai, T. Horiuchi and K. Matsushige and P.S. Weiss, Thin Solid Films, 353, 259 (1999)

3. X. Q. Chen, H. Yamada, T. Horiuchi and K. Matsushige, Jpn. J. Appl. Phys., 38, 3932 (1999)

4. A. V. Bune, V. M. Fridkin, Stephen Ducharme, L. M. Blinov, S. P. Palto, A. V. Sorokin, S. G. Yudin and A. Zlatkin, Nature 391, 874(1998)

5. A. V. Bune, Chuanxing Zhu, Stephen Ducharme, L. M. Blinov, V. M. Fridkin, S. P. Palto, N. G. Petukhova and S. G. Yudin, J. Appl. Phys., 85, 7869(1999)

6. Jaewu Choi, P. A. Dowben, Shawn Pebley, A. V. Bune, Stephen Ducharme, V. M. Fridkin, S. P. Palto and N. Petukhova, Phys. Rev. Lett. 80, 1328(1998)

7. T. Horiuchi, K. Fukao and K. Matsushige : Jpn. J. Appl. Phys. 26, L1839(1987)

8. K. Hayashi, K. Ishida, T. Horiuchi and K. Matsushige : Jpn. J. Appl. Phys. 31, 4081(1992)

9. K. Ishida, K. Hayashi, Y. Yoshida, T. Horiuchi and K. Matsushige : J. Appl. Phys. 73 , 7338 (1993).

10. K.Noda, K. Ishida, A. Kubono, T. Horiuchi and K. Matsushige, J. Appl. Phys., 86, 3688(1999)

NON-ELECTRICAL POLING IN NOVEL FERROELECTRIC POLYMERS

S. Tasaka, O. Furutani and N. Inagaki
Department of Materials Science and Technology, Faculty of Engineering, Shizuoka University,
Hamamatsu 432-8561, JAPAN

Abstract

Non-electrical poling was proposed in novel ferroelectric polymers including such as polythioureas, polycyanophenylenesulfides, and polyvinyfluorides. This poling method utilizing the cooperativity of molecular dipoles can be called "surface energy poling" and takes agvantage of the energy difference in the top and bottom surface of a polar aggregate (dipole glass) to form a remanent polarization. A ferroelectric polymer film sandwiched between a metal with higher surface energy and PTFE film with lower surface energy were heated to Tp=Tc (glass tramsition temperature or ferroelectric transition) x 1.2 and cooled slowly to room temperature. In the thin films less than 10μm, we observed the remanent polarization which gives a large pyro- and piezo-electric constant as well as that obtained by electrical poling. This poling is effective for homogenious structures such as amophous ferroelectric polymers.

Introduction

Recently poled polymers have attracted a great deal of attention for sensors and photonic application. [1] To obtain piezo- and pyro-electric and nonlinear optical activities in polymers, a poling process is necessary to make it noncentrosymmetrical structures for the orientation and stabilization of molecular dipoles in polymers. However, these poling methods have serious problems such as dielectric breakdown, surface thermal damage and the non-uniformity of remanent polarization. To solve these problems, several methods have been proposed.[1-3] In these methods, the possibility of dielectric breakdown and the non-uniformity polarization are not completely avoided.

Generally, surface chemical compositions of polymers or solid materials strongly depend on surface conditions. When a sample comes in contact with a higher surface energy material, the surface energy of the sample will increase if thermal molecular motions are possible. On the other hand, when a sample comes in contact with a lower surface energy material, the surface energy will decrease. If the molecular motion of polymers occurs on the surface or interface, the surface of the polymer in contact with different surface energy materials would be arranged by thermal diffusion so as to reduce the interface energy for energetic stabilization. Recently, the relationship between molecular motions and surface dynamics in polymers has been clarified by atomic force microscopy experiments.[4-6]

We have investigated the polymers containing cyano phenyl groups from the viewpoint of ferroelectricity and surface energy.[7,8,9] The poledpolymer had different surface energy on the film surfaces, because of the anisotoropic orientation of the cyanophenyl group. When the cyanophenyl polymers were sandwiched between a higher and a lower surface energy materials, we could get a poled state without electrical poling. (we named this new poling method, "surface

Mat. Res. Soc. Symp. Proc. Vol. 600 © 2000 Materials Research Society

energy poling".) When the dipoles of the polymers can rotate or move thermally, the molecule only located near both surfaces should orient for attaining the energy minimum. If the polymers show ferroelectric behavior, the dipole orientation of the polymer should be as far as the bulk state because of the high cooperativity of the dipoles. In crystals, the dipole orientation is governed by crystallization factors, such as nucleation and crystal growth, therefore it may be difficult to orient bulk dipoles. On the other hand, in an amorphous material with a polar aggregation, it would be possible to get the dipole orientation up to a certain thickness.

In this paper, we will follow up the possibility of surface energy poling in polymers showing ferroelectric behavior. We has been found recently several ferroelectric polymers, such as polythioureas [10] , polycyanophenylenesulfides,[8] polyurethanes [11,12] and poly(vinylfluoride-trifluoroethylene)[13] , having a large dipole moment , and a large dielectric relaxation strength (more than 40).

Experimental

Ferroelectric polymers ware obtained by polymerization methods previously reported.[7-13] Figure 1 shows the chemical structure of these polymers.

Surface energy was estimated from advancing contact angles of water, glycerol, formamide, diiodidomethane and tricrecyl phosphate on the film surfaces poled. The contact angle was measured at 20°C using an Erma contact anglemeter with a goniometer, model G-I. From the contact angle data the surface energy was estimated according to Kaelble's method [14]

In electrical poling, polymer films (surface energy : 33-45 mJ/m^2) were coated on indium tin oxide(ITO) glass slide or Aluminum coated glass slide with higher surface energy (more than 40mJ/m^2) and were given the upper electrode by aluminum evaporation. (Fig.2 a) In surface energy poling, the upper electrode was changed to 10μm Teflon FEP film with lower surface energy (15 mJ/m^2) as shown in Fig.2 b-1 and b-2. The sample were heated up to Tp(surface poling temperature), usually above Tg (glass transition temperature in the amorphous polymers) or Tc (ferro-paraelectric transition temparature in the crystalline polymers), and cooled down slowly (cooling rate 2-3°C/min) to room temperature.

Dielectric measurements were carried out in vacuum using a Hewlett-Packard HP-4285 LCR meter. Thermally stimulated depolarization current(TSC) and pyroelectric response for poled samples were recorded simultaneously from the current through the electrode irradiated by a pulsed semiconductor laser(670 nm, 3mW, 10Hz) upon heating(3°C/min). The absolute value of the pyroelectric constant was determined from the reversible TSC by heating and cooling near room temperature. Piezoelectric constant (d$_t$) was obtained from the thickness strain (Laser Doppler Vibrometer ONO SOKKI LV3100, laser spot 0.1mm) induced by the AC field (1kHz).

Results and Discussion

1 Thermal and Dielectric properties of Ferroelectric Polymers

Table 1 lists thermal and dielectric properties for ferroelectric polymers obtained. These polymers are amorphous or crystalline with good solubility and transparency, and are very easy

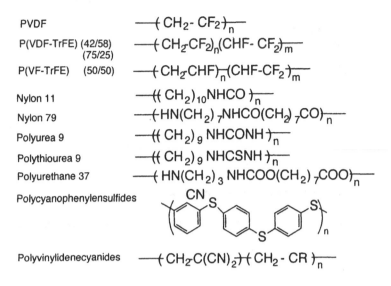

Figure 1. Chemical structure of ferroelectric polymers used in this syudy.

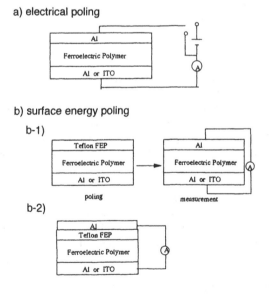

Figure 2. Poling and pyroelectric measurement conditions.

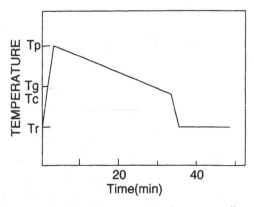

Figure 3. Thermal process of surface enrgy poling.

Table 1. Thermal and electrical properties of novel ferroelectric polymers after electrical poling

polymers	Tg(°C)	Tm	Tc	ε	Pr	Py
polythiourea 9	65	130	-	60	50	15
polyurethane 37	25	145	-	6	50	5
polycyanophenylenesulfides	60-150	-	-	100-150	20-40	5-20
polyvinylidenecyanides	140-180	-	-	60-100	40-60	20-40
P(VF-TrFE)(50/50)	50	200	105	30	40	50
P(VDF-TrFE) (42/58)	-40	160	60	60	50-60	25

Tg: glass transition temperature   Tm: melting temperature
Tc: ferroelectric transition temperature   de: dielectric relaxation strength at tansition region
Pr: remanent polarizarion after poling(mC/m2)   Py: pyroelectric constant (μC/m2K)

to fabricate as thin films.

The dielectric relaxation strength obtained from cole-cole plots of omplex dielectric constants with different frequency in the region of the glass transition(Tg) or the ferro-paralelectric transition (Tc) for these polymer is very large. The dielectric increment from free dipole rotation model[15] is estimated as smaller values than the observed values. Therefore, the dipoles are not independent of each other but undergo thermally cooperative motions. Especially, the hydrogen bonded polymers may be related to the defect motion, which has been observed in aliphatic alcohol by Meakins et al.[16]

All the polymers showed the D-E hysteresis loop. Therefore, the polymer dipoles are able to rotate easily by an external electric field (poling process). The poling process possibly contributes to make the dipole orientation. This process would be able to increase the order of the polar region even in amorphous. The hysteresis confirms that the polymer may be a ferroelectrics. The intersections of the hysteresis loop with y axis give a remanet polarization Pr. These values depend on the temperatures and the muximum applied voltage. The Pr which is obtained from thermally depolarization current for the poled polymers above Tg or Tc is larger than the value from D-E curve. This differnce of the value would be related to methastable structures in the polymers.

If it is possible for certain dipoles in polymer chains to orient in the direction of applied electric field, the ultimate polarization (Nxμ) can be estimated, depending on the conformation of the chain. The value of Pr is smaller than the estimated value.

2 Surface energy of poled ferroelectric polymers

Before discussing surface energy poling, the surface energy of the electrically poled film was measured. Table 2 lists surface energy of the poled films. In electrical poling, corona poling under the condition without surface damage, the surface energy difference between anode and cathode sides was observed. Especially, the difference in cyanopolymers is very large, while little diffrnece in non-polar PVDF. The surface energy difference of cathode side and the anode side is reasonable for the orientation direction of cyanophenyl or fluorocarbon dipoles.

It is very important for this polymer to have a thermally stable surface at room temperature. The surface energy of most polymers strongly depends on the aromosphere and the polymer with especially higher surface energy can be changed to lower value easily because of the absorption of oxygen or other gas. Furthermore, the measurement of surface energy using differnt organic liquids is difficult because of swelling and contamination in the sample. Our samples was fortunatly good for this measurement.

3 Non-electrical poling and Pyroelectricity

For non-electrical poling, Teflon-FEP film and ITO glass are used as a low and high surface energy materials, respectively. The thin film(below 10μm) sandwiched between these materials is heated up to above Tg (1) and then cool down to the room temperature (2). After removing the Teflon-FEP film, aluminum electrode was evaporated on the top of the polymer film.(as shown in Fig.2 b-1) Especially the polymers with hydrogen bonded moiety seem to have a special interaction with metal electrodes, so that the moiety might orient normal to the metal or metal oxide surfaces.

We measured the surface energy of non-electrically poled film. To prevent from contamination, the dry air was used as the lower surface energy material. The tendency of surface energie difference between air side and ITO side in the non-electrically poled film agreed with that between anode side and cathode side in the electrically poled film, respectively. Therefore the dipole orientation of both poled films is similar near the surface, however the remanent polarization of the films should be different in bulk because of the inhomogenity of polymer structure, such as crystal or amorphous.

Table 3 lists pyroelectric activity of ferroelectric polymer films obtained from non-electrical poling. Cyanopolymers and polythioureas, which are perfectly amophous and with large dielectric relaxation, are very effective for non-electrical poling. Little pyroelectric activity is shown in the polymers with the crystallization process. We tried to apply the method to typical ferroelectric crystalline polymers, such as P(VDF-TrFE)s or Nylon 11. We only suceeded in P(VDF-TrFE)(42/58) showing secondary transition in Tc(ferroelectric-paraelectric transition).

Table 2 Surface energy changes with electrical poling (mJ/m^2)

polymer	before poling	after poling -side	+side
polycyanophenylensulfides	42	30	46
p(VDF-TrFE)(42/58)	32	33	30
PVDF	37	36	37

Table 3 pyroelectric activity of polymers obtained from "surface energy poling"
(Al/polymer(5μm)/FEP)

polymers	pyroelectric activity	comment
Polyvinylidenefluoride	n	α crystal
Nylon 11	n	low crystallinity
Nylon 79	m	polar crystal
P(VDF-TrFE)(75/25)	w	Ferro-crystal
P(VDF-TrFE)(42/58)	vs	Ferro-crystal
Polyurea 9	w	low crystallinity
Polythiourea 9	s	amorphous
Polyurethane 37	w	low crystallinity
Polysyanopenylenesulfides	vs	amorphous (with aggregate)
polyvinylidenecyanides	w	amorphous

n: little or non   w: weak   m: middle   s: strong   vs: very strong

The remanent polarization to give pyroelectricity vanished in a critical temperature above Tg in the amophous polymers. This temperature, nearly equals to 1.15xTg, may relate to the molecular motion. In glassy polymers, it is known that intermolecular dissociation involving segment-segment "melting" occurs at 1.2xTg, called as T_{ll} .[17] It is suggested that Tc is a kind of transition about the break-up of the polar aggregation of hydrogen bondings.

To get a larger polarization by nonelectrical poling, we tried to control two poling factors, 1) starting temperatures in cooling and 2) cooling rate.[9] The starting temperature (Tp) in cooling should be higher than 1.15x(Tg or Tc). The cooling rate should be lower to stabilize the remanent polarization. These time dependence of the polarization formation seems to involve a relaxation process in the dipole motion. Therefore, the condition shown in Figure 2 was determined.

Figure 4 show temperature dependence of pyroelectric signals for P(VDF-TrFE)(42/58) as Al/ Teflon-FEP/film / ITO device. This device are already poled by non-electrical poling(the methods in Fig.2 b-2)... The remanent polarization which was formed by the contact with different surface energies can diminish during heating and recover during cooling. Therefore, the remanent polarization would be always kept a certain value at room temperature if the device received thermal stimulations.

Consequently, this device which is stabilized by both surfaces can behave as single crystal ferroelectrics.

Figure 5 shows the relationship between thickness of the film and pyroelectric signal for P(VDF-TrFE)(42/58) , prepared by nonelectrical poling. The thinner films make the larger pyroelectric constants which may be due mainly to surface effects. The surface dipole orientation ocuurs in the region of less than 0.5μm depth from the both sides of film. This is a very interesting phenomenon in the polymer, just like a epitaxial growth of crystal. To get the information of homogenity of remanent polarization in the electrode area, the inverse piezoelectric effect was measured for nonelectrically poled samples. The absolute values of piezoelectric constant (d_p=10-20pm/V), which is smaller than electrically poled one, was fluctuated by the heterogenity (chemically and physically) of the film surface.

Figure 6 shows schematic view of thermally reversible polarization and depolarization for a Al/Teflon-FEP/PTUs/ITO devise. The dipoles show surface anchoring by surface energy minimum. The surface anchoring which is relatively strong in the ferroelectric polymers, but

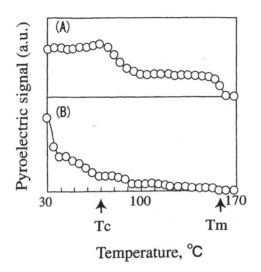

Figure 4. Temperature dependence of pyroelectric signal for Al/FEP/P(VDF-TrFE)(42/58)/Al, polymer film thikness (A) 1μm and (B) 10μm.

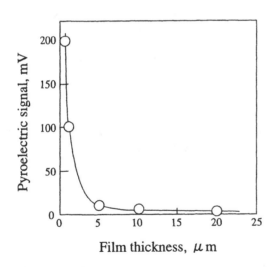

Figure 5. Film thickness dependence of pyroelectric signal for P(VDF-TrFE)(42/58) by surface energy poling.

Amorphous Ferroelectric Polymer
(dipole glass)

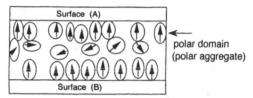

polar domain
(polar aggregate)

Crystalline Ferroelectric Polymer

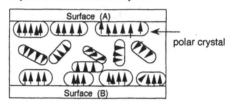

polar crystal

Figure 6. Schematic draw of a ferroelectric thin film sandwiched between different surface enrgy materials.

weak in liquid crystal [18] , depends on the molecular motion of polymers. Surface energy difference induces dipole anchoring but gives no effect in bulk. There are two types of polarization, namely surface and bulk polarizations as shown in this figure. We always have a question about the thickness of the surface polarization region. This strongly depends on the cooperativity of dipoles, which should be related to the polar domain size or polar glass size.

Conclusion

Non-electrical poling utilizing surface energy differences was attained in ferroelectric polymers with high cooperative dipoles. This method, "surface energy poling" is very useful for fabrication of their ferroelectric device.

References

1) S. Bauer, J.Appl.Phys., **80**, 5531(1996).
2) S.Tasaka, K.Shiraishi.K.Murakami and S.Miyata, Sen-i Gakkaishi, **39**,456(1983).
3) T.T.Wang et al. Ed., *Application of Ferroelectri Polymers*,(Blackie, Glasgow ,1987).
4) A.Takahara, in *Modern Approachs to Wettability: Theory and Applications*, M.E.Schrader and G,Loed, eds., (Plemum, New York,1992) p179.
5) T.Kajiyama, K.Tanaka and A.Takahara, Polymer, **39**,4665(1998).
6) T.Kajiyama, K.Tanaka, N.Satomi and A.Takahara, Macromolecules, **31**,5150(1998).

7) S.Tasaka, in *Ferroelectric Polymers* H.S.Nalwa ed.(Mercel Dekker, New York,1995) p325.

8) S.Tasaka, T.Nakamura and N.Inagaki, Jpn.J.Appl.Phys., 33,5838-5841(1994).

9) J.Ide, S.Tasaka and N.Inagaki, Jpn.J.Appl.Phys., 38, 2094-2052(1999).

10) S.Tasaka, K.Ohishi and N.Inagaki, Ferroelectrics, 171, 203-210(1995).

11) A.C.Jayasuriya, S.Tasaka, T. Shouko and N. Inagaki, J. Appl. Phys., 80, 362(1996).

12) S. Tasaka, T. Shouko, K. Asami and N. Inagaki, Jpn. J. Appl. Phys., 33, 1376 (1994)

13) K.Maeda, S.Tasaka, N.Inagaki and T.Kunugi, Polymer, 34,3387(1993).

14) D.H.Kaelble, *Physical Chemistry of Adhesion*, Wiley, New York (1971).

15) H.Froehlich, *Theory of Dielectrics*, Clarendon Press, Oxford (1958).

16) R.J.Meakins and R.A.Sack, Aust. J.Sci.Res., 4, 213-228(1953).

17) R.F.Boyer, in *Polymer year book 2*, Ed. R.A.Pathrick, (Godon & Breach, New York,1985)p233.

18) G.P.Bryan-Brown, E.L.Wood and I.C.Sage, Nature, 399,338-340(1999).

Piezoelectric, Electrostrictive, and Dielectric Elastomer

ULTRA-HIGH STRAIN RESPONSE OF ELASTOMERIC POLYMER DIELECTRICS

ROY KORNBLUH , RONALD PELRINE, JOSE JOSEPH, QIBING PEI, SEIKI CHIBA
SRI International, 333 Ravenswood Avenue, Menlo Park, CA 94025, roy.kornbluh@sri.com

ABSTRACT

The strain response of dielectric elastomers sandwiched between compliant electrodes was studied. These electroactive polymer artificial muscle (EPAM) materials show excellent overall performance and appear more attractive than many competing actuator technologies. Based on the available data, the actuation mechanism is due to the free charge interaction of the compliant electrodes, enhanced by the dielectric properties of the elastomer (Maxwell stress). Strains over 200%, actuation pressures up to 8 MPa, and energy densities up to 3.4 J/cm^3 have been demonstrated with silicone rubber and acrylic elastomers. Response time is rapid, and the potential efficiency is high. The fabrication of EPAM actuators can be simple and low cost. A wide range of small devices have been made, to demonstrate the potential of the technology and reveal more about performance and fabrication issues. These devices include bending beam actuators for scanners and clamps, diaphragm actuators for pumps and valves, stretched-film actuators for electro-optics, and bow actuators for muscle-like actuators for small robots and other micro machines.

INTRODUCTION

Motivation

In small-scale systems such as microrobots and micromachines, conventional electromagnetic technologies generally perform poorly due to physical scaling effects and fabrication difficulties. Such small-scale systems could benefit from improved actuators. There has been much recent interest in electroactive polymers as actuator materials. In general, polymers are attractive as actuator materials because they are lightweight, easily fabricated in various shapes, and low cost; in addition, their properties can often be modified as desired by various chemical means. Within the general category of polymers, the many different possible approaches to actuators include electrostrictive polymers,[1-5] piezoelectric polymers,[6] shape memory polymers,[7] electrochemically actuated conducting polymers,[8-13] and polymer-based air-gap electrostatic devices.[14]

Actuators and actuator materials have several important performance parameters including energy density, specific energy density, strain, actuation pressure, response time, and efficiency. To this list must be added practical considerations such as environmental tolerance, fabrication complexity, and reliability. Given this range of parameters, it is not surprising that individual applications depend more heavily on only one or a few of the parameters. Nonetheless, we consider a useful actuator technology to be one with good overall performance, as opposed to excellent performance in one or two parameters and poor performance in others. This view is supported by the dominance of electromagnetic technology on macro scales, where electromagnetic actuators have good overall performance, as well as by the good overall performance of natural muscle.

This paper describes an approach to electroactive polymer actuators that uses the deformation of elastomeric dielectrics. In studies of this approach since 1992, SRI has demonstrated overall performance similar to or exceeding that of natural muscle in many respects. Because it uses

Mat. Res. Soc. Symp. Proc. Vol. 600 © 2000 Materials Research Society

elastomers and has performance comparable to that of natural muscle, we refer to this approach as electrostrictive polymer artificial muscle (EPAM).

This paper is organized as follows: We first describe the basic principal of operation of the technology. Next, we describe the measured performance of several materials, including a recently identified acrylic that is capable of extremely large strains (more than 200%). We then describe issues related to the fabrication of devices based on this technology. Next, the design, fabrication, and performance of specific actuator embodiments are described. Finally, we summarize and discuss the potential applications of this technology and the research challenges that remain.

Background

The principle of operation of the EPAM technology is shown in Figure 1. An elastomeric polymer is sandwiched between two compliant electrodes. When a voltage difference is placed across the top and bottom electrodes, the polymer is squeezed in thickness and stretched in area. We have previously shown that the principal cause of this stress condition and the resultant deformation of the polymer is the electrostatic forces of the free charges on the electrodes.[15] It is useful to introduce an analytical model that relates the observed stresses and strains to the applied voltage.

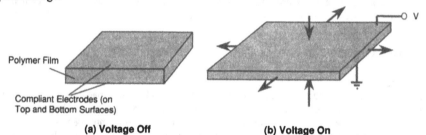

Polymer Film

Compliant Electrodes (on Top and Bottom Surfaces)

(a) Voltage Off   (b) Voltage On

Figure 1. Principle of Operation of EPAM; Film Expands in Area and Contracts in Thickness

The derivation of the electrostatic model is described by Pelrine, Kornbluh, and Joseph.[15] The actuation pressure, p, is given by

$$p = \varepsilon \, \varepsilon_o \, E^2 = \varepsilon \, \varepsilon_o \, (V/z)^2 \quad , \tag{1}$$

where E is the electric field, ε is the dielectric constant, ε_o is the permittivity of free space, V is the voltage, and z is the polymer thickness. Note that this pressure is greater by a factor of 2 than that arising from the commonly used equation for Maxwell's stress in a dielectric of a rigid plate capacitor. The greater pressure is due to the compliance of the electrodes, which allows both the forces of attraction between the oppositely charged electrodes and the forces tending to separate the charges on each electrode to couple into the effective pressure normal to the plane of the film.

For small strains with free boundary conditions, the polymer thickness strain, s_z, is given by

$$s_z = -p/Y = -\varepsilon \, \varepsilon_o \, (V/z)^2/Y \quad , \tag{2}$$

where Y is the modulus of elasticity. The model for large strains with more realistic constrained boundary conditions, such as those required to drive a load, is more complex. However, this simple case illustrates the influence of the electrical and mechanical properties of the polymer on actuation performance. The model also assumes that the elastomer is an ideal rubber, that is, that the rubber is incompressible and has a Poisson's ratio of 0.5.

One of the more useful metrics for comparing actuator materials, independent of size, is the energy densities of the materials. The actuator energy density is the maximum mechanical energy output per cycle and per unit volume of material. For small strains with free boundary conditions, the actuator energy density, e_a, of the material can be written as

$$e_a = Y s_z^2 = (\varepsilon \, \varepsilon_o)^2 \, (V/z)^4/Y \quad . \tag{3}$$

Conventionally, the elastic energy density $e_e = \frac{1}{2} Y s_z^2$ is often used. However, for large strains with a linear stress-strain relation this formula must be modified because as the thickness strain becomes increasingly negative, the film flattens out and the area over which the pressure must be applied increases. Nonlinear moduli, common for elastomers at high strain, further complicate the energy density formula. A detailed derivation for large strains gives a more general approximate formula for the elastic energy density of materials[16] as

$$e_e = -\frac{1}{2} p \ln [1 + s_z] \quad ,$$
(4)

where p is the actuation pressure given by Equation 1. This equation agrees with the more common formula at small strains but is significantly higher for strains greater than 20%.

With electronics that drive at constant voltage, Equation 4 is ambiguous because p changes throughout the stroke as the thickness of the film decreases and the electric pressure increases. This effect is not an issue with small-strain materials such as piezoelectrics because the fields are essentially constant with a constant applied voltage at small strains. With large-strain materials the effect can be dramatic, and we need to distinguish between the fundamental material performance, assuming optimal electronics, and the performance one expects to see with constant voltage. The simplest material assumption is that the breakdown strength of the material does not change throughout the stroke, so that the maximum field pressure should be used in Equation 4 when estimating peak performance with optimal electronics.[*] For constant voltage drivers, however, a more detailed analysis indicates that the $\ln[1 + s_z]$ term in Equation 4 should be replaced by $(s_z + 0.5 \, s_z^2)$.

Various electrostrictive mechanisms can be considered for polymer actuators. While the mechanism for EPAM relies on electrostatic forces, the performance of an EPAM actuator is critically determined by the electrical and mechanical properties of the polymer. In particular, EPAM performance depends on the macroscopic permittivity of the polymer as well as on its modulus of elasticity. Therefore, it is appropriate to consider EPAM as an electrostrictive polymer technology.

EXPERIMENT

Experimental Procedure For Measuring Material Performance

The experiments in our study were designed to measure the response of different polymer materials to applied electric fields.

The measurement of the performance of different polymer materials is complicated by the fact that several of the polymer materials evaluated are relatively soft (have elastic moduli below

---

[*] Real electronic drivers could closely approximate the optimal electronics by servoing the applied voltage according to the sensed strain.

1 MPa). Therefore, the constraints on the polymer film must be carefully controlled. The situation is further complicated by the fact that film samples are often quite thin. We therefore used optical methods to measure the strain condition of the polymer film.

The measurement configuration is shown in Figure 2. A thin film of a polymer is stretched uniformly across a circular hole in a rigid frame. Electrodes are applied to a relatively small circular area at the center of this frame. When a voltage is applied to the electrodes, the film between the electrodes expands in area and contracts in thickness. This expansion in area is measured with an optical microscope, a video camera, video digitizing hardware, and digital measurement software. The software measures the amount of motion of identifiable features on the surface of the electrodes (such as texture features) when a voltage is applied. By comparing the location of the features at a given voltage to the locations at zero voltage, we determine the in-plane strain at a given voltage. The magnification of the microscope and resolution of the video camera are such that a single pixel represents only a small portion of the observed motion. Most measured motions were on the order of tens of pixels. Photographs of one such experiment with 3M's* VHB 4910 acrylic adhesive are shown in Figure 3.

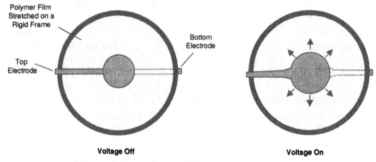

Voltage Off   Voltage On

Figure 2. Experimental Setup (Top View)

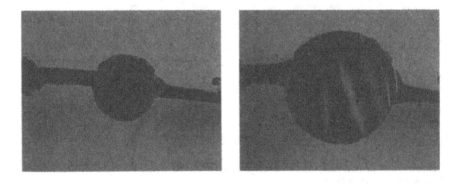

Voltage Off   Voltage On

Figure 2. Photo of VHB 4910 Acrylic Showing Approximately 200% Area Strain (−66% Thickness Strain)

The effects of creep and mechanical hysteresis on the measured strain were minimized by obtaining the strain reading immediately following the application of the voltage. The voltage was brought to zero between each measurement. Any effects of electrical hysteresis on the strain measurement were assumed to be insignificant compared to the total measured strain.

Films of the polymers were prepared by dissolving the uncured polymer in a suitable solvent, such as naphtha for the silicone rubbers (polydimethy siloxane). The polymer solution was then spun onto a disk at a speed appropriate to give a thin uniform coating. This operation was performed in a clean room to minimize the introduction of particulates into the film. The sample was then allowed to dry, to remove the solvent. Certain polymers were further cured at elevated temperatures according to the manufacturers' specifications for a complete cure. Film samples varied from 1 to 100 μm in thickness, depending on the properties of the film. In general, film thickness was chosen to give a maximum sustainable voltage between 1 and 10 kV.

The acrylic and polyurethane films are commercially available in rolls. These films were found to have good uniformity. With the acrylics, however, the thickness of the best performing films is fairly large (typically 0.5 to 1 mm). Thus, it was necessary to stretch these films considerably, by a factor of 3 to 4 in both planar directions. While a large prestrain is necessary in the film, due to the large actuated strain, the large prestrain also changes the elastic properties of this film, bringing it into a stiffer regime with lower viscoelastic losses. Finally, it has been observed in both acrylics and silicones that a large prestrain can actually increase the dielectric strength of the film and therefore increase its performance as an actuator.[16]

The selection of electrode materials is an area of ongoing research. For our measurements, we wanted a thin, extremely low-modulus electrode that provided uniform charge distribution over the surface of the film under the electrodes. In most cases, ultrafine graphite powder and carbon blacks were brushed onto the surface of the film through a stencil. For many of the silicones that are capable of undergoing extremely large strains, it was necessary to coat these electrodes with a mixture of silicone-polymer-based graphite grease and carbon-filled silicone (Chemtronics CW7200 and Stockwell RTV 60-CON, respectively) in order to ensure full coverage at large strains. Pure graphite grease was used as the electrode material on the acrylic.

The tensile elastic modulus was obtained by measuring the force on a thin strip of polymer material at different linear strains. The compressive modulus was assumed to be equal to the tensile modulus.

EPAM Actuators Demonstration

In addition to material measurements, several actuators based on silicone and acrylic polymer films were fabricated. The performance of these actuators was measured by applying a fixed voltage across the electrodes. The applied voltage was below the maximum breakdown strength of the materials.

RESULTS
Performance Of EPAM Materials

As seen from Equations 1–3, the ideal EPAM material, with maximal energy density, has a high dielectric constant ε, a high breakdown strength (V/z), and a relatively low modulus of elasticity Y. For most applications, desirable material properties also include low viscoelastic losses, a wide range of temperature and humidity tolerance, and ease of fabricating thin films.

Table 1 shows the performance of various polymers. Relative strains are used based on the formula (actuated length − unactuated length) / (unactuated length). Table 1 shows measurements of relative area strain, electric field, modulus, and dielectric constant. The relative thickness strain is calculated from the relative area strain and the assumption of constant volume for the polymer. The pressure and elastic energy density are estimated by means of Equations 1 and 4.

Table 1. Maximum Response of Representative Elastomers (Circular Strain Test)

Polymer (Specific type)	Elastic Energy Density, $\frac{1}{2}e$ (J/cm^3)	Pressure (MPa)	Relative Thickness Strain, s_z (%)	Relative Area Strain (%)	Young's Modulus (MPa)	Electric Field (V/μm)	Dielectric Constant (@1 kHz)
Acrylic 3M VHB 4910	3.4	7.2	−61	158	~2	412	4.8
Silicone Nusil CF19-2186	0.75	3.0	−39	63	1.0	350	2.8
Silicone Dow Corning HS3 (centrifuged)	0.098	0.3	−48	93	0.125	110	2.8
Polyurethane Deerfield PT6100S	0.09	1.6	−11	12	17	160	7.0
Silicone Dow Corning Sylgard 186	0.096	0.50	−32	47	0.7	144	2.8
Fluorosilicone Dow Corning 730 (centrifuged)	0.064	0.39	−28	39	0.5	80	6.9
Fluoroelastomer LaurenL143HC	0.027	0.65	−8	9	2.5	32	12.7
Polybutadiene Aldrich PBD	0.013	0.2	−12	14	1.7	76	4.0
Isoprene Natural Rubber Latex	0.006	0.11	−11	12	0.85	67	2.7

Average engineering modulus at the maximum strain.

Several materials listed in Table 1 deserve special mention, two silicones and an acrylic. The acrylic has the greatest strain and energy density.[16] Two silicones, Nusil CF19-2186 and Dow Corning HS3, also are capable of large strains and have among the highest measured energy densities. Other materials such as polyurethane can achieve higher actuation pressures at lower voltages, due to their greater permittivity (the dielectric constant of polyurethane is roughly 7, compared to 3 for silicones). Silicone has good coupling efficiency as well as other properties, including low creep and excellent tolerance to temperatures and humidity, that make it attractive for many actuator applications. Most of our actuator development work has focused on these silicones and the acrylic.

As indicated by the values, the VHB 4910 acrylic elastomer gave the highest performance in terms of strain and actuation pressure. The performance of both the acrylic and the silicones is enhanced using high prestrains. Data indicates high prestrain increases the breakdown strength of the material and thus the actuation pressure.[16] By adjusting the prestrain, we could manipulate the performance of VHB 4910 to increase the strain while lowering the pressure. Area strains of up to 330% and thickness strains of −77% have been demonstrated using lower prestrains. However, qualitatively the acrylic elastomer has relatively high viscoelastic losses that suggest that the response would be significantly reduced at high speeds. Depending on the configuration, the bandwidth of the acrylic (50% of the full 1 Hz strain response) has been measured at 10–40 Hz. This bandwidth is adequate for many mechanical applications, it and may be further enhanced with material modifications in the future. However, it will limit high frequencies applications such as acoustic transducers. By comparison, HS3 silicone has been used for loudspeakers at frequencies of 2–20 kHz.[17,18] The actuation of CF19-2186 silicone, albeit at lower strains and fields than reported here, has been measured directly via laser reflections with full strain response up to 170 Hz (resonance effects prevented measurement at higher speeds).[19] The only apparent fundamental limits on actuation speed are the viscoelastic losses, the speed of sound in the material, and the time to charge the capacitance of the film (electrical response time).

Comparison to Competitive Technologies

Table 2 shows several characteristics of EPAM materials and other electric actuation technologies, including several electroactive polymer technologies. The maximum strain values listed for EPAM in Table 2 are for the maximum linear, planar strain in one direction (note that Table 1 shows the area strain, which includes strain in two planar directions). The maximum linear strain is generally obtained by the use of a high prestrain in one planar direction, which causes the polymer to actuate primarily in the softer, orthogonal planar direction. The coupling efficiency for EPAM is based on treating the material as a variable capacitance.[19] This estimate is expected to be accurate for the silicone but may be an overestimation of the acrylic coupling because of the acrylic's higher viscoelastic losses. Nonetheless, the coupling of the acrylic is expected to be quite good at lower frequencies, and viscoelastic and resistivity measurements indicate that the maximum efficiency can be 60–80%, depending on drive conditions.

Table 2. Comparison of EPAM with Other Actuator Technologies

Actuator Type (specific example)	Maximum Strain (%)	Maximum Pressure (MPa)	Specific Elastic Energy Density (J/g)	Elastic Energy Density (J/cm³)	Coupling Efficiency k^2 (%)	Maximum Efficiency (%)	Specific Density	Relative Speed (full cycle)
Electroactive Polymer Artificial Muscle								
Acrylic	215	7.2	3.4	3.4	85	60–80	1	Medium
Silicone (CF19-2186)	63	3.0	0.75	0.75	63	90	1	Fast
Electrostrictor Polymer (P(VDF-TrFE)[3]	4	15	0.17	0.3	5.5	–	1.8	Fast
Electrostatic Devices (Integrated Force Array)[14]	50	0.03	0.0015	0.0015	~50	> 90	1	Fast
Electromagnetic (Voice Coil)*	50	0.10	0.003	0.025	n/a	> 90	8	Fast
Piezoelectric								
Ceramic (PZT)[†]	0.2	110	0.013	0.10	52	> 90	7.7	Fast
Single Crystal (PZN-PT)[20]	1.7	131	0.13	1.0	81	> 90	7.7	Fast
Polymer(PVDF)[‡]	0.1	4.8	0.0013	0.0024	7	–	1.8	Fast
Shape Memory Alloy (TiNi)[21]	> 5	> 200	> 15	> 100	5	< 10	6.5	Slow
Shape Memory Polymer[7]	100	4	2	2	–	< 10	1	Slow
Thermal (Expansion)**	1	78	0.15	0.4	–	< 10	2.7	Slow
Electrochemo-mechanical Conducting Polymer (Polyaniline)[8]	10	450	23	23	< 1	< 1%	~1	Slow
Mechano-chemical Polymer/Gels (polyelectrolyte)[11]	> 40	0.3	0.06	0.06	–	30	~1	Slow
Magnetostrictive (Terfenol-D, Etrema Products)	0.2	70	0.0027	0.025	–	60	9	Fast
Natural Muscle (Human Skeletal)[22]	> 40	0.35	0.07	0.07	n/a	> 35	1	Medium

*These values are based on an array of 0.01 m thick voice coils, 50% conductor, 50% permanent magnet, 1 T magnetic field, 2 ohm-cm resistivity, and 40,000 W/m2 power dissipation.
[†]PZT B, at a maximum electric field of 4 V/μm.
[‡]PVDF, at a maximum electric field of 30 V/μm.
*Aluminum, with a temperature change of 500°C.

As can be seen from Table 2, EPAM has good overall performance compared to that of competitive technologies. Its high strain and its very high specific energy density, together with its potential for fast response at good efficiencies, are particularly attractive for small robot and other mobile applications.

Performance of EPAM Actuators

A wide variety of EPAM actuator configurations have been demonstrated. Several are analogous to well-known piezoelectric configurations; others take advantage of the unique capabilities of high-strain polymer actuators. Here we briefly describe the fabrication and performance of four EPAM actuators that illustrate the range of possible designs and potential applications.

Unimorph and bimorph actuators are similar to their piezoelectric counterparts and work well with EPAM materials. Applications for unimorph and bimorph EPAM actuators include oscillating microfans, displays, and low-force robotic elements such as grippers. Strains are higher with EPAM than with competing piezoelectric materials, so higher bending angles can be achieved with shorter devices and without resorting to submicron film thicknesses. For example, bending angles approaching 360° have been achieved with 5-mm-long unimorphs. Figure 3 shows an EPAM unimorph made from silicone. The electrodes are sputtered gold on the bonded surface and carbon black on the free surface. The sputtered gold is sufficiently stiff, compared to the silicone, that the strain is much less at the gold electrode than at the carbon black electrode. The result is that the film bends toward the carbon black electrode when a voltage is applied.

Figure 3. An Array of EPAM Unimorphs Undergoing Actuation

One possible use of unimorphs or bimorphs is to deflect light for scanner applications or optical switches. Figure 4 shows optical scanning with a 2-mm unimorph microscanner.

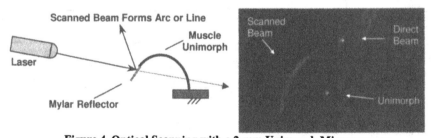

Figure 4. Optical Scanning with a 2-mm Unimorph Microscanner

The unimorph or bimorph construction is simple enough for large arrays that could, for example, be used to alter the physical properties of a surface such as in aerodynamic flow control applications.

Diaphragm EPAM actuators work very well, in part because a diaphragm can easily exploit both directions of planar expansion of the film. Figure 5 illustrates the structure of an EPAM diaphragm actuator. Diaphragms are particularly well suited for pumps but could also be used for adaptive optics, loudspeakers,[17,18] or controllable surface roughness (for example, on an aerodynamic surface). For pumps, single EPAM diaphragm actuators with up to 20 kPa (3 psi) pressure with 3-mm-diameter diaphragms have been demonstrated with silicone films. Single-layer acrylic diaphragms with diameters of up to 17 mm have produced pressures of 10 kPa. We have also demonstrated small proof-of-principle EPAM pumps using single-layer acrylic film diaphragms and one-way valves. These pumps produced flow rates of roughly 30-40 ml per minute and pressures up to 2500 Pa. Multiple cascaded pumps or thicker diaphragms could be used to increase pressure. An attractive feature of EPAM diaphragms, as opposed to piezoelectric diaphragms, is that the displacement can be relatively large without the sacrifice of other performance parameters.

Elastomer with Electrodes Bonded to Frame

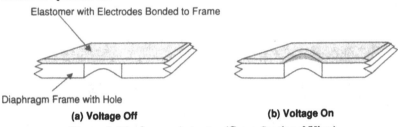

Diaphragm Frame with Hole

(a) **Voltage Off**   (b) **Voltage On**

Figure 5. Diaphragm Actuator (Cross-Sectional View)

Our highest-performing films allow for out-of-plane deflection equal to 50% or more of the diaphragm diameter. Figure 6 shows an acrylic diaphragm actuator undergoing large out-of-plane deformation in which the diaphragm changes shape from flat to hemispherical. In principle, piezoelectrics can achieve large diaphragm strokes, but in practice only very thin piezoelectric diaphragms can achieve similar strokes, because the intrinsic strain of piezoelectrics is so much smaller than that of EPAM. The use of very thin piezoelectric diaphragms, however, sacrifices other parameters such as pumping pressure or packaging density, and in most cases significantly reduces the size of piezoelectric diaphragm strokes.

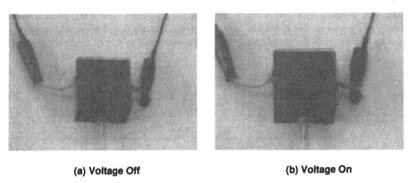

(a) **Voltage Off**   (b) **Voltage On**

Figure 6. An Acrylic Diaphragm Undergoing Actuation

Diaphragm actuators are well suited to microdevices because they can easily be fabricated in small sizes. Arrays of diaphragms with 150 μm diameters have been demonstrated, as well as in situ fabrication of diaphragms on silicon wafers. Linear actuators can also be made in a configuration by adding a flexure that presses down on the diaphragm. We used linear diaphragm actuators to demonstrate micro light scanners, in a configuration alternative to that of the unimorph light scanner in Figure 4. Driving voltages of 190–300 V were used to demonstrate scanning similar to that in Figure 4, but at 60–200 Hz.

A **stretched film** actuator consists of a polymer film stretched over a rigid frame. The stretch can be uniform in both plane directions (as in the configuration used for strain measurement described in Section 2 and shown in Figure 3), or can be stretched much more in one direction than the other (anistropically). In the latter case, the film tends to actuate primarily in the direction with lower strain. This directional compliance effect can be combined with geometric effects to produce long, thin actuation areas that are capable of extremely large strains in the direction orthogonal to the long axis of the active area. Such an effect is shown in Figure 7. We have used this approach to produce linear strains of more than 100% in silicones and more than 200% in acrylics.

| (a) Voltage Off | (b) Voltage On |

Figure 7. Linear-Motion-Stretched Film Actuator Showing Approximately 200% Strain

Such actuators can be used for many applications where small motions are needed, such as an optical switch in which an opaque electrode area interrupts a light beam when actuated. The advantage of this approach is that the structure is extremely simple and therefore low cost. The apparatus is basically solid state and has just one moving part, the artificial muscle film.

A binary switch of this type can be useful, but in some cases it may be desirable to continuously modulate the amount of light transmitted. This modulation can also be done with a stretched film actuator that uses carbon fibrils as an electrode. The fibril electrode gradually becomes less opaque as the electroded area increases. Figure 8 shows an example of such an actuator. Its advantage is the simplicity of its design compared to that of mechanical apertures.

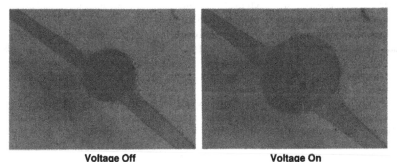

| Voltage Off | Voltage On |

Figure 8. Solid-State Optical Aperture Based on a Stretched Film Actuator

Bow actuators are a simple and efficient means of coupling the energy of deformation of a polymer film to linear motion.. While stretched film actuators are capable of linear motion, they do require frames. It is often desirable to include a discrete element that changes length, much like a muscle. The bow actuator uses flexures to allow the film to expand as shown in Figure 9, with suitable prestrains so that the actuation in the desired direction is enhanced.

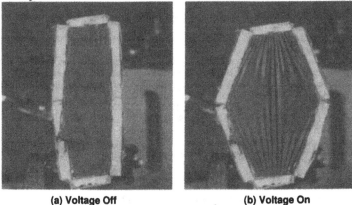

| (a) Voltage Off | (b) Voltage On |

Figure 9. Bow Actuator Undergoing Approximately 100% Peak Strain

The bow actuator is an example of an EPAM actuator design that specifically exploits the properties of EPAM as an actuator material. Just as optimal piezoelectric designs differ from optimal magnetic actuator designs, it is expected that optimal EPAM designs will be different from those in which other actuator materials are used, particularly because of the large intrinsic strains of EPAM.

CONCLUSIONS

Electroactive polymers based on an elastomeric dielectric between compliant electrodes show great promise for actuation. These EPAM materials show excellent overall performance and appear more attractive than many competitive actuator technologies. Strains of over 200%, actuation pressures of 8 MPa, and energy densities of 3 J/cm^3 have been demonstrated with silicone rubbers and acrylics. Their response is fast and their efficiency is potentially high. A wide range of small devices have been made to demonstrate the potential of the technology and clarify the performance, fabrication, and actuator design issues.

The key technical issues for EPAM technology include the selection of elastomer and compliant electrode materials, the fabrication of integrated elastomer-electrode structures, and actuator design. Improved elastomer materials are an ongoing area of research, though their current performance is already attractive for many applications. Compliant electrodes have been based primarily on carbon particle materials (e.g., graphite, carbon black). Excellent-quality single-layer elastomer films can be fabricated by spin coating or commercial processes. In the fabrication area, some progress has been made in demonstrating in-situ multilayer fabrication, but at present the performance of these films is below that of films made by single-layer fabrication.

ACKNOWLEDGEMENTS

Much of this work was performed under the management of the Micromachine Center as the Industrial Science and Technology Frontier Program, Research and Development of Micromachine Technology of the Ministry of International Trade and Industry, Japan, and supported by the New Energy and Industrial Technology Development Organization.

REFERENCES

1. R. Pelrine, J. Eckerle, and S. Chiba, *Proc. Third Intl. Symposium on Micro Machine and Human Science*, Nagoya, Japan (1992).
2. R. Pelrine, R, R. Kornbluh, J. Joseph, and S. Chiba, *Proc. IEEE Tenth Annual Intl. Workshop on Micro Electro Mechanical Systems*, Nagoya, Japan, 238–243 (1997).
3. Q. Zhang, V. Bharti, and X. Zhao, "Giant Electrostriction and Relaxor Ferroelectric Behavior in Electron-Irradiated Poly(vinylidene fluoride-trifluoroethylene) Copolymer," Science, 280, 2101–2104 (1998).
4. M. Zhenyl, J.I. Scheinbeim, J.W. Lee, and B.A. Newman, "High Field Electrostrictive Response of Polymers," J. Polymer Sciences, Part B—Polymer Physics, 32, 2721–2731 (1994).
5. Y. Shkel, and D. Klingenberg, J. Applied Physics, 80(8), 4566–4572 (1996).
6. T. Furukawa, and N. Seo, Japanese J. Applied Physics, 29 (4), 675–680 (1990).
7. H. Tobushi, S. Hayashi, and S. Kojima, in JSME International J., Series I, 35 (3) (1992).
8. R. Baughman, L. Shacklette, R. Elsenbaumer, E. Pichta, and C. Becht, in Conjugated Polymeric Materials: Opportunities in Electronics, Optoelectronics and Molecular Electronics, eds. J.L. Bredas and R.R. Chance, Kluwer Academic Publishers, The Netherlands, 559–582 (1990).
9. D. De Rossi, and P. Chiarelli, Macro-Ion Characterization, American Chemical Society Symposium Series, 548, 40, 517–530 (1994).
10. K. Oguro, Y. Kawami, and H. Takenaka, *J. Micromachine Society*, 5, 27–30 (1992).
11. M. Shahinpoor, J. Intelligent Material Systems and Structures, 6, 307–314 (1995).
12. E. Smela, O. Inganas, and Q. Pei, Advanced Materials, Communications Section, 5, 9, 630–632 (1993).
13. Q. Pei, O. Inganas, and I. Lundstrom, Smart Materials and Structures, 2, 1–6 (1993).
14. S. Bobbio, M. Kellam, B. Dudley, S. Goodwin Johansson, S. Jones, J. Jacobson, F. Tranjan, and T. DuBois, in Proc. IEEE Micro Electro Mechanical Systems Workshop, February, Fort Lauderdale, Florida (1993).
15. R. Pelrine, R. Kornbluh, and J. Joseph, Sensor and Actuators A: Physical 64, 77–85 (1998).
16. R. Pelrine, R. Kornbluh, Q. Pei, and J. Joseph, "High Speed Electrically Actuated Elastomers With Over 100% Strain," to be published in Science.
17. R. Heydt, R. Kornbluh, R. Pelrine, and B. Mason, J. Sound and Vibration, 215(2), 297–311 (1998).
18. R. Heydt, R. Pelrine, J. Joseph, J. Eckerle, and R. Kornbluh, "Acoustical Performance of an Electrostrictive Polymer Film Loudspeaker," to be published in J. Acoustical Society of America (1999).
19. R. Kornbluh, R. Pelrine, J. Joseph, R. Heydt, Q. Pei, and S. Chiba, Proc. SPIE Sixth Intl. Symposium on Smart Structures and Materials: Electro-Active Polymer Actuators and Devices, 149–161(1999).
20. S. Park, and T. Shrout, J. Applied Physics, 82, 1804–1811 (1997).
21. I. Hunter, S. Lafontaine, J. Hollerbach, and P. Hunter, Proc. 1991 IEEE Micro Electro Mechanical Systems—MEMS '91, 166–170 (1991).
22. I. Hunter, and S. Lafontaine, Technical Digest of the IEEE Solid-State Sensor and Actuator Workshop, 178–185 (1992).

ELECTROSTRICTIVE GRAFT ELASTOMERS AND APPLICATIONS

J. SU*, J. S. HARRISON**, T. L. St. CLAIR**, Y. BAR-COHEN***, and S. LEARY***,
*National Research Council, NASA-Langley Research Center, Hampton, VA 23681, USA
**NASA-Langley Research Center, Hampton, VA 23681, USA
***Jet Propulsion Laboratory/CalTech, Pasadena, CA 91109, USA

ABSTRACT

Efficient actuators that are lightweight, high performance and compact are needed to support telerobotic requirements for future NASA missions. In this work, we present a new class of electromechanically active polymers that can potentially be used as actuators to meet many NASA needs. The materials are graft elastomers that offer high strain under an applied electric field. Due to its higher mechanical modulus, this elastomer also has a higher strain energy density as compared to previously reported electrostrictive polyurethane elastomers. The dielectric, mechanical and electromechanical properties of this new electrostrictive elastomer have been studied as a function of temperature and frequency. Combined with structural analysis using x-ray diffraction and differential scanning calorimetry on the new elastomer, structure-property interrelationship and mechanisms of the electric field induced strain in the graft elastomer have also been investigated. This electroactive polymer (EAP) has demonstrated high actuation strain and high mechanical energy density. The combination of these properties with its tailorable molecular composition and excellent processability makes it attractive for a variety of actuation tasks. The experimental results and applications will be presented.

INTRODUCTION

Materials that sustain mechanical displacement under controlled electrical excitation are needed as actuators for many applications. For aerospace applications, there is also a need for low mass, high performance, and ease of processability which are inherent characteristics of electroactive materials. Electroactive polymeric elastomers that show electromechanical activities, especially the large electric field induced strain, are being thought of potential candidates for such applications. Existing materials include polyurethane elastomers, [1-3] and silicon rubber [4,5]. Recently, we have demonstrated an electrostrictive response in graft elastomers. The elastomers offer large electric field induced strain and significant high mechanical modulus. Therefore, a high electromechanical output power, or high strain energy density is achieved. In addition to the high performance as a new class of electromechanically active polymeric materials, the electrostrictive graft elastomers also offer advantages such as excellent processability, and electrical and mechanical toughness.

EXPERIMENT

An electrostrictive graft elastomer consists of two components, a flexible backbone polymer and grafted crystalline groups. The schematics in Figures 1a and 1b show the structure and molecular morphology, respectively, of the graft elastomer. The graft crystalline phase provides the polarizable moieties and serves as cross-linking sites for the elastomer system.

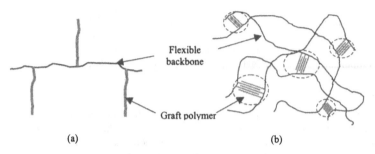

Flexible backbone

Graft polymer

(a)   (b)

Figure 1.   Structure and morphology of the graft elastomer.

The graft elastomer films were prepared by solution casting. Five grams of graft elastomer powder was added to N,N-dimethylformamide (as received from Aldrich) and was heated to 60 °C while stirring for 2 hours to make a 5 wt% solution. The solution was cooled to room temperature and cast on glass substrates and placed in a vacuum chamber. After drying overnight under vacuum, at room temperature, 20 micrometer thick films of the graft elastomer films were obtained. Gold electrodes were sputtered on opposing sides of the films using a plasma deposition device (Hummer III, Technics).

Dynamic mechanical and piezoelectric properties of the graft elastomer films were measured as a function of temperature and frequency using a Rheovibron DDV-II-C mechanical analyzer that has been modified to collect electric charge data as a function of applied stress. Dielectric data of the films was measured as a function of temperature and frequency using an HP4192A Impedance Analyzer.

The electric field induced strain was tested as a function of the applied electric field. The deflection of a bilayer bending actuator was measured and the strain was calculated using the relationship

$$-S_L = 2S_T = 2d(1/R) = 4dL/(L^2+l^2) \qquad \bullet \qquad (1)$$

where S_L is the longitudinal strain of the active elastomer layer, S_T is the strain of the active layer in the transversal direction, $1/R$ is the bending curvature of the actuator, d is the distance between the central layers of active and non-active films, and the L and l are the deflection of the tip of the actuator in in-surface and off-surface directions.[6,7] The low temperature response of the bending actuator was tested in a Satec TCS1200 temperature controller with cryovac chamber (Satec Systems Inc.). Voltage was applied using a Trek 10/10A amplifier. Deflection was measured as a function of temperature (-55 °C to 25 °C) and voltage (up to 2.75 kV) using a Sony XC-55 progressive scan CCD and IMAQ 1407 PCI image acquisition board (national Instruments) controlled by LabView software (National Instruments).

The results of the x-ray diffraction (XRD) test of the films showed that the crystallinity is about 22%. Since the graft elastomer contains 30 wt% of the graft polymer, approximately, 70% of the graft polymer crystallized for the present processing condition. Differential scanning calorimetry (DSC) analysis showed that the melting temperature of the graft elastomer is about 160 °C with a melting range from 145 °C to 185 °C. DSC results also indicate that a glass transition occurs around 50 °C (from 40 °C to 60 °C). This transition should be the secondary glass transition, which is related to the cross linking grafted units.

RESULTS AND DISCUSSION

Figure 2 shows the relationship between the electric field induced longitudinal strain, S, and the applied electric field at room temperature. The strain exhibits a quadratic dependence with the applied electric field. The electric field induced strain was observed to be as large as 3.9% at an applied electric field of 120MV/m.

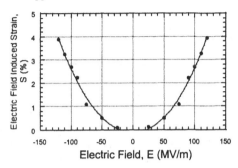

Figure 2.   Electric field dependence of the induced strain on the graft elastomers.

After subjecting the graft elastomer film to a high electric field, 130MV/m, the film was tested for piezoelectric activity. The result is shown in Figure 3. The piezoelectric strain coefficient, d_{31}, is only about 0.1 pC/N in the temperature range from 25 °C to around 100 °C at frequencies of 1 Hz, 10 Hz, and 100 Hz. This indicates that there is no significant remanent polarization in the graft elastomer after being treated under a high electric field.

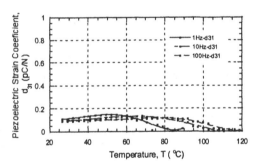

Figure 3.   Piezoelectric activity of the graft elastomer after poling treatment.

In order to understand the mechanisms of the electrostrictive response of the graft elastomer, dynamic mechanical and dielectric properties of the elastomer were investigated as a function of frequency and temperature. The results of the dynamic mechanical analysis are shown in Figure 4. The temperature range for the test was from room temperature to 100 °C at frequencies of 0.1, 1, 10, 100 Hz. As can be seen, at room temperature, the mechanical modulus of the graft elastomer is in the range from 550 MPa to 700 MPa for the measured frequencies.

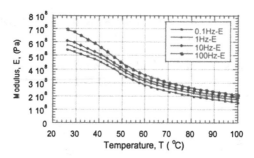

Figure 4.   Temperature and frequency dependence of dynamic
            mechanical properties of the graft elastomer films.

The temperature and frequency dependence of the dielectric constant of the graft elastomer is shown in Figure 5. At room temperature, there is no significant disparity in the dielectric constant with frequency; however, as temperature increases, the dielectric constant varies for the range of frequencies measured.

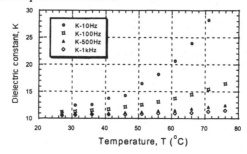

Figure 5. Temperature and frequency dependence of dielectric constant of the graft
          elastomer.

In the polymeric elastomers exhibiting a large electric field induced strain, two intrinsic mechanisms are considered as primary contributors. They are electrostriction and the Maxwell stress effect.[7]   The contributions from both mechanisms exhibit a quadratic dependence with an applied electric field.   The electrostrictive effect is due to the reorientation of the polar phase in response to an applied electric field while the Maxwell stress effect is attributed to the force generated by the accumulated charges on the opposing surface of films under an applied electric field. The strain response of an elastomer can be contributed by either one of them or both of them. The following equations give the relationships of the stain and the applied electric field through the two mechanisms:

$$S_{electrostriction} = -Q\varepsilon_0^2(K-1)^2E^2 \quad \text{and} \quad S_{Maxwell} = -s\varepsilon_0^2KE^2 \tag{2}$$

where Q is the electrostrictive coefficient, K is the dielectric constant, s is the mechanical compliance, and $\mathbf{E}$ is the applied electric field. The total strain should be

$$S = S_{electrostriction} + S_{Maxwell} = RE^2 \qquad (3)$$

where R is the electric field induced strain coefficient.

According to the experimental results, a comparison can be made relating the contributions of the different mechanisms to the strain response of polymeric elastomers, which is tabulated in Table 1. For silicon rubber, 100% of the strain response resulted from

Table 1. Comparison of the contribution of mechanisms to the strain response

Materials	Electrostriction	Maxwell Stress
Silicon rubber	0	100
Polyurethane	65	35
Graft elastomer	95	5

the Maxwell stress, or electrostatic, contribution.[5] For electrostrictive polyurethane, the majority of the strain response is due to electrostriction; however, the Maxwell stress contribution is still significant due to the low mechanical modulus of polyurethane.[3] For the newly developed electrostrictive graft elastomers, more than 95% of the strain response is contributed by electrostriction mechanism while the contribution from the Maxwell stress is less than 5%.

Table 2 gives a comparison of some key properties of electrostrictive polyurethane and graft elastomers as electroactive polymeric materials. As can be seen,

Table 2. Comparison of electromechanical properties of electrostrictive elastomers

Materials	Strain S, (%)	Modulus Y, (MPa)	Output power P, (MPa)	Energy density E_{strain}, (J/kg)	Dielectric Constant, K
Graft elastomer	4	550	22	247	11
Polyurethane	11	17	2	87	6

the graft elastomers offers about 4% induced strain which is smaller than that of polyurethane. However, the graft elastomer exhibits a significantly higher mechanical modulus, which results in the high mechanical output power and high specific strain energy density. In addition to the good electromechanical properties, the electrostrictive graft elastomers also has a higher dielectric constant than that of the polyurethane elastomers. This is an advantage of the graft elastomer over the polyurethane in applications.

APPLICATIONS

Two types of actuators were fabricated using the electrostrictive graft elastomer: a unimorph actuator and a bimorph actuator. The unimorph is fabricated by adhering a layer of the graft elastomer film with electrodes and a layer of the graft elastomer film without electrode together. The adhesive used was a room temperature curable epoxy resin. Figures 6a and 6b show the unimorph in the unexcited state and excited state, respectively. The unimorph actuator bends in one direction and its response frequency is two times of the frequency of the driving power supply, or f (bending) = $2f$ (driving).

Figure 6. The unimorph actuator (a) unexcited and (b) electrically excited state.

Figure 7 shows the bimorph actuator, which can bend in both direction when controlled by the power supply. The response frequency of the bimorph (two direction bending actuator) is the same as the frequency of the power supply, f (bending) $= f$ (driving).

Figure 7. The bimorph actuator in the state of unexcited (middle), one direction excited (left), and opposite direction excited.

In the low temperature test, it was observed that the actuators still function at $-50\ ^\circ C$.

CONCLUSIONS

A new class of electromechanically active polymer was developed using electrostrictive graft elastomers. These elastomers offer a unique combination of desirable promising properties including: light weight, large electric field induced strain, high performance, and excellent processability. Optimization of electromechanical properties of the graft elastomers can be realized by molecular design, composition adjustment, and processing to meet requirements of various applications

REFERENCES

1. M. Zhenyi, J. I. Scheinbeim, J. W. Lee, and B. A. Newman, *J. Polym. Sci., Part B: Polym. Phys.,* **32**, 2721 (1994).
2. Q. M. Zhang, J. Su, C. H. Kim, R. Ting, and R. Capps, *J. Appl. Phys.,* **81**, 2770 (1997).
3. J. Su, Q. M. Zhang, C. H. Kim, R. Y. Ting, and R. Capps, *J. Appl. Polym. Sci.,* **65**, 1363 (1997).
4. R. Pelrine, R. Kornbluh, and J. Joseph, *Sensor and Actuators A: Physical,* **64**, 77 (1998).
5. R. Kornbluh, R. Pelrine, J. Joseph, R. Heydt, Q. Pei, and S. Chiba, *Proceedings of SPIE,* **3669**, 149 (1999).
6. W. Takashima, M. Kaneko, K. Kaneto, and A. G. MacDiarmid, *Synthetic Metals,* 483 (1995).
7. H. Wang, Q. M. Zhang, L. E. Cross, R. Ting, C. Coughlin, and K. Rittenmyer, *Proc. Int. Symp. Appl. Ferro.,* **9**, 182 (1994).

DEHYDRATION TIME DEPENDENCE ON PIEZOELECTRIC AND MECHANICAL PROPERTIES OF BOVINE CORNEA

A. C. Jayasuriya*, J. I. Scheinbeim*, V. Lubkin**, G. Bennett** and P. Kramer**
* Polymer Electroprocessing Laboratory, Department of Chemical and Bio-Chemical Engineering, Rutgers University, New Jersey, 98 Brett Road, Piscataway, NJ 08854
** Aborn Laboratory, New York Eye and Ear Infirmary, 310 East 14th Street, New York, NY 1000

ABSTRACT

The Young's Modulus (E) and piezoelectric coefficient (d_{31}) have been investigated as a function of dehydration time for bovine cornea at room temperature. The piezoelectric and mechanical responses observed were anisotropic for bovine cornea and d_{31} decreased, while E increased with dehydration. In addition, water molecules appear to increase the crystallinity (of collagen) in the cornea. With dehydration of the cornea, reduction of crystallinity and changes in hydrogen bonding were observed by Fourier Transform Infra Red (FTIR) and Wide Angle X-ray Diffracion (WAXD) measurements.

INTRODUCTION

The existence of piezoelectricity has been studied for a number of biological polymers such as bone [1], tendon [2], muscle [3], and DNA [4]. Piezoelectric and related properties in synthetic organic polymers such as poly(vinylidene fluoride) (PVF_2) [5] and its copolymers [6] have been extensively studied. These polar materials exhibited piezoelectricity due to orientation of dipoles towards the applied electric field i.e. these materials are ferroelectric.

Biological polymers can often be represented as a water-filled proteinous matrix, unlike commercial organic polymers. Collagen is a fibrous protein, which provides associated tissue with the majority of its tensile strength. Collagen molecules arrange in triple helix polypeptide chains [7]. The collagen structure contains water molecules which are described in three ways: tightly bound water, slightly bound water and free water [8]. These water molecules are bound by hydrogen bonds.

It has been reported that bovine corneas contain approximately 66% (or more) water according to X-ray diffraction measurements [9]. The collagen fibrils swell very little with water, and most of the water goes to the interfibrillar spaces. It has been reported that the piezoelectric and dielectric properties of biological polymers are greatly influenced by small amounts of absorbed water [10]. In addition, the mechanical behavior of collagen also strongly depends on the degree of hydration [11]. In this paper, we report the dehydration time dependence of the piezoelectric and mechanical properties of bovine corneas. The structural changes of bovine cornea during dehydration were analyzed using FTIR and WAXD.

MATERIALS AND METHODS

Bovine cornea samples, supplied by the New York Eye Bank, were stored in normal saline solution. Samples were kept in a refrigerator at 4 °C until starting the measurements. The corneas were cut into rectangular strips in different directions such as horizontal, vertical and diagonal to study anisotropic effects. The roughly circular corneas were marked as the right hand direction on the face of a clock. The positions for 12 and 3 o'clock were determined by the eye muscles.

Mat. Res. Soc. Symp. Proc. Vol. 600 © 2000 Materials Research Society

Then, rectangular strips were cut in different directions using the following method: 12-6 o'clock strip as vertical; 3-9 o'clock strip as horizontal; 1:30-7:30 or 10:30-5:30 as diagonal (see figure 1). The opposite surfaces of the strips were painted with conductive silver print to make electrodes. The piezoelectric constant (d_{31}) and Young's modulus (E) were measured simultaneously using a TOYOSEIKI Rheolograph® at room temperature. FTIR spectra were obtained for very thin cornea tissue samples using a NICOLE® spectrometer. Wide angle X-ray diffraction patterns (WAXD) were obtained using a SIEMENS diffractometer with Cu Kα filtered radiation.

RESULTS AND DISCUSSION

(i) Mechanical and piezoelectric anisotropy in bovine cornea.

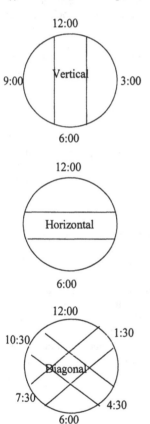

Fig. 1 Sample preparation of
bovine cornea

Figure 2 shows the dehydration time dependence of Young's Modulus, E, averaged for (a) five bovine corneas vertically cut (b) averaged for five corneas horizontally cut and (c) averaged for nine corneas diagonally cut. It can be seen that E increases with dehydration time for all samples. The diagonally cut samples show lower E values, 0.09 MPa, at the beginning of dehydration and then increase to around 0.16 MPa within 80 min. The horizontally cut samples show higher E values than the diagonally cut samples and lower values than the vertically cut samples which shows the highest initial E value, 0.15 MPa, which reaches to 0.23 MPa after 80 min. These data illustrate the anisotropic behavior of the mechanical properties of bovine cornea.

Figure 3 shows the piezoelectric constant d_{31} averaged for (a) five bovine corneas vertically cut (b) averaged for five corneas horizontally cut and (c) averaged for nine corneas diagonally cut. Initially, the diagonally cut samples show much higher d_{31} values than the horizontally and vertically cut samples. The initial d_{31} value of 2500 pC/N for the diagonally cut samples is approximately 2.5 times larger than the d_{31} of the other two directional samples. All three kinds of samples exhibit a decrease in d_{31} with dehydration time. The horizontally cut samples exhibit a higher d_{31} value than the vertically cut samples. After 80 min. of dehydration, d_{31} is about 300 pC/N for all types of samples but the values maintain their relative order: diagonal > horizontal > vertical.

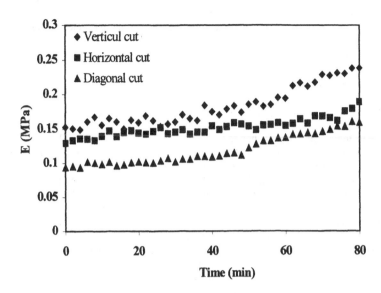

Fig. 2 Dehydration time dependence of Young's modulus (a) averaged five bovine corneas vertical cut (b) averaged five bovine corneas horizontal cut (c) averaged nine bovine corneas diagonal cut.

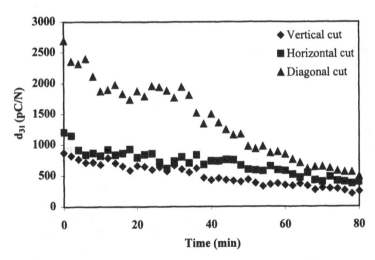

Fig. 3 Dehydration time dependence piezoelectric constant d_{31} (a) averaged five bovine corneas vertical cut (b) averaged five bovine corneas horizontal cut (c) averaged nine bovine corneas diagonal cut.

Similar anisotropic mechanical and piezoelectric behavior were reported for human corneas as a function of dehydration time [12]. In that case, diagonally cut samples also exhibit much higher piezoelectric response than other two directionally cut samples. The initial piezoelectric response of 2250 pC/N decreases to 400 pC/N for diagonally cut samples while it decreases to 100 pC/N for vertically and horizontally cut samples with 30 min. of dehydration time.

(ii) FTIR and WAXD Measurements

Bovine corneas contain approximately 66% or more water, the fibrils themselves swell very little and most of the additional water goes into the interfibrillar spaces [9]. The water molecules in collagen appear in three ways: tightly bonded water, slightly bonded water and free water. Some of the tightly bound water can be definitely associated with structure stabilizing roles for the macromolecules [7]. From X-ray diffraction studies of collagen, a very specific arrangement for the bound water molecules has been deduced, where two water molecules are hydrogen bonded to the collagen triple helix for every three amino-acid residues [13]. By bridging across two different polypeptide chains, these bound water molecules stabilize the triple helix structure of the collagen macromolecule [7].

Therefore, different types of hydrogen bonded O-H peaks can be expected in the IR spectra. According to our FTIR results, we observed several different peaks in the range of 3300 to 3600 cm^{-1}. These peaks are interpreted as the OH and N-H stretching vibration bands. We obtained IR spectra every 5 min. over a 35 min. time period and observed significant changes in the peaks in the region of 3300-3700 cm^{-1} wavenumbers. These are shown in Fig. 4. Changes in absorbance intensity of the peaks can be seen between the initial spectra and those obtained during dehydration. At longer dehydration times, the intensity difference between subsequent peaks gradually decreases. The intensity reduction of all the peaks can be explained due to the change in the amount of water present in the sample. There are several O-H stretching hydrogen bond vibration bands appearing in the first and second spectra due to the differently bonded states in the collagen fiber and then those bands merged beyond the second spectra. This significant change observed in the O-H hydrogen bonded vibration is related to the breaking or loosing of hydrogen bonds by disoriented dipoles between the peptide and water molecules in collagen during dehydration.

Figure 5 shows the X-ray diffraction patterns for bovine cornea samples (a) stored in saline (b) air dried for 1 h and (c) air dried for 2h. Three main WAXD peaks can be seen in the region of 2θ=15° to 60°. The first small peak appears at 2θ=18.8°, the second and larger peak at 2θ=28.3° and the third peak at 2θ= 42°. When the samples become more dehydrated, the intensity of all the peaks decrease significantly. The largest change is observed in the second peak.

According to Fig. 10, it seems that the water molecules play a major roll in determining the crystallinity of these bovine cornea collagen samples. The 3.4 Å d-spacing of the second peak (2θ=28.3°) is similar to the hydrogen bond distance observed in other collagen studies [14]. The large decrease in the second WAXD peak is interpreted as resulting from the breaking of hydrogen bonds between the loosely bound water molecules and the peptide groups in collagen with dehydration. It is obvious that the crystallinity of these collagen samples is reduced with dehydration. When collagen fibers are thoroughly dried, they become highly disoriented. In addition, X-ray diffraction patterns (both low angle and wide angle) show disorientation when the fibers are dry, and that the orientation improves remarkably, even without stretching, when the fiber is rewetted [14]. The piezoelectric response observed is most likely attributed to the crystalline regions in the bovine cornea.

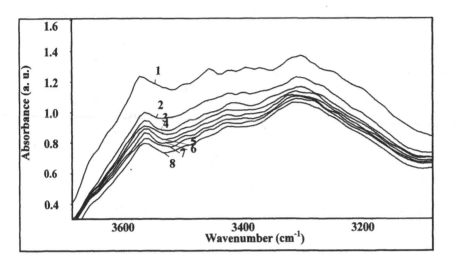

Fig. 4 FTIR spectra obtained for every 5 min. in the range of 3000-4000 cm^{-1} for bovine cornea tissue.

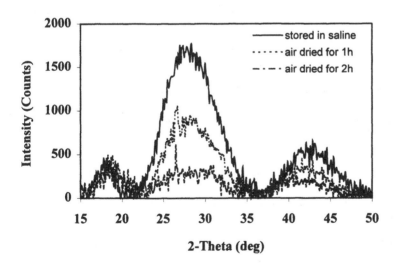

Fig. 5 WAXD pattern for bovine cornea sample

CONCLUSIONS

Bovine cornea exhibits a significant piezoelectric response. The piezoelectric coefficient d_{31} decreases with continued dehydration. On the other hand, Young's modulus exhibits an increase with dehydration. It was observed that both d_{31} and E exhibit anisotropic behavior: samples cut at 45° to the horizontal have an initial d_{31} value of approximately 2500pC/N while for the horizontal and vertical cut samples, d_{31} is approximately 1000pC/N. It is also seen that water molecules in collagen plays a major roll in both mechanical and piezoelectric properties in bovine collagen. According to WAXD and FTIR measurements, the loss of water molecules decreases the crystallinity of collagen. The piezoelectric response observed in bovine cornea is attributed primarily to the N-H and C=O dipole polarization in the crystal phase in the collagen. The anisotropic mechanical and piezoelectric behavior observed in bovine cornea is not well understood but is likely due to oriented crystalline collagen fibrils.

REFERENCES

1. E. Fukada and I. Yasuda, J. Phy. Soc. Jpn, 12(10), pp. 1158-1162 (1957).

2. E. Fukada and I. Yasuda, Jpn. J. App. Phys, 3 (2), pp. 117-121 (1964).

3. E. Fukada and H. Ueda, Jpn. J. App. Phys. 9, p. 844 (1970).

4. J. Duchesne, J. Depirevx, A. Bertinchamps, N. Covnet and J. M. Van Der Kaa, Nature, 188, p. 405 (1960).

5. H. Kawai, Jpn. J. Appl. Phys. 8, p. 975 (1969).

6. T. Furukawa, A. J. Lovinger, G. T. Davis and M. G. Broadhurst, Macromolecules, 16, p.1885 (1983).

7. R. Pethig, *Dielectric and Electronic Properties of Biological Materials*, John Wiley & Sons, New York, 1979.

8. N. Sasaki, S. Shiwa, S. Yagihara, and K. Hikichi, Biopolymers. 22, pp. 2539-2547 (1983).

9. K. M. Meek and D. W. Leonard, Biophys. J. 64 pp. 273-280 (1993).

10. E. Fukada, Ann. N. Y. Acad. Sci. pp. 238:7-25 (1974).

11. I. V. Yannas, J. Macromol, Sci. Rev. Makromol. Chem. C7 pp. 49-104 (1972).

12. S. Ghosh, J.I. Scheinbeim, V. Lubkin, G. Bennett, P. Kramer and Daena Ricketts, (submitted).

13. G. N. Ramachandran, *Treatise on Collagen*, Vol. 1. Academic Press, London, 1967.

14. G. N. Ramachandran and R. Chandrasekharan, Bioplymers, 6, pp. 1649-1658 (1968).

NOVEL POLYMER ELECTRETS

G. M. SESSLER and J. HILLENBRAND
Institute for Communications Technology, Darmstadt University of Technology, Merckstrasse 25, D-64283 Darmstadt (Germany)

ABSTRACT

Permanently charged films with a cellular or porous structure represent a new family of polymer electrets. These materials show piezoelectric properties with high transducer constants. The electromechanical response equations of such films are derived for their operation as sensors and as actuators. Experimental results are also presented for cellular polypropylene. In particular, measurements of the direct and inverse transducer constants, the thermal stability of the charge, and Young's modulus are discussed. Assuming reasonable charge distributions and charge densities, the calculated transducer constants are in good agreement with the measured values. Both the theoretical model and the measurements show the reciprocity of the transducer constants.

1. INTRODUCTION

There has been considerable recent interest in the permanent charging of films with a cellular or porous structure [1-15]. It was shown that such films of polypropylene, polytetrafluoroethylene, and silicon dioxide, when charged with corona or other methods, show relatively good electret behavior. These films can be used, either as single layers or together with another, less compliant layer, as reversible electromechanical or electroacoustic transducers. They exhibit piezoelectric properties and possess transducer constants or sensitivities comparable to those of piezoelectric ceramics, but are mechanically better matched to air or water. This makes such materials attractive for a variety of applications [1].

In the present paper, the electromechanical response equations of such films are derived for their operation as sensors and as actuators and the reciprocity of the transduction is shown. Following this, various experimental results are reported on cellular films of polypropylene. These results demonstrate the operation of such films in both directions and allow one to determine the transducer constants. Finally, a comparison of the theoretical and experimental results yields information on the amount of charge and on its distribution in the cellular structure of the films.

2. RESPONSE EQUATIONS

2.1 Model of Cellular Film

Typical cross sections of a cellular film are shown in Fig. 1. To allow for a ready analysis of the electromechanical operation of such films, a simplified structure, as shown in Fig. 2 is considered. The charged material, electroded on top and bottom, comprises plane parallel solid layers and air layers of thicknesses d_{1n} and d_{2m}, respectively, with $n = 1, 2, .. N$ and $m = 1, 2, .. N-1$, where N is the total number of solid layers. It is further assumed that the two solid surfaces confining the m-th air layer carry a total planar charge density of σ_m and $-\sigma_m$, respectively, and that no volume charges exist. The quantity σ_m includes all permanent charges (surface charges and the ends of polarization chains, see [16], p. 13 ff). The permanent charges on the two sides of the air gap are taken to be equal in magnitude since it is assumed that they originate from discharges in the air gap during poling (see Sect. 3). Fig. 2 also shows the denomination of the electric fields E_{1n} and E_{2m} in the solid and air layers, respectively.

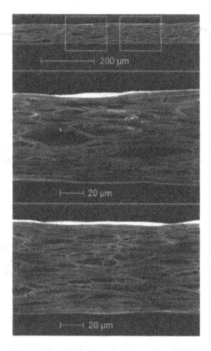

Figure 1: SEM images of the cross section of a larger (top) and two smaller (middle and bottom) segments of a 70-µm-thick cellular polypropylene film.

2.2 Field Equations

The electric fields in the solid layers and air layers may be obtained from Gauss' and Kirchhoff's laws. For the uppermost solid-air interface, Gauss' law can be written as

$$-\varepsilon E_{11} + E_{21} = \sigma_1 / \varepsilon_0. \tag{1}$$

Similar relations hold for the other interfaces. Kirchhoff's second law is for short-circuit conditions

$$\sum_i d_{1i} E_{1i} + \sum_i d_{2i} E_{2i} = 0 \tag{2}$$

These equations yield $E_{11} = E_{12} = ... = E_1$ with

$$E_1 = -\left[\varepsilon_0 (d_1 + \varepsilon d_2)\right]^{-1} \cdot \sum_j d_{2j} \sigma_j, \tag{3}$$

and

$$E_{2i} = (\sigma_i / \varepsilon_0) - [\varepsilon_0(d_1 + \varepsilon d_2)]^{-1} \cdot \varepsilon \sum_j d_{2j}\sigma_j \qquad (4)$$

where $d_1 = \sum_i d_{1i}$ and $d_2 = \sum_i d_{2i}$.

2.3 Sensor Response for Electrical Short Circuit

The charge on the top electrode is given by $\sigma_0 = -\varepsilon_0\varepsilon_1 E_1$. In short circuit, it depends on

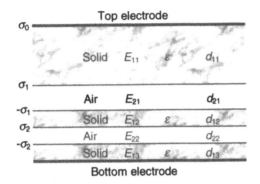

Figure 2: Simplified model of cellular film with $N = 3$.

thickness changes of the film, caused by an applied force. Since the thickness changes are primarily due to the compression of the air layers, the electrode charge is controlled by $\partial\sigma_0 / \partial d_2$. If $\partial d_{2i} / \partial d_2 = d_{2i} / d_2$ is assumed, $\partial\sigma_0 / \partial d_2$ follows from Eq. (3) as

$$\frac{\partial\sigma_0}{\partial d_2} = \varepsilon \frac{d_1 \sum_i d_{2i}\sigma_i}{d_2(d_1 + \varepsilon d_2)^2}. \qquad (5)$$

In the quasistatic case (well below any resonance), a strain relation $\Delta d_2/d = p/Y$ holds, where d equals $d_1 + d_2$, p is the applied pressure and Y is Young's modulus for the film, one obtains the transducer constant $t_{33} = \Delta\sigma_0/p$ from Eq. (5) in finite-difference form as

$$t_{33} = \frac{\Delta\sigma_0}{p} = \frac{\varepsilon d}{Y} \frac{d_1 \sum_i d_{2i}\sigma_i}{d_2(d_1 + \varepsilon d_2)^2} \left[\frac{C}{N}\right]. \qquad (6)$$

A similar equation was derived before for a two-layer system [17].

2.4 Sensor Response for Electrical Open Circuit

Under open-circuit conditions, the electrode charge σ_0 and thus the fields E_1 and E_{2i} remain constant. In particular, they do not change with sudden thickness changes of the air layers, but rather maintain their quiescent values which are assumed to be the short-circuit values. Thus, such thickness changes induce a voltage

$$\Delta V = \sum_i \Delta d_{2i} E_{2i}. \qquad (7)$$

Using again the above strain relation, one obtains the open-circuit transducer constant

$$m_{33} = \frac{\Delta V}{p} = \frac{d}{\varepsilon_0 Y} \cdot \frac{d_1 \sum_i d_{2i} \sigma_i}{d_2(d_1 + \varepsilon d_2)} \left[\frac{\text{Vm}^2}{\text{N}} \right]. \tag{8}$$

In this open-circuit mode of operation, the sensor reduces to a conventional condenser microphone if one assumes just a single layer of dielectric and a single layer of air (thicknesses d_1 and d_2, respectively). Thus Eq. (8) yields the microphone sensitivity below resonance

$$m_{33} = \frac{dd_1\sigma}{\varepsilon_0 Y(d_1 + \varepsilon d_2)}. \tag{9}$$

For condenser microphones, d/Y is frequently replaced by $d_0/\gamma p_0$, where d_0 is the thickness of the air cavity, γ is the ratio of the specific heats and p_0 is the air pressure ([16], p. 349).

2.5 Actuator Response for Mechanically Free-Running System

An applied ac- or dc-voltage V generates additional fields e_1 and e_2 in the solid and air layers, respectively, of the film such that

$$V = e_1 d_1 + e_2 d_2. \tag{10}$$

The field e_2 in the air gaps causes an additional force per unit area between any two dielectric layers given by

$$\Delta F_{2i} = \varepsilon_0 e_2 E_{2i} \tag{11}$$

which, in turn, causes in a mechanically free-running system a thickness change of the air layer Δd_{2i}. The sum of these thickness changes is

$$\Delta d = \sum_i \Delta d_{2i} = \frac{d}{Y d_2} \sum_i d_{2i} \Delta F_{2i}. \tag{12}$$

With Eqs. (4), (10), and (11), one obtains the inverse transducer constant from Eq. (12)

$$\frac{\Delta d}{V} = \frac{\varepsilon d}{Y} \frac{d_1 \sum_i d_{2i} \sigma_i}{d_2(d_1 + \varepsilon d_2)^2}. \tag{13}$$

From this relation, the ratio $\Delta u/I$, where Δu is the volume velocity of the free surface and I is the current into the film capacitance C, may be obtained from

$$\frac{\Delta u}{I} = \frac{S \partial(\Delta d)/\partial t}{C \partial V / \partial t} = \frac{S \Delta d}{CV}, \tag{14}$$

where S is the sample area. Substituting $\Delta d/V$ from Eq. (13), this yields

$$\frac{\Delta u}{I} = \frac{d}{\varepsilon_0 Y} \cdot \frac{d_1 \sum_i d_{2i} \sigma_i}{d_2(d_1 + \varepsilon d_2)}. \tag{15}$$

2.6 Reciprocity Relations

The above equations obey, as expected, the reciprocity relations for reversible systems [18]. Using the four-pole equations in admittance form and applying the impedance analogy to one side of the four pole, the reciprocity relation can be written as

$$\frac{\Delta \sigma_0}{p} = \frac{\Delta d}{V} = t_{33},$$ (16)

which is also found from Eqs. (6) and (13). If the four-pole equations are expressed in impedance form and if the mobility analogy is used, the reciprocity relation reads

$$\frac{\Delta V}{p} = \frac{\Delta u}{I} = m_{33},$$ (17)

which, in turn, follows from Eqs. (8) and (15). The expressions in Eq. (16) correspond to the piezoelectric charge constant [19] while the expressions in Eq. (17) differ from the piezoelectric voltage constants by a replacement of the electric field by the voltage, but correspond to the definition of microphone sensitivity.

3. MEASUREMENTS

All measurements presented were performed with polypropylene EMF foils [1,20] (HS 01, from VTT, Finland) of 70 µm thickness, covered on both sides with aluminium electrodes. The corona charging process was carried out by the manufacturer of the foils. Measurements corresponding to the direct as well as the inverse longitudinal piezoelectric effect are presented in this section.

3.1 Measuring Methods

3.1.1 Direct transducer constant

Mechanical stress was applied to, and removed from, the foils and the generated charges were measured with a charge amplifier (Brüel & Kjaer 2635). The output signal of the amplifier was recorded with a digital storage oscilloscope (Philips PM3350) and a subsequent PC.

3.1.2 Inverse transducer constant

For these measurements a dynamic method was used: A sinusoidal voltage ($V_{rms} < 350$ V, $f < 30$ Hz) was applied to the EMF samples while the surface vibration was recorded by a profilometer (Veeco Instruments Dektak 8000). Profile scans with this instrument are made electromechanically by moving the foil beneath a diamond-tipped stylus which rides over the surface of the sample. The relevant quantity, i. e. the amplitude of the periodic thickness change as function of the surface position, was obtained by applying a digital bandpass filter to the raw data.

Measurements of the inverse constant were also performed with an accelerometer bonded to the film surface. These measurements will not be discussed in the following.

3.2 Experimental Results

3.2.1 Direct transducer constant

An example of a measurement of the direct effect is presented in Fig. 3. The times of applying and removing the force of 3.8 N are marked in the figure. The slow variation of the charge during times of constant force is due to the charge amplifier. The analysis of Fig. 3 yields a transducer constant of 220 pC/N. Further measurements, performed with other HS 01 foils using different stress, yielded comparable mean transducer constants.

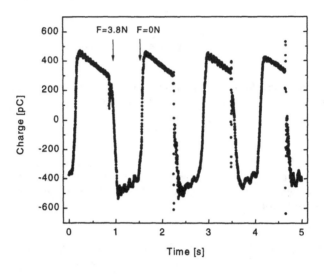

Figure 3: Change of output signal of the charge amplifier upon periodic application and removal of a force of 3.8 N.

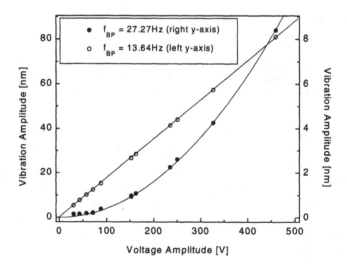

Figure 4: Mean amplitudes of the surface vibration after bandpass filtering at 13.64 Hz and 27.27 Hz. The EMF foil was excited with different voltages at 13.64 Hz.

3.2.2 Inverse transducer constant

In Fig. 4 measurements of the inverse transducer effect are presented. All measurements were performed exactly at the same surface position of the foil with a laterally fixed stylus. The foil was excited with a frequency of $f = 13.64$ Hz and different voltages. Bandpass filtering at 13.64 Hz and 27.27 Hz was performed and the mean vibration amplitudes were calculated. At 13.64 Hz a linear relation between these amplitudes and the applied voltages can be found, rendering an in-

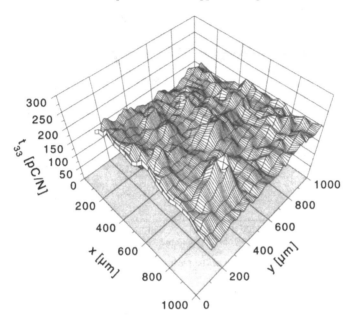

Figure 5: Inverse transducer constant t_{33} of en EMF foil measured within an area of 1 mm^2.

verse transducer constant of 176 pm/V = 176 pC/N. For the second harmonic at 27.27 Hz, a quadratic dependence of the vibration amplitudes versus the voltage was observed in our measurements (closed symbols in Fig. 4).

In the case of EMF foils, this electrostrictive effect is due to the electrostatic forces between the two electrodes and therefore should be independent of the electret charges. This independence could be shown by measurements on thermally depolarized foils. For Young's modulus a value of $Y = 9.5 \times 10^5$ Pa for a foil of 70 μm thickness was calculated from the data for 27.27 Hz in Fig. 4. To eliminate the influence of a locally varying transducer constant the preceding measurements were performed on a single point of the surface.

In contrast, Fig. 5 presents the variation of the transducer constant t_{33} of an EMF foil within an area of 1 mm^2. For this measurement 20 scans of 1000 μm length and 50 s duration were performed. A voltage of $V_{rms} = 280$ V at $f = 27.27$ Hz was applied to the EMF foil during the measurement. The first and the last of these scans are shown again in Fig. 6. The non-uniformity of t_{33} is due to the cellular structure, which has typical dimensions of 30 to 50 μm (see Fig. 1).

The relationship between t_{33} and Young's modulus is presented in Fig. 7. In this measurement the foil was excited with a voltage of $V_{rms} = 280$ V at $f = 13.64$ Hz and a scan of 600 µm length was performed. It can be seen that the vibration amplitude of the fundamental (proportional to t_{33}) and the second harmonic (inversely proportional to Young's modulus) are approximately proportional. This indicates that the variations in t_{33}, as seen in Fig. 5, are to a large degree due to fluctuations in Young's modulus.

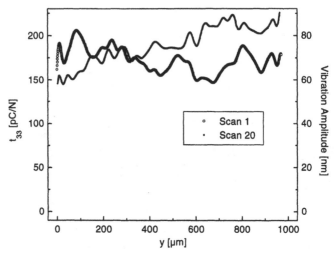

Figure 6: First and last scan of the piezoelectric profiles shown in Figure 5.

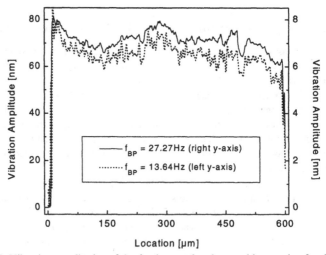

Figure 7: Vibration amplitudes of the fundamental and second harmonic of an EMF foil, excited with a voltage of $V_{rms} = 280$ V at $f = 13.64$ Hz.

150

3.2.3 Thermal stability

Another series of measurements was taken to investigate the thermal stability of the transducer constants. The foils were heated at a constant temperature and the transducer constants were measured at room temperature. As seen in Fig. 8, the transducer constant decreased to approximately half of its original value in five hours at 80 °C.

3.2.4 Charge distribution

In order to obtain information about the polarity of the surface charges within an EMF foil, a voltage pulse was applied to the foil and the direction of the resulting thickness change was measured. Positive voltage pulses applied to a positively charged foil increased the thickness of the foil. This result can be explained, if discharges occur in the air gaps during poling. This mechanism has been assumed to be responsible for the charging in previous studies.

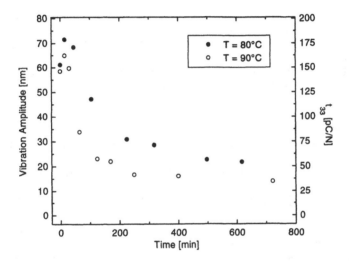

Figure 8: Decrease of the transducer constant t_{33} at temperatures of 80 and 90° C.

4. DISCUSSION AND CONCLUSIONS

For a comparison of the experimental results with theory, an evaluation of Eqs. (6) and (13) is necessary. From the constants in these equations, $\varepsilon = 2.35$, $Y = 9.5 \times 10^5$ Pa (see above), $d = 70$ μm, $d_1 = 26$ μm, $d_2 = 44$ μm (d_1 and d_2 are determined from the densities) are known. The remaining quantities d_{2i} and σ_i have to be estimated. From micro-photographs such as Fig. 1 it can be inferred that there are about 10 voids of more than 2 μm thickness and a larger number of smaller voids across a 70 μm thick film. Assuming further that only the surfaces of the large voids are charged and that $\sigma_i = 10^{-3}$ C/m², one obtains $t_{33} \approx 200$ pC/N (or pm/V) in agreement with the above data and [1,8,9]. The assumed charge density yields from Eq. (4) a voltage of less than 150 V even across large voids, which is below the Paschen voltage of ≈ 400 V for such gaps. This indicates that the above model for the piezoelectric behavior of the cellular material is based on reasonable assumptions.

With the above charge density, microphone or hydrophone sensitivities of about 1.7 mV/Pa, corresponding to −55 dB vs 1V/Pa, are expected from Eq. (8) below resonance. This constant and relatively high sensitivity shows that cellular electrets are also potentially useful in such applications.

ACKNOWLEDGMENTS

The authors are grateful to Prof. Siegfried Bauer (Linz) and Prof. Reimund Gerhard-Multhaupt (Potsdam) for stimulating discussions. They are further indebted to VTT Chemical Technology for providing the EMF material and to the Deutsche Forschungsgemeinschaft (DFG) for financial support.

REFERENCES

1. J. Lekkala, R. Poramo, K. Nyholm, T. Kaikkonen, Med. & Bio. Eng. & Comp., **34** Suppl. 1, Pt. 1, pp. 67 – 68 (1996).
2. Z. Xia, J. Jian, Y. Zhang, Y. Cao, Z. Wang, 1997 Annual Report, Conference on Electrical Insulation and Dielectric Phenomena, pp. 471 – 474.
3. Y. Cao, Z. Xia, Q. Li, L. Chen, B. Zhou, Proc., 9th International Symposium on Electrets, 1996, pp. 40 – 45.
4. Z. Xia, Proc., 10th International Symposium on Electrets, 1999, pp. 23 – 26.
5. G. M. Sessler, J. Hillenbrand, Proc., 10th International Symposium on Electrets, 1999, pp. 261 – 264.
6. R. Gerhard-Multhaupt, Z. Xia, W. Künstler, A. Pucher, Proc., 10th International Symposium on Electrets, 1999, pp. 273 – 276.
7. R. Schwödiauer, G. Neugschwandtner, S. Bauer-Gogonea, S. Bauer, J. Heitz, D. Bäuerle, Proc., 10th International Symposium on Electrets, 1999, pp. 313 – 316.
8. M. Paajanen, H. Välimäki, J. Lekkala, Proc. 10th International Symposium on Electrets, 1999, pp. 735 – 738.
9. J. Lekkala, M. Paajanen, Proc., 10th International Symposium on Electrets, 1999, pp. 743 – 746.
10. J. van Turnhout, R. E. Staal, M. Wübbenhorst, P. H. de Haan, Proc., 10th International Symposium on Electrets, 1999, pp. 785 – 788.
11. J. Hillenbrand, G. M. Sessler, 1999 Annual Report, Conference on Electrical Insulation and Dielectric Phenomena, pp. 43 – 46.
12. Z. Xia, R. Gerhard-Multhaupt, W. Künstler, A. Wedel, R. Danz, J. Phys. D: Appl. Phys. **32**, L83 – L85 (1999).
13. G. M. Sessler, J. Hillenbrand, Appl. Phys. Lett. **75**, 3405-3407 (1999)
14. W. Künstler, Z. Xia, T. Weinhold, A. Pucher, R. Gerhard-Multhaupt, "Piezoelectricity of porous polytetrafluoroethylene single- and multiple-film electrets containing high charge densities of both polarities", to be published in Applied Physics A.
15. G. S. Neugschwandtner, R. Schwödiauer, S. Bauer-Gogonea, S. Bauer, "Large piezoelectric effects in charged, heterogeneous fluoropolymer electrets", to be published in Applied Physics A.
16. G. M. Sessler (ed), Electrets (3rd Edition), Laplacian Press, 1999.
17. R. Kacprzyk, A. Dobrucki, J. B. Gajewski, J. Electrostat. **39**, 33-40 (1997).
18. P. M. Chirlian, "Basic Network Theory" McGran Hill, 1969.
19. D. A. Berlincourt, D. Curran, H. Jaffe, "Piezoelectric and Piezomagnetic Materials and their Function in Transducers", in: W. P. Mason "Physical Acoustics I, Part A", 1964, p. 169-270.
20. A. Savolainen, K. Kirjavainen, J. Macromol. Sci., **A 26**, 583 – 591 (1989).

POLARIZATION STABILITY OF AMORPHOUS PIEZOELECTRIC POLYIMIDES

C. PARK*, Z. OUNAIES**, J. SU*, J.G. SMITH JR. AND J.S. HARRISON
Advanced Materials and Processing Branch, NASA Langley Research Center, Hampton VA, 23681-2199.
*National Research Council
**ICASE

ABSTRACT

Amorphous polyimides containing polar functional groups have been synthesized and investigated for potential use as high temperature piezoelectric sensors. The thermal stability of the piezoelectric effect of one polyimide was evaluated as a function of various curing and poling conditions under dynamic and static thermal stimuli. First, the polymer samples were thermally cycled under strain by systematically increasing the maximum temperature from 50°C to 200°C while the piezoelectric strain coefficient was being measured. Second, the samples were isothermally aged at an elevated temperature in air, and the isothermal decay of the remanent polarization was measured at room temperature as a function of time. Both conventional and corona poling methods were evaluated. This material exhibited good thermal stability of the piezoelectric properties up to 100°C.

INTRODUCTION

Aromatic polyimides have been used in a wide variety of aerospace applications because of their chemical and radiation resistance, with excellent thermal, mechanical, and dielectric properties. Recently, a series of novel piezoelectric polyimides containing pendant, polar groups have been synthesized and evaluated at NASA-LaRC for potential use in micro-electro-mechanical systems (MEMS) devices [1,2]. The initial investigations have shown that they exhibited a piezoelectric response at temperatures in excess of 150°C [3].

Experimental studies of the thermal stability of a piezoelectric amorphous polyimide, (β-CN)APB/ODPA, are presented here as a function of various curing and poling conditions under dynamic and static thermal stimuli. Both conventional and corona poling methods were employed for this study. Corona poling was used to maximize the degree of dipolar orientation and minimize localized arcing during in-situ imidization and poling. The results of both poling methods are discussed.

EXPERIMENTAL

Film preparation
The polyimide evaluated was (β-CN)APB/ODPA, which was prepared from 2,6-bis(3-aminophenoxy) benzonitrile ((β-CN)APB) and 4,4' oxidiphthalic anhydride (ODPA) via a polyamic acid solution in N,N-dimethylacetamide and subsequent thermal imidization. The synthesis was reported in detail elsewhere [3]. The polyamic acid solution was cast to form approximately 30μm thick films. The tack-free films were imidized under various cure cycles to produce samples having different degrees of imidization. The exact cure cycles are summarized in Table I. A silver layer, approximately 200nm thick, was evaporated on both sides of the films for conventional poling and only one side of the films for corona poling.

Table I. Processing parameters and properties of partially-cured, corona poled (β-CN)APB/ODPA: cure cycle, T_g, degree of imidization ($A1780cm^{-1}/A1500cm^{-1}$), and P_r.

Sample	Cure cycle (ßC) 1 hour each, N_2	T_g before poling (ßC)	$A1780cm^{-1}/A1500cm^{-1}$	P_r (mC/m^2)
P100	50, 100	97	0.18	N/A
P150	50, 150	142	0.69	26
P200	50, 150, 200	166	0.82	9
P240	50, 150, 200, 240	218	1.00	4

Poling

The film specimens were poled using either a conventional or a positive corona poling procedure. For the conventional poling, each sample was polarized by the application of a DC electric field (80MV/m) at an elevated temperature (T_g + 5°C) in a silicone oil bath. For corona poling, a DC field of 20 kV was applied to generate a positive corona using a single tungsten wire for four hours at 223ßC and one hour at 212ßC. The distance between the corona tip and the specimen was approximately 30mm. An argon gas was maintained during the poling process. For both poling processes, the dipoles were oriented with the applied field at a temperature above T_g, with subsequent cooling to below T_g in the presence of the applied field. The resulting remanent polarization (P_r) is directly proportional to the material s piezoelectric response, and estimated from the following equation [4],

$$P_r = \varepsilon_0 \Delta \varepsilon E_p \qquad (1)$$

where ε_0 is the permittivity of free space (8.854 pF/m), $\Delta\varepsilon$ is the dielectric relaxation strength, and E_p is the poling field.

Characterization

Degree of imidization. The glass transition temperatures (T_g) of the films were measured by differential scanning calorimetry using a Shimadzu DSC-50 at a heating rate of 20ßC/min in air. The T_g was taken as an inflection point of the shift of the baseline of the DSC thermogram. The degree of imidization was determined by a Nicolet FTIR spectrometer in an ATR mode using a Nicolet Contin m IR microscope. The absorption peak at 1780cm^{-1} (sym. carbonyl stretch) was used to determine the degree of imidization and that at 1500cm^{-1} (ring modes of aromatic moieties) was used as an internal standard [5]. The results are presented in Table I.

Thermally stimulated current (TSC) measurement. After poling, P_r was measured as a function of temperature. As the sample was heated through its T_g at a heating rate of 1.5¡C/min, the depolarization current was measured using a Keithly 6517 electrometer. The P_r is equal to the charge per unit area, which is obtained from the data by integrating the current with respect to time and plotting it as a function of temperature.

Piezoelectric Measurement. The piezoelectric strain coefficient (d_{31}) was measured using a Rheovibron DDV-II-C mechanical analyzer as a function of temperature for a range of frequencies. As the polymer is strained along the direction of applied stress, a charge Q is generated on the surface of the electrodes. A geometric factor is used to produce a geometry independent parameter, namely, surface charge density per unit applied stress.

Thermal stability Measurements. Thermal stability of the piezoelectricity in the (β-CN) APB/ODPA system was carried out under both dynamic and isothermal conditions. First, the polymer sample was thermally cycled under strain by systematically increasing the maximum temperature from 50 to 200°C at a heating rate of 2°C/min. The d_{31} was measured as a function of temperature for four runs, where the maximum temperature for each run was 50, 100, 150, and 200°C. The effect of the dynamic temperature cycling on the d_{31} was assessed as % retention of the piezoelectric response. Second, the as-poled samples were isothermally aged at 50, 100, 150, and 200°C in static air ovens. The aged samples at each temperature were tested periodically and the P_r was measured and compared to that of the as-poled sample. The % retention of P_r was plotted as a function of aging temperature and time.

RESULTS

Fully-cured, conventionally poled (β-CN)APB/ODPA   The remanent polarization of the fully-cured, conventionally poled (β-CN)APB/ODPA was approximately 20mC/m^2 when poled at 80MV/m for one hour above T_g. Figure 1 summarizes the results of the long-term thermal stability of this polymer. The retained P_r after isothermal aging is presented as a function of aging time at the various temperatures. Excellent thermal stability was observed up to 100°C, and no loss of the piezoelectric response was seen after aging at 50°C and 100°C up to 500 hrs. After aging at 150°C, 50% of the initial P_r was retained after 100 hrs, and remained constant until 450 hrs, while at 200°C, 13% of the P_r was retained after 100 hrs.

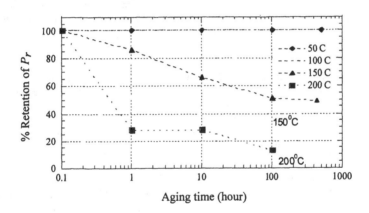

Figure 1. Thermal stability of P_r of (β-CN)APB/ODPA after aging at various temperatures.

Thermal stability of the piezoelectric strain coefficient (d_{31}) under dynamic conditions was also studied and the results are shown in Figure 2. After cycling up to 50°C, no loss of the initial d_{31} was observed on the 2nd run, as seen in the inset of Figure 2. The d_{31} of this polymer rapidly increased with temperature above 90°. A d_{31} value of 5pC/N at the end of the 3rd run at 150°C and that of 10pC/N at 200°C at the end of the 4th run were obtained. These results are encouraging and portend use of these polyimides as sensors in high temperature aerospace applications.

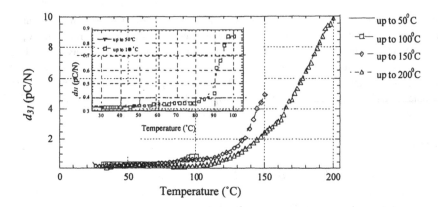

Figure 2. Dynamic stability of d_{31} of fully-cured, conventionally poled polyimide film (1 Hz).

Partially-cured, corona poled (β-CN)APB/ODPA. A positive corona poling was used for the partially-cured polymers in an attempt to maximize the degree of dipolar orientation and minimize localized arcing during in-situ imidization and poling. The aligned polar groups should be immobilized by additional imidization and subsequent cooling in the presence of an electric field. Table I summarizes the cure cycle, T_g, and degree of imidization of the polyimides. Both the T_g and the degree of imidization increased almost linearly with the final cure temperature. Thus, higher mobility of the dipoles should be expected for the polymers cured at a lower temperature. The remanent polarization of these specimens was measured, and listed in Table 1. The value of P_r appeared higher when cured at lower temperature. Since the mobility of the molecules of the partially-cured polyimide should be much higher than that of the fully-cured one, the polar groups of the former are expected to orient parallel to the field direction more efficiently than the latter. Therefore, poling in a partially imidized state may produce a higher degree of dipole orientation than poling in a fully imidized state, thereby generating higher P_r. The P150 partially-cured polyimide exhibited six times higher P_r than the fully-cured one under the same corona poling condition, which was greater than expected from the preliminary dielectric measurement.

Figure 3 shows an example of the P_r of a partially-cured, corona poled (β-CN) APB/ODPA (P150). The partially-cured polymers tend to show broad double depolarizing current peaks during TSC measurement while the fully-cured ones depolarize with one sharp peak near T_g. The double peaks imply incomplete imidization during poling since the peak represent the T_g and the dipoles poled in a partially imidized state are thermally less stable than those poled in a fully imidized state. Since the poling temperature for the partially-cured polymers was selected 5°C above T_g of the fully-cured one, the dipoles aligned in the early stage (unimidized state) may not have had enough time to relax their excessive free volume during the in-situ imidization and poling process. A gradual increase of the poling temperature may be desirable to shift the lower peak maximum to the higher temperature, providing more thermal stability to the polarized dipoles.

Figures 4(a) and (b) show the normalized retention of the piezoelectric strain coefficient, d_{31} of the dynamic stability tests for both fully-cured, conventionally poled and the partially-cured, corona poled (β-CN)APB/ODPA (P150), respectively. In both cases, the films exhibit

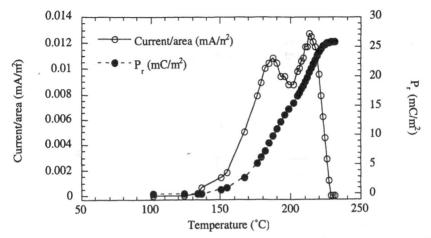

Figure 3. Thermally stimulated current measurement: current/area and P_r versus temperature for partially-cured (β-CN) APB/ODPA (P150).

very good thermal stability and virtually no loss of d_{31} was observed prior to reaching 150°C. After 150°C, both films exhibited noticeable loss of d_{31}, however, the fully-cured specimen still retained about 60% of the initial value while the partially-cured one retained less than 20%. The lower value of the partially-cured one may be an artifact due in part to incomplete imidization (Figure 3). Further investigations in this matter are underway. The thermal stability of the poled amorphous polyimide films did not show a significant change in the range of the measurement frequencies, from 1Hz to 100Hz for both poling processes.

CONCLUSIONS

Thermal stability of the amorphous piezoelectric polyimide was evaluated as a function of curing and poling conditions. The piezoelectricity of this polymer was stable under both dynamic and static thermal stimuli and statistically no loss was observed up to 100°C. The partially-cured, corona poled polymers exhibited improved remanent polarization. Since this amorphous piezoelectric polyimide can generate a piezoelectric response at elevated temperatures, its thermal stability should be beneficial for high temperature aerospace applications.

ACKNOWLEDGEMENTS

The authors acknowledge Dr. Terry L. St. Clair of NASA-LaRC for his technical insight. We also acknowledge Mr. Bill White of Wyle Laboratories for development of the characterization measurement software.

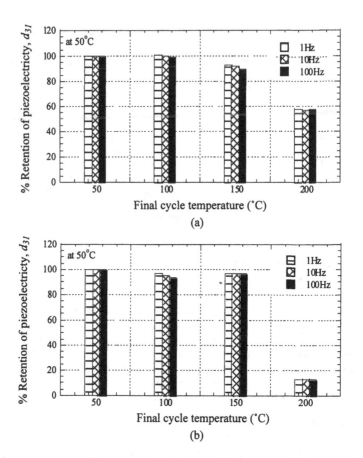

Figure 4. Percent retention of the piezoelectric strain coefficient, d_{31} at 50°C after cycling up to 50, 100, 150, and 200°C, (a) fully-cured sample (conventional poling: 80MV/m, 220°C for 60min), (b) partial cured sample (corona poling)

REFERENCES

1. J.O. Simpson, S.S. Welch, and T.L. St. Clair, in *Materials for Smart Systems II*, edited by E.P. George, et al. (Mater. Res. Soc. Proc. **459**, Boston, MA, 1997) pp. 351–356.
2. J.O. Simpson, Z. Ounaies, and C. Fay, in *Materials for Smart Systems II*, edited by E.P. George, et al. (Mater. Res. Soc. Proc. **459**, Boston, MA 1997) pp. 59–64.
3. Z. Ounaies, C. Park, J.S. Harrison, J.G. Smith, and J. Hinkley, in Smart Structures and Materials: Electroactive Polymer Actuators and Devices, edited by Y. Bar-Cohen (Proc. SPIE **3669**, Newport Beach, CA 1999) pp. 171–178.
4. B. Hilczer and J. Malecki, *Electrets; Studies in Electrical and Electronic Engineering 14*, Elsevier, New York, 1986, p. 19.
5. C.A. Pryde, J. Polym. Sci.: Part A: Polym. Chem. **27**, p. 711 (1989).

A STUDY OF LIQUID CRYSTALLINE ELASTOMERS AS PIEZOELECTRIC DEVICES©

A.G. BIGGS[*], K.M. BLACKWOOD[*], A. BOWLES[**], S. DAILEY[*], ALISON MAY[†]
*DERA Malvern, St Andrews Road, Malvern, Worcestershire, WR14 3PS, UK.
**DERA Farnborough, Ively Road, Farnborough, Hampshire, GU14 0LX, UK.
†Merck R&D UK, Building 35, University of Southampton, Southampton, SO17 1BJ, UK.
Email : kmblackwood@dera.gov.uk

ABSTRACT

Ferroelectric liquid crystalline polymer films exhibit a number of unique physical properties and have been extensively studied during the last few years. Ferroelectric polymer films exhibiting a high level of piezoelectric activity would provide a new route to piezo- and pyro-electric sensors.

We report the fabrication of free-standing ferroelectric liquid crystal films which exhibit piezoelectric coefficients up to 4.4pC/N. We also report the investigation of the physical properties of these films by dynamic mechanical analysis.

INTRODUCTION

In the smectic phases of liquid crystals, the mesogens arrange themselves into distinct layers, with increasing positional ordering within the layers. In the smectic C phase, the mesogens within each layer are tilted to one side at a constant tilt angle. As Meyer showed by an elegant symmetry argument, if the mesogen has a chiral centre near the core then the chirality serves to break the symmetry of the unit cell [1]. Thus each layer will have a degree of polarisation even when no external field is present. This polarisation is responsible for the appearance of ferroelectricity and is referred to as the spontaneous polarisation (P_s). There is a complication, however. The chirality causes each layer to become twisted with respect to its neighbouring layers, producing a helical superstructure when considered over a large number of layers. This cancels out the P_s vector from each layer, producing a sample with an overall P_s of zero. Clark and Lagerwall [2] first succeeded in unwinding this helix and producing a low molecular mass liquid crystal sample with a non-zero P_s. Side chain ferroelectric liquid crystal polymers (FLCPs) were initially fabricated [3] in an effort to address the mechanical stability problems associated with the alignment of low molecular mass ferroelectric liquid crystals. Similar problems in unwinding the helical superstructure of the polymers were also encountered, but eventually a non-zero P_s in a FLCP was reported by Uchida et al in 1988 [4]. Much of this work has been extensively reviewed elsewhere [5].

Organic piezoelectrics based on poly(vinylidene fluoride) (PVDF) have been known for some time [6]. Piezoelectricity in PVDF is thought to arise from changes in the dipole density per unit area, brought about by mechanical deformation of the sample. The dipole density in FLCPs is lower than that in PVDF due to the size of the molecule contributing to the unit cell, larger FLCP molecules give a larger unit cell volume. Hence, the magnitude of the polarisation in FLCPs will, be consequently lower than in PVDF. However, the increased order parameter and the coupling of the mechanical strain with the order parameter in FLCPs may offer increased figures of merit. Also, FLCP elastomers are of inherently lower modulus than PVDF, hence the mechanical coupling will be much improved.

Although PVDF exhibits a large value of polarisation it has relatively poor ordering, compared to FLCPs which can exhibit a high degree of self-ordering. Hence, while the piezoelectric activity of some FLCP systems is similar in magnitude to PVDF, others may have the potential to surpass it. Further, their self ordering nature means that FLCPs do not require large thermal and electric poling fields to impose order unlike the case of PVDF, facilitating processing and fabrication.

In general, organic systems are considerably less active than electro-ceramics such as lead zirconic titanate (PZT) [7,8], due to the molecular scale dipoles in the organics compared to the atomic scale dipoles in the ceramics. However, in certain applications, the lower modulus and ease of production offered by polymer systems offsets this reduction in sensitivity.

An advantage which FLCPs have over ceramics and PVDF is their relatively low modulus which will produce a larger piezoelectric response. The most widely recognised measure of electromechanical activity in piezoelectrics is the d coefficient, defined as the ratio of dielectric displacement to applied mechanical stress. It is clear to see that this coefficient is a good indication of an electromechanical materials efficacy when used as a sensor. Due to the anisotropic nature of most piezoelectrics, the d term is described with reference to an orthogonal crystallographic system, via two subscripts defining, in this case, the direction of stress and resultant charge generated. They are referenced to the direction of net polarisation, designated as the z-axis or 3 direction. Hence the coefficient d_{33} refers to a force applied in and a charge generated in a plane perpendicular to the polarisation direction (z axis).

It is also informative to note that the d coefficient can be approximated as illustrated in equation 1.

$$d \approx \frac{Polarisation}{Modulus} \tag{1}$$

Since the modulus of FLCPs can be considerably lower than the corresponding value for PVDF, they clearly offer the possibility of increased piezoelectric sensitivity.

To achieve a high order parameter, a highly ordered, planar smectic C* phase is required. When a planar monodomain smectic C* phase is produced, the dipoles on the mesogenic chains will be unidirectional, thus maximising the magnitude of the spontaneous polarisation. Any subsequent perturbation to this idealised system will alter the P_s and therefore produce an electrical response. In order to achieve this it is usual to cool an appropriate liquid crystal from the isotropic state, via the nematic and smectic A states to the smectic C*. This phase sequence gives a more gradual increase in order parameter and hence leads to improved alignment and larger domain sizes. The helical superstructure can be unwound using modest surface; mechanical or electrical forces. In order to realise all of the above parameters a mixture of liquid crystal monomers is typically used. This provides for a system with a chiral dipole, the desired phase sequence and a highly ordered smectic C* phase over the required operating temperature range.

EXPERIMENT

Three liquid crystal monomers were used in this study, LCP138, LCP145 and RM22. The structures of these materials are shown in Figure 1. The synthesis of LCP138 and LCP145 is reported by Ebbutt et al [9,10].

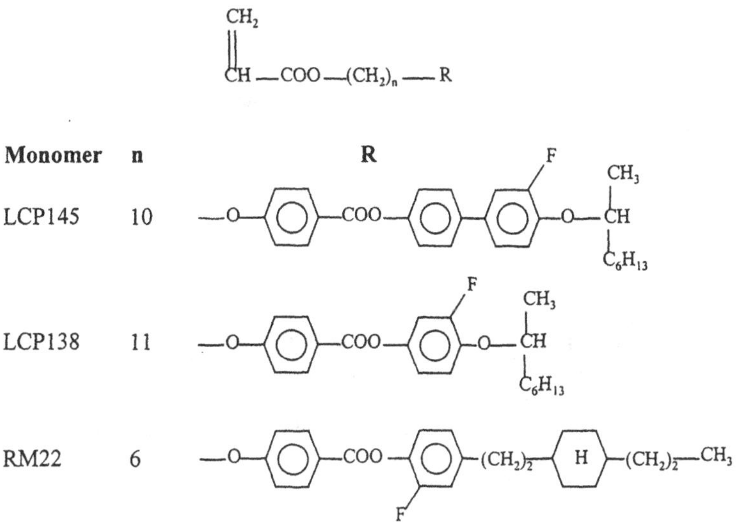

Figure 1. *Structures of the liquid crystal monomers*

LCP145 has a chiral dipole and exhibits isotropic, smectic A and smectic C* phases. LCP138 also has a chiral dipole and is used to reduce the smectic C temperature range of LCP145. RM22 was added in an attempt to improve the smectic C* ordering by inducing a nematic phase.

Free standing FLCP films were prepared by mixing the liquid crystal monomers above in various ratios with the commercially available cross-linker, pentaerythritol tetraacrylate (Aldrich). Irgacure 184 (Ciba Specialities) was used as a photoinitiator. The cross-linker content of the mixtures was 2% by weight and the initiator concentration was 4% by weight. This mixture was placed between two glass substrates separated by Mylar strips with a thickness of 25µm, 50µm or 100µm. Polymerisation was carried out following cooling on a Mettler FP82HT Hot Stage at a typical rate of less than 0.5°C/min. A Teklite ELC400 was used to provide ultra-violet light, and the polymerisation step took approximately 30 seconds. Once polymerised, the substrates could be separated and the free-standing films removed.

The phase behaviour of the mixtures of these materials was measured using differential scanning calorimetry on a TA Instruments DSC 2910. Phases were identified using polarising optical microscopy. The Storage Modulus (G') of the films was determined using a TA Instruments DMA 2980 Dynamic Mechanical Analyser at a frequency of 1Hz. Dielectric measurements of the materials have been reported elsewhere [11].

The d_{33} coefficient of the free standing films was measured using a proprietary system, as conventional techniques are not suitable for the measurement of the piezoelectric parameters of low modulus materials such as FLCPs. The system consists of a force transducer and an electromechanical shaker mounted on an optical stage, with the sample under test lightly clamped between two conducting probes (see Figure 2).

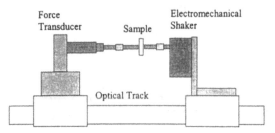

Figure 2. *Test rig used for piezoelectric measurements of FLCP films*

The experiment operates by applying a controllable force to the material under test and measuring the resultant charge by feeding the signal into an amplifier. The system is highly sensitive to small charge outputs and, more importantly, is able to control the dynamic stress levels such that the material remains within its linear piezoelectric regime.

RESULTS

Figure 3 shows the phase diagrams for the samples examined in this study. Figure 3(a) shows the thermal behaviour of LCP145 doped with LCP138 and figure 3(b) illustrates the thermal behaviour of a 50:50 mixture of 145/138 doped with RM22. It can be seen that doping LCP138 into LCP145 can reduce the temperature range of the smectic C phase from 125°C to 140°C to around room temperature. Also doping in RM22 to this systems introduces a nematic phase above the smectic A phase between the concentration ranges of 50% to 100%.

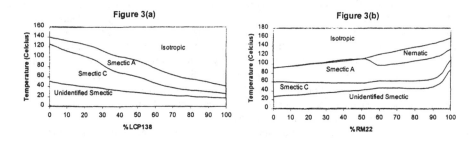

Figure 3. *Phase diagrams of a) LCP138/LCP145 and b)LCP138/LCP145/RM22 elastomers*

The introduction of RM22 monomer at percentages less than 50% results in a nematic phase in the monomer mixture, but this phase does not appear in the final polymerised elastomer, as shown in figure 3(b). However, the ability to induce a nematic phase in the monomer mixture during the cooling sequence may still offer improved alignment in the final elastomer.

Table 1 below, shows a summary of the results obtained on the elastomer films prepared using the process outlined above. Although d_{33} results are still below those expected of PVDF (12 pC/N) the storage modulus is extremely low and we believe there is much scope for improvement in terms of the d_{33} figures. The net polarisation, or product of d_{33} and G' gives an

Monomer Concentration Weight % (4% Cross Linker + 2% Initiator)			Phase Behaviour	d_{33} (25°C)	G' (25°C)	d_{33} X G'
LCP138	LCP145	RM22		pC/N	Mpa	nC/cm^2
30	70	0	I–S$_A$–S$_C$– S$_?$	0.51	141.2	7.50
50	50	0	I–S$_A$–S$_C$– S$_?$	3.50	19.3	6.75
45	45	10	I–S$_A$–S$_C$– S$_?$	0.76	27.8	2.11
37.5	37.5	25	I–S$_A$–S$_C$– S$_?$	4.40	21.4	9.41
20	20	60	I–N–S$_A$–S$_C$– S$_?$	1.4	10.31	1.44

Table I. Results summary for materials studied. Phase Behaviour shows I-isotropic liquid, N – nematic phase, S$_A$–smectic A phase S$_C$– chiral smectic C phase, S$_?$ – unidentified higher order smectic phase. d_{33} – piezoelectric coefficient (measured at room temperature), G' – storage modulus (measured at room temperature).

indication of the degree of alignment in the system. Previous studies have shown that the maximum polarisation achievable in these types of liquid crystal mesogens is about 35nC/cm^2 [9,10].

The largest d_{33} result is for a mixture of 25% by weight RM22 with the balance made up equally of LCP145 and LCP 138. This suggests that introducing the nematic phase during the cooling step does indeed lead to improved alignment and hence higher d_{33} results compared to 0% RM22. However, going to 60% RM22, while further reducing the modulus, dilutes the active components to the extent that the d_{33} figures deteriorate due to the lack of active dipoles. Adding small amounts of RM22 (10%), is clearly not sufficiently nematogenic to give the required increase in alignment, but instead serves to increase crystallisation and hence increase the modulus. This is typical of an anti-plasticising effect observed in other polymeric systems [12]. Hence both a reduction in net polaristaion and an increase in modulus give a smaller d_{33} result. Varying the concentrations of LCP145 and LCP138 also varies the results. Increasing the concentration of LCP145 gives rise to the presence of the higher order smectic phase at room temperature. While it is entirely possible that this may also be ferroelectric, and therefore produce a higher net polarisation, the much higher modulus results in a lower d_{33} value.

CONCLUSIONS

We have studied the electromechanical properties of various ferroelectric liquid crystalline elastomers, based on three liquid crystal mesogens. We have found a wide range of piezoelectric activity from 0.51pC/N rising to 4.4 pC/N. By looking at the product of this figure with measured storage modulus we derive a figure for the apparent net polarisation in the system. It would appear that the LCP145 monomer makes the largest contribution to Ps, but the storage modulus increases rapidly upon further addition of LCP145. Introduction of RM22 induces a nematic phase in the monomer mixture at concentrations above 10%, but no nematic phase is observed in the polymerised mixtures below 50% RM22. It also appears that the addition of RM22 at levels of 10% leads to an increase in modulus and a disruption to the dipolar alignment. We believe this may be caused by some degree of phase separation or crystallisation upon polymerisation. Going to higher concentrations of RM22 leads to marked improvements in the d_{33} results up to about 25% from then on the RM22 serves only to dilute the active components and lead to lower values of activity. Our previous studies [9,10] have indicated that we should be measuring much higher polarisation values, this suggests that there is still much scope for improvement in the alignment of these systems and also, that d_{33} values of about 15 pC/N are possible with this type of molecular archietecture.

In summary it appears that these liquid crystalline elastomers offer the potential for novel piezoelectric devices of very low modulus which can be easily fabricated without the long and arduous poling processes usually associated with organic piezoelectrics. However, more work has to be done to optimise the system discussed in this work and it is to be hoped that this will lead to materials which can offer similar performance figures to PVDF.

REFERENCES

[1] R.B. Meyer et al; Fifth International LC Conf., Stockholm, (1974): R.B. Meyer, L. Liebert, L. Strzelecki and P. Keller, *Mol. Cryst. Liq. Cryst.*, **38**, L-69, (1975).
[2] N.A. Clark and S.T. Lagerwall, Appl. Phys. Lett. **36**, p. 899 (1980)
[3] V. Shibaev, M. Kozlovsky, L. Beresnev, L. Blinov, A. Platé, Polym. Bull. **12**, p. 299 (1984)
[4] S. Uchida, K. Morita, K. Miyoshi, K. Hashimoto, K. Kawasaki, Mol. Cryst. Liq. Cryst. **155**, p. 93 (1988)
[5] See for example: "Side Chain Liquid Crystal Polymers", ed. C.B. McArdle, Blackie, Glasgow, (1989). And references therein
[6] J.F. Tressler, S. Alkoy, R.E. Newnham, Journal of Electroceramics **2** p.257 (1998)
[7] R.G. Kepler, R.A. Anderson, J. Appl. Phys. **49** p. 4490 (1978)
[8] C.J. Dias, D.K. Das-Gupta, Ferroelectric Polymers and Ceramic-Polymer Composites **92** p.217 (1994)
[9] J. Ebbutt, R.M. Richardson, J. Blackmore, D.G. McDonnell, M. Verrall, Mol. Cryst. Liq. Cryst. **261** p.549 (1995)
[10] I.C. Sage, D.G. McDonnell, K.P. Lymer, K.M. Blackwood, J.M. Blackmore, D. Coates, D.R. Beattie, M. Verrall, J.W. Goodby, M. Watson, Conference Proceedings, FLC '95, Cambridge, UK, (1995).
[11] K.M. Blackwood, R. Sabral, M. Jones, D.K. Das-Gupta, *J. Phys. D: Cond. Matter*, In Press
[12] Blackwood, K.M., Pethrick, R.A., Simpson, F.I., Day, R.E. and Watson, C.L.; *J. Mat. Sci.*, **30**, 4435, (1995).

Conductive Polymers

POLY(SNS) ELECTROCHEMICAL SYNTHESIS AND PROCESSABILITY

T.F. Otero*, S. Villanueva*, E. Brillas**, J. Carrasco**.
* Laboratorio de Electroquímica, Facultad de Químicas, Universidad del País Vasco, P.O. Box 1072, 20080 San Sebastián, SPAIN, qppfeott@sq.ehu.es
** Departament de Química Física; Departament d'Enginyeria Química i Metal.lúrgia, Facultad de Química, Universitat de Barcelona, Martí i Franqués, 1. 08028 Barcelona, SPAIN.

ABSTRACT

The flow of an anodic current density of 0.5 mA. cm^{-2} through a 5 mmol/l solution of 2,5-di-(2-thienyl)pyrrole (SNS), in acetonitrile, gives dark-violet and uniform polymer films coating the platinum electrode in presence of different electrolytes. The process is named either electropolymerization or electrogeneration.

The new material was insoluble in acetonitrile solutions of different salts. When a cathodic current density of 0.2 mA. cm^{-2} was applied to the coated electrode, a yellow-green cloud was observed around it. The electrodissolution occurs due to the high, and fast, solubility of the reduced poly(SNS) in the electrolyte. This electrodissolu. on also followed a faradaic process.

Deposits of oxidized poly(SNS) films have been further obtained on a Pt electrode by flow of an anodic constant current through a solution containing a salt and the reduced poly(SNS) in acetonitrile. The electrodeposition occurs in presence of different electrolytes. Electrodeposited weights are proportional to the consumed charge: the electrodeposition is a faradaic process. Electrodeposited films are partially electrosoluble in their background solutions. Electrodissolution and electrodeposition from ammonium salts are reverse processes showing the same productivities and mimicking similar processes with inorganic metals.

Electrodissolution and electrodeposition processes are being applied to the processability of new actuators and artificial muscles.

INTRODUCTION

Conducting polymers are considered in literature as one of the most important components of organic metals[1-7]. Many of the studied conducting polymers seem to lack some of the most specific polymeric properties: solubility and fusibility.

The interest in our laboratories was centered in a main question: if conducting polymers reproduce most of the electrochemical properties of inorganic metals, like oxidation and reduction processes, it could be possible to generate a polymer able to mimic electrodissolution and electrodeposition process. Those processes are the origin of many industrial processes: electroerosion, electromachining, electroforming, electroplating, electrophotography, electro-litography, etc.

In order to produce polymers able to be electrochemically solved, our aim was to find conducting polymers, which oxidized and reduces states show different solubilities. The studied was centered in different trimers such as: SOS, SSS[8,9] and SNS[10-13]. However, only the last one, 2,5-di-(2-thienyl)pyrrole, (SNS), had different solubilities between his oxidized and reduced state. In this way the method of synthesis of 2,5-di-(2-thienyl)pyrrole, (SNS), was improved.

The electrochemical oxidation of monomer from $LiClO_4$ acetonitrile solutions gives a film of poly-conjugated and lineal oxidized chains, which solve in most of the organic solvents.

167

The kinetics of the electrosynthesis of this polyconjugated material was followed by ex situ ultramicrogravimetry.

The electrochemical reduction of the film in aqueous solution gives an insoluble film of the reduced material. On this way both, oxidized and reduced states of the material, are available as solid films.

The flow of an anodic current through a solution of reduced material gives a faradaic electrodeposition of the oxidized state. The process was also followed by microgravimetric determination of the deposited dry films. Checking different electrolyte productivities similar to those obtained during electrodissolution were obtained. The resulting films were also soluble in organic solvents.

The influence of the different chemical and electrical variables in the electrogeneration, electrodissolution and electrodeposition processes were studied.

The final aim of this paper is to find new ways of processability for conducting polymers, which can be applied to manufacture new actuators and artificial muscles.

EXPERIMENTAL

The availability of important quantities of monomer was possible by improving the method of synthesis described by Wynberg and Matselaar[14]. The intermediate 1,4-di-(2-thienyl)-1,4-butanedione was kept under reflux with ammonium acetate, glacial acetic acid and acetic anhidric overnight under nitrogen atmosphere. The reaction mixture was then poured into 250 ml of distilled water and the resulting dark-green solid was chromatographed through silica gel columns using a mixture of dichloromethane and hexane (3:2) for elution. A 75% of SNS, as pale yellow crystals, was obtained. The synthesized monomer, several grams can be obtained every time, was stored under nitrogen.

Acetonitrile (Lab Scan, HPLC grade) was used as solvent in the electrochemical experiments. Lithium perchlorate (Aldrich, A.C.S. reagent), tetraethyl-ammonium perchlorate (TEAClO$_4$), tetrabutyl-ammonium perchlorate (TBAClO$_4$) and thetrahexyl-ammonium perchlorate (THAClO$_4$) (Fluka purum reagent) were dried in an oven at 80 °C before use.

All the electrochemical studies were performed in a one-compartment thermostat three-electrode cell. The working and counter electrodes were two platinum sheets of 1 and 2 cm^2 of surface area, respectively. Both electrodes were immersed in sulfochromic mixture and rinsed with water and acetonitrile before each electrogeneration or electrodeposition. The reference electrode was an Ag/AgCl (3MKCl) electrode. Chronoamperometric, chonopotentiometric and voltammetric measurements were carried out using a PAR 273 potentiostat-galvanostat connected to a personal computer and controlled through a PAR M270 program.

Each solution was deoxygenated by nitrogen bubbling, presaturated with acetonitrile, for 20 min before each experiment. The polymer weight was obtained by "ex situ" microgravimetry. Each film was dried in hot air, rinsed several times in acetonitrile, dried again and weighed until constant weight. Weights of electrodes or polymer films were determined using a Sartorious 4504 MPS ultramicrobalance having a precision of 10^{-7} g. Very reproducible polymeric weights (higher than 95 %) were obtained following this procedure.

RESULTS AND DISCUSSION

Electrogeneration process.

Until now, we have studied the electrogeneration processes from LiClO$_4$ solutions in acetonitrile[10-13]. The obtained films became soluble by electrochemical reduction, but electrodeposition processes from the reduced solutions took place with a very low productivity. In order to improve the reversibility of both processes now we try to electrogenerate the soluble polymer using different ammonium salts: TEAClO$_4$, TBAClO$_4$ and THAClO$_4$.

The flow of an anodic constant current density of 0.5 mA cm^{-2} through 5 mmol/l SNS acetonitrile solutions containing 0.1 mol/l of different ammonium salts (TEAClO$_4$, TBAClO$_4$ or THAClO$_4$), at room temperature gives, in every electrolytes, adherent and uniform polymer films coating the platinum electrode.

The kinetics of the electrogenerated poly(SNS) films were studied following the weights of the dry poly(SNS) electrogenerated at different times of current flow. Once electrogenerated, each film, was dried in hot air, rinsed in acetonitrile and weighed until constant weight, using an ultramicrobalance. The procedure allows the obtention of very reproducible films ($\pm$ 0.0005 mg). The polymer weight was calculated by weight difference between coated and uncoated electrodes. A linear correlation was obtained in every electrolytes (table I) between the electrogenerated polymer weight and the electrogeneration time.

Table I: *Amount of poly(SNS) electrogenerated on the electrode when a current density of 0.5 mA. cm^{-2} is applied through different acetonitrile solutions containing 5 mmol/l of SNS and 0.1 mol/l of different ammonium salts.*

	LiClO$_4$	TEAClO$_4$	TBAClO$_4$	THAClO$_4$
10 s	0.0149	0.0084	0.0070	0.0055
20 s	0.0259	0.0182	0.0137	0.0096
30 s	0.0350	0.0262	0.0200	0.0167
40 s	0.0436	0.0344	0.0264	0.0227
50 s	0.0582	0.0414	0.0320	0.0274
60 s	0.0682	0.0490	0.0393	0.0327
	Electrogenerated poly(SNS) weight/ mg			

The productivities (defined as the polymer weight generated on the electrode per consumed unit of charge, mg. mC^{-1}) of the electrogeneration processes were: 2.1 x 10^{-3}, 1.6 x 10^{-3}, 1.3 x 10^{-3} y 1.1 x 10^{-3} mg. mC^{-1} from LiClO$_4$, TEAClO$_4$, TBAClO$_4$ or THAClO$_4$ solutions, respectively.

Electrodissolution process.

The low solubility of the poly(SNS) films in acetonitrile drops to zero when different salts, like LiClO$_4$, TEAClO$_4$, TBAClO$_4$ and THAClO$_4$, are solved in this solvent. Films containing the same polymer weight were electrogenerated, rinsed and transferred into 0.1mol/l acetonitrile solutions of LiClO$_4$, TEAClO$_4$, TBAClO$_4$ or THAClO$_4$ respectively. There, they were reduced by flow of a cathodic current density of 0.2 mA. cm^{-2}. When the current was applied, a yellow-green cloud was formed around the electrode and the potential of the electrode keeps a constant value. When the black polymer film is solved a cathodic potential jump greater than 500 mV was presented in each chronopotentiogram. Times of current flow before the potential jump were proportional to the mass present on the electrode. After each electrodissolution, the electrode was dried in hot air, rinsed with acetonitrile, dried again and weighted until constant weight. The polymeric weight remaining after each electrodissolution was very low.

Films containing the same weight of poly(SNS) were electrogenerated in order to follow electrodissolution kinetics. A constant cathodic current density of 0.2 mA cm^{-2} was applied to each electrode during different times in their background solutions (the electrolytic solution where every film was electrogenerated). After that, every film was rinsed, dried, and weighed. The electrodissolved weight was obtained by weight difference before and after current flow. The solved mass increases linearly at increasing times of current flow (table II).

Table II: Weight of electrodissolved poly(SNS) when a cathodic current density of 0.2 mA. cm^{-2} was applied to a Pt electrode, during different times in the background solutions (0.1 mol/l acetonitrile solutions of: LiClO$_4$, TEAClO$_4$, TBAClO$_4$, or THAClO$_4$)

	LiClO$_4$	TEAClO$_4$	TBAClO$_4$	THAClO$_4$
4 s	0.0078	0.0050	0.0032	0.0023
8 s	0.0144	0.0097	0.0104	0.0071
12 s	0.0172	0.0147	0.0167	0.0137
16 s	0.0229	0.0191	0.0237	0.0172
20 s	0.0283	0.0250	0.0302	0.0224
24 s	0.0339	0.0305	0.0357	0.0255
Electrodissolved poly(SNS) weight/ mg				

The productivities of the electrodissolution process (defined as the polymer weight electrodissolved per unit of consumed charge, in mg. mC^{-1}) were: 7.4 x 10^{-3}, 7.2 x 10^{-3}, 6.4 x 10^{-3} y 5.7 x 10^{-3} mg. mC^{-1} in THAClO$_4$, LiClO$_4$, TEAClO$_4$ and TBAClO$_4$ respectively.

Electrodeposition process.

Different productivities were obtained for electropolymerization (2.1 x 10^{-3} mg. mC^{-1}) and electrodissolution (7.2 x 10^{-3} mg. mC^{-1}) processes of poly(SNS) in LiClO$_4$ acetonitrile solutions, showing that they are not reverse processes. We are interested to extend the processability of the polymeric material checking whether conducting polymers can mimic both electrodissolution and electrodeposition of metals. This possibility has never been described in literature. Taking into account that by electrodissolution only the oxidation state of the material is changed without formation of covalent bonds between new atoms in the chain, we expect that the oxidation of the solved neutral polymeric molecules $(SNS)_n$ must give an oxidized and insoluble polymer film on the electrode (electrodeposition).

The electrodeposition process only can be studied using very low anodic current densities and long times of current flow, because of the low solubility (low concentration) of the reduced poly(SNS) in the electrolytic solutions.

First of all we studied the electrodeposition process from acetonitrile solution using $LiClO_4$ as electrolyte. The best productivity (1.44×10^{-3} mg. mC^{-1}) was obtained by applying an anodic current density of 8 μA. cm^{-2} through 6 ml of 0.5 mol/l $LiClO_4$ acetonitrile solution containing 0.75 mg of reduced poly(SNS). This productivity is much lower than that of the electrodissolution process. Up to a 30 % of the electrodeposited films was solubilized again by cathodic electrodissolution. Both, the non-soluble fraction of the film and the low productivity indicate the irreversibility of the process, with presence of parallel cross-linking reactions[15].

In order to check the influence of the electrolyte on the electrodeposition productivity, we prepared acetonitrile solutions containing $TEAClO_4$, $TBAClO_4$ and $THAClO_4$. Anodic current densities of 4 μA cm^{-2} were applied during 1800 s, through 6 ml of 0.1 mol/l acetonitrile solutions of $TEAClO_4$, $TBAClO_4$ or $THAClO_4$, where 1.25 mg of poly(SNS) were electrodissolved by reduction previously. The obtained results are shown in table III.

Table III: *Poly(SNS) electrodeposited weight (mg) and productivity (mg. mC^{-1}) obtained when an anodic current density of 4 μA. cm^{-2} was applied, during 1800 s, through 0.1 mol/l acetonitrile (6 ml) solutions of $TEAClO_4$, $TBAClO_4$ or $THAClO_4$ containing 1.25 mg of reduced poly(SNS).*

	$TEAClO_4$	$TBAClO_4$	$THAClO_4$
Deposited weight/ mg	0.0409	0.0370	0.0311
Productivity/ mg. mC^{-1}	5.68×10^{-3}	5.14×10^{-3}	4.32×10^{-3}

Those productivities are higher than that obtained in $LiClO_4$ solutions (1.44×10^{-3} mg. mC^{-1}) and very close to the productivity of the electrodissolution process. Under those conditions we can consider that electrodissolution and electrodeposition, using $TEAClO_4$ or $TBAClO_4$, are reverse processes. On this way we have a polymeric material able to mimic electrodissolution and electrodeposition of inorganic metals, as well as the concomitant applications. The processability of the polymer open new ways to develop actuators and artificial muscles if we are able to control the cross-linking degree of the material once the device was constructed. This is the new field we are working now trying to develop an electrochemical control of the degree of cross-linking.

CONCLUSIONS

The monomer SNS, gives adherent and uniform films by electropolymerization from different electrolytes ($LiClO_4$, $TEAClO_4$, $TBAClO_4$ and $THAClO_4$), following a faradaic process: the electrogenerated polymer weight is proportional to the polymerization charge. Those films when submitted to a cathodic current density in their background solutions also follow a faradaic electrodissolution process: the weight of solved polymer is proportional to the cathodic charge. Electropolymerization and electrodissolution are not reverse processes as show the different productivities. We had been able to obtain electrodeposited films from solutions of the reduced polymer by flow of anodic currents, the electrodeposition being a faradaic process. However the productivity of the electrodeposition process, using LiClO$_4$ as electrolyte, is very low and only a fraction of the obtained film was soluble. That indicates the presence of electrochemically induced cross-linking processes and parallel reactions consuming charge in parallel to the polymeric oxidation-deposition. However, using ammonium salts as TEAClO$_4$ or TBAClO$_4$, productivities of both electrodeposition and the electrodissolution processes are much closer. Electrodeposition and electrodissolution are reverse faradaic processes.

ACKNOWLEDGEMENTS

The authors are grateful for the financial support from the Ministerio de Educación y Cultura of the Spanish Government, the Diputación Foral de Guipuzcoa and the Basque Government.

REFERENCES

1. J. Roncali, F. Garnier, M. Lemaire, R. Garreau, Synth. Met., **15**, 323, (1986).

2. G.G. McLeod, M.G.B. Mahboubian-Jones, R.A. Oethrick, S.D. Watson, N.D. Truong, J.C. Galin, J. François, Polymer, **27**, 455, (1986).

3. M.T. Zhao, M. Samoc, B.P. Singh, P.N. Prasai, J. Phys. Chem., **93**, 7916, (1989).

4. J.P. Ferraris, T.R. Hanlon, Polymer, **30**, 1319, (1989).

5. J.R. Reynolds, M.Pomerants, in *Electroresponsive Molecular and Polymeric Systems*, edited by Marcel Dekker (New York, 1991), pp. 187-256.

6. A.R. Sørensen, L. Overgaad, T. Johannsen, Synth. Met., **55-57**, 1626, (1993).

7. M.V. Soshi, C. Henler, M.P. Cava, J.L. Cain, M.G. Bakker, A.J. Mc-Kinley, R.M. Metzeger, J. Chem. Soc. Perkin Trans., **2**, 1081, (1993).

8. J. Carrasco, T.F. Otero, E. Brillas, M. Montilla, J. Electroanal. Chem., **418**, 115, (1996).

9. E. Brillas, M. Montilla, J. Carrasco, T.F. Otero, J. Electroanal. Chem., **418**, 123, (1996).

10. J. Carrasco, A. Figueras, T.F. Otero, E. Brillas, Synth. Met., **61**, 253, (1993).

11. T.F. Otero, J. Carrasco, A. Figueras, E. Brillas, J. Electroanal. Chem., **370**, 231, (1994).

12. T.F. Otero, E. Brillas, J. Carrasco, A. Figueras, Mat. Res. Symp. Proc., **328**, 799, (1994).

13. E. Brillas, J. Carrasco, A Figueras, F. Urpí, T.F. Otero, J. Electroanal. Chem., **392**, 55, (1995).

14. H. Wynberg, J. Metselaar, Synth. Commun., **14(1)**, 1, (1984).

15. T.F. Otero, S. Villanueva, E. Brillas, J. Carrasco, Acta Polym., **49**, 433, (1998).

IN SITU CONDUCTIVITY OF POLY(3,4-ETHYLENEDIOXYTHIOPHENE) ELECTROSYNTHESIZED IN AQUEOUS SOLUTIONS IN THE PRESENCE OF LARGE ANIONS

H. J. AHONEN, J. LUKKARI, J. KANKARE
Department of Chemistry, University of Turku, FIN-20014 Turku, Finland

ABSTRACT

Thick poly(3,4-ethylenedioxythiophene) films incorporating various anion dopants were anodically deposited from aqueous solutions. *In situ* AC conductometry was applied during the electropolymerisation and the conductivity values obtained in different electrolyte systems were compared. The results indicate that 3,4-ethylenedioxythiophene can be polymerised also from relatively dilute aqueous solutions containing various types of electrolytes and that the most conducting films are obtained using large organic anions.

INTRODUCTION

Organic conducting polymers are widely studied materials in physics, chemistry and materials science due to numerous potential applications of these materials, e.g., in the area of electronics and molecular electronics. Usually, these materials are synthesized chemically or electrochemically from monomer solutions in organic solvents (acetonitrile, propylenecarbonate, dichloroethane, etc.). Contrary to some other monomers like pyrrole and aniline, thiophene is not easily polymerised from aqueous medium. Perhaps the most prohibitive factor is the low solubility of thiophene in water, which leads to a low radical cation concentration during the electrosynthesis. The few radical cations can react instantly and irreversibly with the nucleophilic water molecules instead of each other and, consequently, only soluble oligomers are obtained. The solubility of the thiophene monomer can be enhanced by a methoxy substitution at the 3-position with some sacrifice of the conductivity of the corresponding polymer due to the introduction of steric effects and loss of the effective conjugation in the polymer backbone. The high oxidation potential of thiophene, which is beyond the electrochemical stability window of water is another limiting factor. On the other hand, for practical, economical and enviromental reasons it would be desirable if high quality materials could be synthesised in a non-toxic and non-expensive medium like water. Indeed, there has been some success in obtaining polythiophene from highly acidic aqueous solutions or methanol/water mixtures, but the properties of the resulting polymers have been inferior to the materials polymerised in organic solvents [1,2].

Due to its high conductivity and electrochemical stability in oxidised state poly(3,4-ethylenedioxythiophene) (PEDOT) is presently under intense research in several academic and industrial laboratories. EDOT monomer can be dissolved in water to a sufficiently high concentration (up to 14 mM) and its oxidation potential is relatively low (ca. 1 V vs. SSCE). Furthermore, there have been some recent reports about the improvement of the physicochemical properties of poly(3,4-ethylenedioxythiophene) using sodium dodecylsulfate (SDS) micellar aqueous medium, where regular and well ordered films have been obtained [3]. Molecular order in the film is usually connected to the increased conductivity of the material. Therefore, we electrosynthesized potentiostatically PEDOT films from various aqueous solutions including the micellar medium in order to compare their properties with the films

173

obtained from organic solutions. The *in situ* AC conductimetric method used in this work gives more reproducible and reliable conductivity values than the conventional four-point measurements, in which the polymerised film must be peeled off from the electrode surface before measuring the resistance of the film. Often the resulting polymer is mechanically weak and/or powdery, and does not allow detachment from the electrode without breaking. With our method such a mechanical weakness of the film is not a problem. Furthermore, depending on the gap size between the electrodes, this technique allows the conductivity of thinner polymeric films to be measured than with the conventional four-probe method.

EXPERIMENT

The electropolymerisation of 3,4-ethylenedioxythiophene (Bayer AG) in aqueous solutions was carried out in a one-compartment cell in the presence of various organic or inorganic electrolytes: 0.07 M sodium dodecylsulfate (SDS, Aldrich), 0.01 M picric acid (Merck) and 0.1 M lithium perchlorate (LiClO₄, Aldrich). During the electropolymerisation in organic solution, 0.1 M LiClO₄ was used as an electrolyte salt. The aqueous polymerisation solutions were unbuffered in order to avoid any effects resulting from the buffer components. The monomer concentration was kept constant (5 mM). The working electrode was a double band platinum microelectrode, with the surface area 0.005 cm^2 and the gap between the bands 15 μm. A platinum wire was used as a counterelectrode. The reference electrode in aqueous solutions was a sodium saturated calomel electrode (SSCE) and in acetonitrile solutions an Ag/AgCl electrode (Cypress Systems) was used. In voltammetric experiments platinum disk (Cypress Systems), was used as working electrode (area = 0.785 mm^2). The mathematical model which gives the specific conductivity has been published previously [4]. In addition to the *in situ* conductivity, the shape of the curve of conductance vs. logarithm of the consumed charge gives information on the homogeneity of the growing layer and current efficiency of the growth process, i.e., the volume yield.

RESULTS

The cyclic voltammograms of the EDOT monomer in different electrolyte systems are presented in figure 1. All the voltammograms show an irreversible behaviour and present a shoulder or a prepeak followed by a second peak during the anodic scan. The existence of a prepeak is connected to the oxidation of adsorbed species on the electrode, whereas the following oxidation process is assigned to the oxidation of the bulk species [3]. On the other hand, the shape of the second peak (non-symmetric) in some cases implies that the overoxidation of the forming polymer layer coincides with the oxidation of the bulk monomer species. Further support to the interpretation of the second oxidation peak being at least partly due to overoxidation is provided by cycling the monomer solution beyond the second oxidation peak potential. Very little or no electroactive polymer was obtained. Therefore, in order to prevent the overoxidation to occur simultaneously during the electropolymerisation, a low polymerisation potential (E_p = 0.85 V) was selected, i.e., from the foot of the first oxidation peak. If polymerisation potentials were higher by 150 mV, the conductivity values of the polymers were considerably lower in all electrolyte systems (vide infra). This emphasizes the importance of the polymerisation potential as one of the significant parameters affecting the quality of the resulting polymer film.

The oxidation potential of the adsorbed monomer is relatively constant (1.01-1.06 V) in different electrolytes but the potential of the second peak is affected to some extent by the electrolyte system. The shift relative to LiClO₄ electrolyte is largest in picrate solutions (ca.

250 mV) and smaller in SDS solutions (ca. 70 mV). Such shifts in the oxidation potential have been attributed to interactions between the anion and the monomer [3].

Figure 2 shows the recorded conductance during the potentiostatic polymerization of 3,4-ethylenedioxythiophene in aqueous medium in the precence of picric acid, SDS and LiClO$_4$.

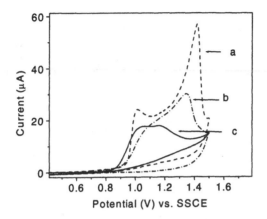

Fig. 1. Cyclic voltammograms of EDOT (5 mM) obtained in a) 0.1 M LiClO$_4$ b) 0.07 M SDS c) 0.01 M picric acid aqueous solutions on platinum electrode (area = 0.785 mm^2). Scan rate 100 mV/s.

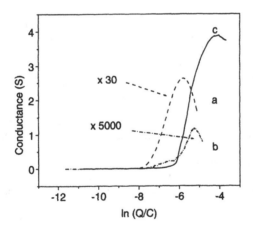

Fig. 2. Potentiostatic polymerization of 5 mM EDOT in aqueous solutions in the presence of a) 0.1 M LiClO$_4$ (multiplied by 30) b) 0.07 M SDS (multiplied by 5000) and c) 0.01 M picric acid. Polymerisation potential +0.85 V vs. SSCE.

The shape of the conductance vs. logarithm of the charge consumed during the polymerization is similar in all electrolyte systems.

The curves show clear maxima after a certain amount of charge has passed. Similar behaviour has been observed with pyrrole polymerized in aqueous solutions containing perchlorates [5] and has been attributed either to the gradual deterioration of the polymer or detachment of the film from the surface.

Using higher polymerization potentials the conductance maximum was reached earlier and the conductivity of polymers was lower in all electrolyte systems, which is attributed to the deterioration of the films by simultaneous overoxidation processes. However, the obtained polymer films were thicker by a factor of ca. 2. Table I summarises the parameters obtained from the conductance measurements in the different electrolyte systems using two different polymerization potentials.

Table I. Parameters from the potentiostatic polymerization of 5 mM EDOT in various aqueous electrolyte solutions on a double band platinum microelectrode with different polymerization potentials (E_p).

electrolyte	$E_p = 0.85$ V vs. SSCE		$E_p = 1.0$ V vs. SSCE	
	conductivity σ S/cm	volume yield cm^{3}C^{-1}	conductivity σ S/cm	volume yield cm^{3}C^{-1}
Picric acid	49.5	9.5×10^{-4}	0.002	$1.7*10^{-3}$
LiClO$_4$	1.0	3.6×10^{-4}	0.3	$7.5*10^{-4}$
NaClO$_4$ (pH 2)	1.79	6.6×10^{-4}		
SDS	0.004	1.1×10^{-3}	0.003	$2.1*10^{-3}$
LiClO$_4$/MeCN	104*	1.1×10^{-3}		

* $E_p = 1.25$ V vs. Ag/AgCl

The *in situ* conductivities of the polymer films obtained in various aqueous electrolytes show remarkable differences. Polymerising EDOT with low oxidation potential in aqueous solutions, the picrate electrolyte system gives the highest conductivity, which is of the same order as with films polymerised from organic solvents, e.g., from acetonitrile. In literature the reported conductivity values for PEDOT polymerised electrochemically in organic solutions vary between 100 and 400 S/cm [6,7]. These values have been obtained by four probe measurements of dry samples and the conductivity of the swollen "wet" films may be different. Nevertheless, the values are in good agreement. The high conductivity of PEDOT has been previously proposed to originate from the structure of the polymer. The PEDOT chains are supposed to lie flat on the electrode surface when polymerised in acetonitrile solution [3]. This kind of "flat" structure would show high conductivity. In the case of picric acid, which is a planar molecule itself, a similar flat structure of the polymer can be assumed. Since the pH of picric acid solution is ca. 2 ($pK_a = 0.29$), the polymerisation was conducted in sodium perchlorate solution with the same pH adjusted with perchloric acid. The polymer obtained from perchloric acid exhibits markedly lower conductivity than the polymer obtained in picric acid. Therefore, it is suggested that a specific interaction with the acid and the polymer is needed to obtain high conductivity and that pH has no effect in general. Interestingly, the volume yield, i.e., the compactness of the film, does not seem to correlate with the conductivity.

It is interesting to notice the very poor conductivity of the film polymerised in SDS micellar medium. The EDOT monomer is preferentially dissolved into the micellar assembly of dodecylsulfate ions because of the hydrophobic nature of EDOT monomers. Therefore, the free monomer concentration is low, keeping also the radical cation concentration low during electrosynthesis in the vicinity of the electrode and coupling of two radical cations is inhibited. These few radicals may react with impurities or with nucleophilic water molecules leading to shorter conjugation in the growing polymer chains, with drastic effect on the conductivity. Attempts to polymerize from more concentrated monomer solutions (50 mM), while keeping the SDS concentration constant led to films with higher conductivities (up to 60 S/cm). In these solutions the free monomer concentration is higher. Therefore, the mass transport to the electrode is sufficient for the effective polymerisation reaction to proceed. Furthermore, in this case the micelles also contain a high concentration of EDOT monomers and taking into account the diffusion constant of the SDS micelles (ca. 10^{-6} cm^{2}s^{-1} [8]) they also contribute to the mass transport. Upon opening, a micelle can deliver a large amount of monomers and increase their effective concentration at the vicinity of the electrode surface. When the monomer concentration is low the micelles contain also less monomer and their contribution to the mass transport is not so important. Therefore, in dilute monomer solutions in micellar electrolytes both the "free" monomer and the micelle-bound concentrations are crucial factors affecting the polymerization process.

It has been suggested that the growth mode of the growing polymer chains would be perpendicular to the surface of the electrode when polymerisation solution contains SDS [3,9]. The growth mode of the polymer can be estimated during the electrosynthesis on double band platinum electrode from the time required for bridging the two platinum bands. If this takes place rapidly the lateral growth is fast. Usually, the first contact between the two growing polymer layers is established after about 100 s. On the other hand, during the polymerisation in the micellar medium, the first appearance of conductance is observed after a considerably longer time (ca. 1000 s). This supports the assumption of the perpendicular growth mode over lateral growth.

CONCLUSIONS

In situ conductivity measurements during electropolymerization of 3,4-ethylenedioxythiophene in different electrolyte solutions were performed. The previous results from the polymerization attempts of thiophene in aqueous solutions have shown that the resulting material is inferior to the films obtained from organic medium. On the contrary, 3,4-ethylenedioxythiophene polymerizes well in aqueous solutions. The polymerization potential and the anion composition of the polymerisation solution have an important role on the structure and electrical properties of the resulting film. It was demonstrated in this work that selecting the right conditions for electropolymerisation highly conducting PEDOT films can be prepared even from dilute aqueous solutions.

ACKNOWLEDGMENT

The financial support of the Academy of Finland (Grant #30579) is gratefully acknowledged.

REFERENCES

1. S. Mu, and S. Park., Synth. Met., **69** 309 (1995).
2. S. Dong, and W. Zhang, Synth. Met., **30** 359 (1989).
3. N. Sakmeche, S. Aeiyach, J.-J. Aaron, M. Jouini, J. C. Lacroix, and P.-C. Lacaze, Langmuir, **15**, 2566 (1999).

4. J. Kankare, and E.-L. Kupila, J. Electroanal. Chem., **322**, 167 (1992).
5. E.-L. Kupila, and J. Kankare. Synth. Met., **82**, 89 (1996).
6. G. Heywang, and F. Jonas, Adv. Mater., **4**, 116, (1992).
7. Q. Pei, G. Zuccarello, M. Ahlskog, and O. Inganäs, Polymer, **35**, 1347 (1994).
8. B. Lindman, and H. Wennerström, Top. Curr. Chem. **87**, 1 (1980).
9. K. Naoi, Y Oura, M. Maeda, and S. Nakamura, J. Electrochem. Soc., **142**, 417 (1995).

HYPERBRANCHED CONDUCTIVE POLYMERS
CONSTITUTED OF TRIPHENYLAMINE

S.TANAKA*, K.TAKEUCHI*, M.ASAI*, T.ISO**, M.UEDA***
*National Institute of Materials and Chemical Research, 1-1 Higashi, Tsukuba 305-8565
JAPAN, sutanaka @ home.nimc.go.jp
**Ibaraki Prefectural Industry and Technology Center, 189 Kanakubo, Yuhki 307-0015
JAPAN
***Tokyo Institute of Technology, Ohokayama, Meguro-ku, Tokyo 152-8552 JAPAN

ABSTRACT

A hyperbranched conjugated polymer containing triphenylamine was prepared by the
Grignard reaction of tris(4-bromophenyl)amine 1, via the coupling of N,N-bis(4-
bromophenyl)-N-(4-bromomagnesiophenyl)amine 2 with the catalytic amount of Ni(acac)$_2$.
Grignard reagent 2 reacted as an AB$_2$-type monomer to give hyperbranched conjugated
polymer 3 in a one-step process. Polymer 3 was also obtained via the Pd-catalyzed coupling
of N,N-bis(4-bromophenyl)-4-animobenzeneboronic acid 4. Polymer 3 had an average
molecular weight of 4.0-6.3x10^3 and was found to be soluble in organic solvents such as
THF and CHCl$_3$. A cast film had an anodic peak at 0.95-1.20 V vs. Ag wire. It was dark blue
above the oxidation potential and brown-yellow in the neutral state. When polymer 3 was
doped with iodine, its conductivity rose to 0.8-3.0 S/cm

INTRODUCTION

Dendritic polymers can be prepared through repetitive monomer addition sequences
using either a convergent method or a divergent method. Every approach requires a stepwise
growth process[1-4], where a protection-deprotection step is involved at every step of the
growth process. Although this approach yields monodisperse polymers with well-defined
structure, it usually requires extensive intermediate purification of the products at every
protection-deprotection step of the growth process. This makes dendritic polymers expensive
and difficult to produce on a large scale. Hyperbranched polymers can be obtained by a one-

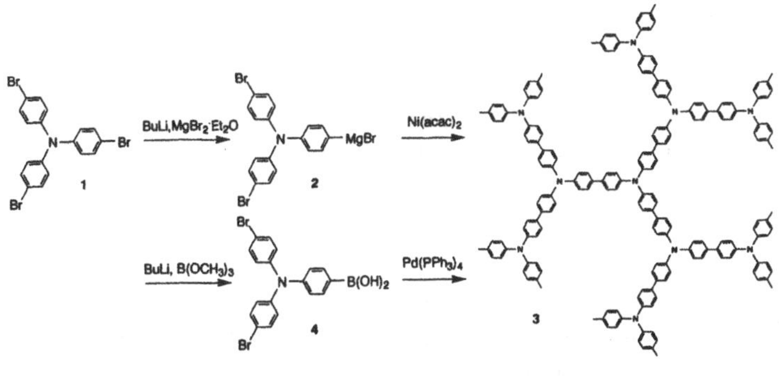

Scheme 1

179

step process, where an AB_x-type monomer undergoes self-condensation polymerization as first discussed theoretically by Flory in 1952[5]. The advantage of this one-pot AB_x approach lies in its potential for greater general applicability. It is important to produce hyperbranched polymers on a large scale at a reasonable cost. But this method involves the loss of control in molecular weight, accompanied by a broad molecular weight distribution.

We report here the preparation of a hyperbranched conjugated polymer containing triphenylamine by the Grignard reaction of tris(4-bromophenyl)amine 1 (Scheme 1). Polymer 3 was also obtained via the Pd-catalyzed coupling of N,N-bis(4-bromophenyl)-4-animobenzeneboronic acid 4. Oligomeric and polymeric materials based on triphenylamine have been attracted much attention because of their stability and useful properties such as electrical conductivity and electroluminesence[6-8].

EXPERIMENTAL

Polymerization of tris(4-bromophenyl)amine 1 to polymer 3 via a Grignard route

To tris(4-bromophenyl)amine 1 in dry THF was added a 1.6 M n-BuLi solution at -78°C under argon. The mixture was stirred for 10 min. To the lithiate was added $MgBr_2 \cdot Et_2O$ in dry diethyl ether at -78°C. The yellow solution was warmed to room temperature and was added to $Ni(acac)_2$ in THF. The dark green mixture was refluxed for 24 h, and then water was added at room temperature. The solution was concentrated in vacuo, and the polymer was precipitated from petroleum ether. After washing with a dilute HCl solution, water, and methanol and drying, a light yellow powder was obtained. M.p. 192-199°C.

N,N-Bis(4-bromophenyl)-4-aminobenzeneboronic acid 4

To a solution of amine 1 in dry diethyl ether was added n-BuLi at -78°C under argon. The mixture was stirred for 15 min and then added to trimethyl borate in diethyl ether at -78°C. The solution was stirred for 15 min and then allowed to warm to room temperature. After stirred for 20 h, the solution was treated with 2 N HCl. The ether layer was extracted with a KOH solution. The combined aqueous solution was washed with ether and then acidified with 6 N HCl to pH 3 at 0°C. A gray powder was obtained by filtration, washed with water and dried in vacuo. M.p. 186-198°C.

Preparation of hyperbranched polymer 3 via Pd-catalyzed coupling

To a mixture of benzene, a 2 M Na_2CO_3 solution, and $Pd(PPh_3)_4$ was added acid 4 under argon. The mixture was refluxed for 20 h and then cooled to room temperature. The whole mixture was poured into acetone. The solid material was collected by filtration, washed with 1 N HCl. After drying in vavuo, the hyperbranched polymer 3 was obtained. M.p.> 500°C.

Measurements

Cyclic voltammograms were measured in 0.1 mol/l Bu_4NBF_4 solution in PC at a sweep rate of 100 mV/s by using a Hokuto-Denko HA-501 potentio-galvanostat and a HB-104 function generator. A platinum plate and a Ag wire were used as counter and reference electrodes, respectively. Working electrode was a platinum plate or indium-tin oxide glass (ITO). FTIR and UV-VIS spectra were recorded on a Perkin-Elmer 1720-X spectrometer and a Shimadzu UV-3100PC spectrometer, respectively. GPC was performed in THF at 40°C using a TOSOH TSKgel G3000H$_{XL}$ column which was calibrated with polystyrene stardards.

RESULTS and DISCUSSION

FTIR and UV-VIS spectra

A FTIR spectrum of monomer **1** showed two bands at 817 and 1269 cm^{-1}, assignable to the C-H deformation vibration of 1,4-disubstituted benzene rings and the C-N stretching vibration of tertiary amines, respectively (Fig. 1a). Polymer **3** had these two bands and there was a weak band at 510 m^{-1}, due to the C-Br stretching vibration of terminal benzene rings (Fig. 1b), although monomer **1** had a strong band at 510 cm^{-1}. Fig. 2 shows the UV-VIS spectra of monomer **1** and polymer **3**. Monomer **1** had a band at 307 nm assignable to the conjugation between benzene units and nitrogen atoms. In polymer **3** a new $\pi\text{-}\pi^*$ transition band was observed clearly at 360 nm, possibly brought about by the conjugation between biphenylene units and nitrogen atoms. These results indicate that a C-C bond was formed at the carbon atoms coupled with bromine originally and a highly conjugated system is formed. The ^{13}C-NMR spectrum of monomer **1** had four bands at 116.0, 125.6, 132.5 and 146.0 ppm. In addition to these bands, polymer **3** had three bands at 124.7, 127.8 and 135.5 ppm, which would be assignable to biphenylene carbon resonances.

Molecular weight

Polymer **3** had an average molecular weight of 4.0-6.3X10^3 (Fig. 3) and was found to be soluble in organic solvents such as THF and CHCl$_3$. The solubility is in striking contrast to that reported for the polymer prepared by the conventional polycondensation of tris(4-bromophenyl)amine **1** with magnesium[9]. In this case tris(4-bromophenyl)amine **1** reacted as a

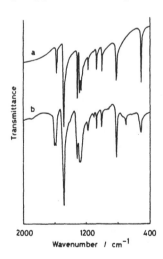

Fig.1. FTIR spectra of (a) monomer **1** and (b) polymer **3**.

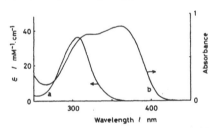

Fig.2. UV-VIS spectra of (a) monomer **1** in CH$_3$CN and (b) polymer **3** in THF.

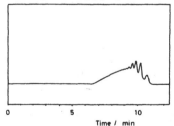

Fig.3. GPC of polymer **3** in THF.

crosslinking agent, and poly(triphenylamine) obtained was insoluble in organic solvents. The hyperbranched polymer prepared by us was soluble, which implies that the polymer is processable.

The results of GPC indicated that the degree of polymerization was about 16-26. Considering that the repeating unit of polymer 3, $C_{18}H_{12}N$, contains three benzene rings, this degree of polymerization signifies that the polymer had about 50-80 benzene rings, which was sufficient to show electroactivity as will be described below.

Electrochemistry and conductivity

A cast film deposited on a platinum plate was used for cyclic voltammetry. The polymer showed a cathode-activity at applied voltages higher than 0 V; an anodic peak was observed at 0.95-1.20 V (Fig. 4). The film was dark blue above the oxidation potential and brown-yellow in the neutral state. The change in the VIS-near IR spectra during electrochemical doping was observed by using a cast film deposited on ITO. When the applied voltage was 0 V, a polymer film had only one band at 360 nm (3.44 eV) assigned to a π-π^* transition. At 1.0 and 1.2 V, three new bands were observed around 500, 750 and 1500 nm (2.48, 1.65 and 0.83 eV). The first and last bands were considered to be assigned to the transition from valence band to higher and lower polaron bands[10], respectively. The second band would be due to the transition from lower polaron to higher polaron band. Electrochemically created charge carrier would be identified as polaron species in this case. The stability of the charge carriers would be brought about by the resonance effect of charged triphenylamine moiety. Polymer 3 was insulator without doping. When it was doped with iodine, its conductivity rose to 0.8-3.0 S/cm (Fig. 5).

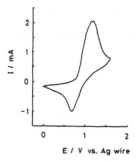

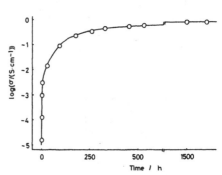

Fig.4. Cyclic voltammogram of polymer 3 in PC with 0.1 mol/l Bu4NBF4. Sweep rate : 100 mV/sec.

Fig.5. Conductivity vs. doping time of polymer 3 with iodine.

CONCLUSIONS

A hyperbranched conjugated polymer 3 containing triphenylamine was prepared by using an AB2-type monomer. Polymer 3 was found to be soluble in organic solvents such as THF and CHCl3. A cast film had an anodic peak at 0.95-1.20 V vs. Ag wire. When polymer 3 was doped with iodine, its conductivity rose to 0.8-3.0 S/cm

Many conductive polymers are insoluble in organic solvents due to the rigid main chains. Introducing a long alkyl or alkoxy group to the main chains is one of the way to enhance solubility. But these groups do not contribute to increase in π-conjugation length. A new hyperbranched conjugated polymer containing triphenylamine has been designed to have both hyperbranched and conjugated backbone, and was found to be soluble in organic solvents without a long alkyl or alkoxy group. It showed a distinct anode-activity, indicating that it has a highly developed conjugated system.

REFERENCES

1 D.A. Tomalia, A.M. Naylor, and W.A. Goodard, Angew. Chem., Int. Ed. Engl., **1990**, *29*, 138

2 C.J. Hawker and J.M.J. Frechet, J. Am. Chem. Soc., **1990**, *112*, 7638

3 G.R. Newkome, V.V. Narayanan, L. Echegoyen, E.Perez-Cordero and H. Luftmann, Macromolecules, **1997**, *30*, 5187

4 E.C. Constable, P. Harverson and M. Oberholzer, J. Chem. Soc., Chem.Commun., **1996**, 1821

5 P.J. Flory, J. Am. Chem. Soc., **1952**, *74*, 2718

6 Y. Kuwayama, H. Ogawa, H. Inada, N. Noma and Y. Shirota, Adv. Mater., **1994**, *6*, 677

7 Y. Shirota, T. Kobata and N. Noma, Chem. Lett., **1989**, 1145

8 H. Tanaka, S. Tokito, Y. Taga and A. Okada, J. Chem. Soc., Chem. Commun., **1996**, 2175

9 M. Ishikawa, M. Kawai, and Y. Ohsawa, Synth. Met., **1991**, *40*, 231

10 J.C. Scott, J.L. Bredas, K. Yakushi, P. Pfluger and G.B. Street, Synth. Met., **1984**, *9*, 165

CONDUCTION BEHAVIOR OF DOPED POLYANILINE UNDER HIGH CURRENT DENSITY AND THE PERFORMANCE OF AN ALL POLYMER ELECTROMECHANICAL SYSTEM

Haisheng Xu, V. Bharti, Z.-Y. Cheng, Q. M. Zhang, Materials Research Laboratory, The Pennsylvania State University, University Park, PA 16802; Pen-Cheng Wang, A. G. MacDiarmid, Chemistry Department, University of Pennsylvania, Philadelphia, PA 19104

ABSTRACT

In many device applications, such as electro-acoustic transducers and actuators based on high strain electroactive polymers, there are many advantages to utilize conductive polymers as electrodes. However, in these applications, a high electric power usually is required which translates to high voltage and high current in the system. Hence, the maximum current density which a conducting polymer can carry is of great interest and importance. In this paper, the conduction behavior at high current density of doped polyaniline(PANI) is reported. It was found that the current density deviates strongly from the ohmic relation with the electric field in high current density region and a saturation of the current density was observed. The maximum current density J_m observed is proportional to the conductivity of the samples and for PANI doped with HCSA, J_m can reach as high as 1200 A/cm^2. Making use of the conducting polymer as the electrodes for the electrostrictive P(VDF-TrFE) copolymer, an all-polymer electromechanical system was fabricated. The all-polymer films exhibit similar or larger electric field induced strain responses than those from films with gold electrodes, presumably due to reduced mechanical clamping from the electrodes. In addition, the all-polymer system also exhibits comparable dielectric and polarization properties to those of gold-electroded P(VDF-TrFE) films in a wide temperature (from -50°C to 120°C) and frequency range (from 1Hz to 1MHz). These results demonstrate that polyaniline can be used for many electro-acoustic devices and provide improved performance.

INTRODUCTION

Recently, we reported that electron-irradiated P(VDF-TrFE) copolymers exhibit an exceptionally high electrostrictive response[1], which will have a great impact in transducer, sensor, and actuator technologies[2]. Because of a high elastic modulus, the commonly used metal electrodes, such as Au and Al, may impose mechanical clamping on the polymer which can reduce the electric field induced strain level and the efficiency of the electromechanical transduction. In addition, at high strain level, thin metal electrodes will crack and cause failure in the devices. Hence, a new electrode material which can lower the clamping effect and withstand high strain is highly desirable. It is believed that a conducting polymer electrode will meet these requirements[3,4]. Due to the flexibility, low acoustic impedance, and elastic modulus of conductive polymer electrodes, such all-polymer electrostrictive systems may improve the performance of electromechanical polymer materials in acoustic and electromechanical applications. In many applications, a high electric power and high electric field will be delivered through polymer electrodes in electromechanical devices, and hence, the current density under high voltage of the conductive polymer electrode is also of great concern.

In this article, we report the behavior of conducting polyaniline films doped with HCl and HCSA in the high current density regimes at room temperature and above. An all-polymer system, consisting of electrostrictive P(VDF-TrFE) copolymer and polyaniline doped with HCSA and casting from m-cresol, was fabricated. The dielectric properties of this all-polymer system were characterized over a wide temperature and frequency range. The polarization and electric field induced strain were also evaluated. The results demonstrate that the all-polymer films exhibit comparable dielectric properties to gold-electroded P(VDF-TrFE) films in a wide temperature (from -50°C to 120°C) and frequency range (from 1Hz to 1MHz). In addition, the all-polymer films seem to show similar or even larger electric field induced strain responses than those from films with gold electrodes under identical measurement conditions.

Experiments

Polyaniline in salt form was prepared by chemical oxidation of aniline according to the reference[5]. The base form of PANI was obtained by treating the salt form with 3% NH4OH for 2h. The free-standing films of PANI were directly cast from N-methyl pyrrolidinone (NMP) solution in emeraldine base (EB) form. For stretched film, the EB film was stretched to ca. 200% of its original length at 180°C by a zone-drawing method[6]. Both stretched and unstretched films were doped by HCl solution. To prepare PANI/HCSA films, EB form of PANI was mixed with HCSA in the molar ratio of 0.5 HCSA per repeat unit of PANI[7], the mixture was ground under N2 atmosphere to fine powder which was then dissolved in m-cresol. The film of PANI doped with HCSA was prepared by casting the resulting PANI-HCSA/m-cresol solution on a glass substrate and then evaporating the solvent at 60°C.

The P(VDF-TrFE) copolymer films were prepared by melt-pressing powder at 225°C and then slowly cooling it to room temperature. For stretched films, films were uniaxially stretched at a temperature between 25 ~ 50 °C with a stretching ratio of 5 times. Both stretched and unstretched films were then annealed at 140°C under vacuum for 24h and cooled down slowly to room temperature. The irradiation was carried out at either 95°C or 120°C under nitrogen atmosphere by electrons at 3 MeV energy with different doses.

To prepare conductive polymer electrodes, the solution of PANI/HCSA was coated on both sides of P(VDF-TrFE) film with a mask. The composite films were dried with an infrared lamp in the hood for 10 minutes to remove the solvent. The composite films are soft and flexible. Gold electroded P(VDF-TrFE) films were also prepared by sputting Au on opposing faces of the films. For conductivity measurement, free standing PANI/HCSA films were obtained by casting the above-mentioned solution on a glass slide and dried on a hot plate at 50°C.

Four-termination setup was used to measure the current density. The DC current (I) was applied and measured between electrodes 1 and 4, while the voltage (V) between electrodes 2 and 3 was recorded concomitantly. Hence, current density (J) is given by

$$J = \frac{I}{t * d} \qquad (1)$$

t and d are the thickness and width of the sample, relatively. The conductivity can be calculated from the data of J and V. The conductivity was also measured by the traditional four-probe method. The results obtained from the current density measurement method in the low current range (will be discussed) are the same as those from the four-probe method which measures conductivity at low current density.

The polarization hysteresis loops of all-polymer systems were measured by a Sawyer Tower circuit[8] at the frequency of 1 Hz under different electric fields. Dielectric properties of the P(VDF-TrFE) films with conductive polymer electrodes were characterized and compared with those of the P(VDF-TrFE) films with gold electrodes. The temperature dependence of dielectric properties was measured in the temperature range from -50°C to 130°C. The heating rate employed was 2°C/min. The frequency dependence of dielectric properties was measured using an HP 4192A Impedance Analyzer in the frequency range between 100 Hz and 13 MHz. The electric field induced strain was characterized with bimorph-based strain sensors designed specially for polymer film strain measurement in our lab[9].

Results and Discussions
Conduction Behavior of Doped Polyaniline

Fig.1 presents the current density J as a function of voltage measured at room temperature for four polyaniline films. All the J-V curves exhibit a linear current-voltage relationship at low current density range. However, in high current ranges, the J-V curves deviate from the linear relationship and the current density shows a saturation. To illustrate this saturation effect more clearly, the field dependent conductivities of these films are presented in Fig.2.

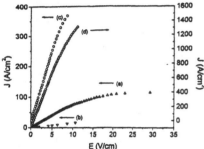

Figure 1. Current density of different polyaniline (a) polyaniline doped with 1M HCl and dry at room temperature, (b) polyaniline doped with 1M HCl and dry at 100°C under vacuum, (c) stretched polyaniline doped with 1M HCl, and (d) polyaniline doped with HCSA and casting from m-cresol solution.

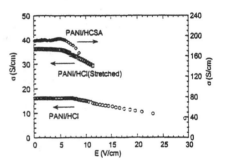

Figure 2. dc conductivity (J/E) as a function of the electric field across the sample. At high fields, σ decreases with field.

Evidently, the films possess an ohmic region where the conductivity is approximately a constant at field. As the field increases, the conductivity of these films drops. For the film doped with HCl (film (a)), the threshold field is about 7.5 V/cm and for the film doped with HCl and stretched, the threshold field is reduced to about 5 V/cm. For the film doped with HCSA, it is at about 5 V/cm. These behaviors are quite different from what has been observed at low temperatures (below 100 K) where the conductivity increases with field[10]. The difference could be caused by temperature related effects and processes in the polymer such as thermal activation of trapped dopants. When the current density at the threshold field (J_m, maximum current density) is plotted against the low field conductivity for the films shown in Fig.1, the data reveals a linear relationship between the two as shown in Fig.3, the maximum current density increases with the conductivity. The result suggests that in analogous to the conductivity measured at low field, the maximum current density is also an intrinsic property of conducting polymers.

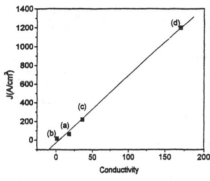

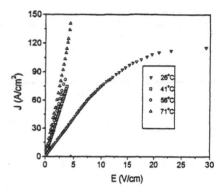

Figure 3. Relation between the low field conductivity and maximum current density for the four samples

Figure 4. The current density of polyaniline doped with 1M HCl at different temperatures

For conducting polymers, it is well known that the conductivity increases with temperature. It is interesting to follow how the nonlinear behavior observed at room temperature changes with temperature. Here, the film (a) was examined and the data is presented in Fig.4 and an increase in the conductivity with temperature was observed. However, unlike the J-V curve at room temperature which contains a relatively large nonlinear region, the J-V curves at elevated temperatures show a linear region between the current and voltage which stops at a maximum current density and voltage beyond which no increase in the voltage and current density can be made even if one tried to increase the current in the external power supplier. Such a behavior suggests that polyaniline here acts as a power source against the external current source.

Polarization Hysteresis Loops and Field Induced Strain Responses

The typical hysteresis loops for P(VDF-TrFE)/PANI and P(VDF-TrFE)/gold films are shown in Fig.5. The films are stretched 65/35 films irradiated at 120°C with 60 Mrad doses.

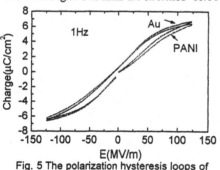

Fig. 5 The polarization hysteresis loops of P(VDF-TrFE) 65/35 copolymer

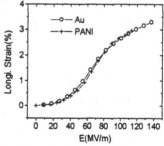

Fig. 6 Longitudinal strain induced by electric fields

The polarization loops measured from the conducting polymer electrode and gold electrode show a quite similar polarization loop. The coercive fields and the remnant polarization of P(VDF-TrFE) films with two different electrode systems measured at different electric fields also exhibit closed values. The results indicate that electrically, PANI can sustain an electric field more than 120 MV/m (the voltage limit of the polarization measurement setup) with a similar performance as that of the gold electrode.

The electrostrictive strains of the irradiated P(VDF-TrFE) 65/35 copolymers were measured and Fig.6 compares the longitudinal strain induced by external electric fields in the P(VDF-TrFE) films with polyaniline electrodes and with gold electrodes, respectively. The two yield nearly identical results for applied electric fields up to 140 MV/m.

For electromechanical applications, the transverse strain response is also of great importance. Fig 7 illustrates the electric field induced transverse strains of stretched (PVDF-TrFE) 65/35 films measured along the stretching direction and the comparison between the films with PANI and with gold electrodes. The transverse strain measured from PANI electroded films is higher than that from gold electroded films. Part of the reasons for this difference could be due to the reduced mechanical clamping from the electrodes. In addition, due to the match of the acoustic impedance of the conductive polymer electrode with that of the electrostrictive polymers, all-polymer systems may improve the performance of electrostrictive polymer systems in acoustic and ultrasonic applications.

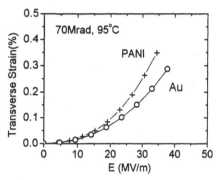

Fig. 7 Transverse strain induced by electric fields

Dielectric Properties

Temperature dependence of the dielectric constant and dielectric loss measured at 100 kHz for the electron irradiated P(VDF-TrFE) 50/50 copolymer films with PANI/HCSA electrodes

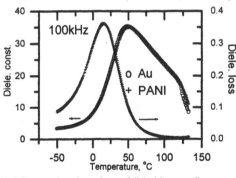

Fig. 8 Temperature dependence of dielectric properties.

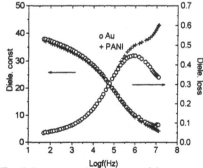

Fig. 9 Frequency dependence of the dielectric properties

are shown in Fig.8, which are nearly the same as the dielectric properties of the same copolymer with gold electrodes shown in the same figure. This demonstrates that the polymer electrode

189

developed here works well in the temperature range at least from -50°C to +120°C, which is more than enough for most applications using the P(VDF-TrFE) copolymers.

The frequency dependence of the dielectric properties of copolymer films with different electrodes measured at room temperature is shown in Fig.9. It is observed that the two systems show very similar dielectric properties over a wide frequency range. The dielectric loss is nearly identical in frequencies from 100Hz to 300kHz and the dielectric constant exhibits nearly the same value in frequencies from 100 Hz to 2 MHz. When the frequency is higher than 300 kHz, the dielectric loss of P(VDF-TrFE) film with conductive polymer electrodes starts increasing and becomes higher than that measured from the film with gold electrodes. The dielectric constant of PANI electroded film is lower than that of gold electroded film above 2 MHz.

Conclusions

An investigation has been carried out to study the conduction behavior of doped polyaniline in the high current density regime. It is found that there exists an ohmic region when the current density is not high. At high current density, the J-V curve deviates from the linear relation and the current density approaches saturation. For PANI doped with HCSA, J_m can reach more than 1200 A/cm^2. All-polymer electromechanical systems were fabricated with electrostrictive poly(vinylidene fluoride-trifluoroethylene) copolymer and conductive polyaniline doped with camphor sulfonic acid. The P(VDF-TrFE)/PANI composite films are soft and flexible with strong coherent interfaces. These all-polymer systems exhibit similar dielectric properties as those from P(VDF-TrFE) films with the gold electrode in a wide temperature (from -50°C to 120°) and frequency range (from 100Hz to 1MHz). The P(VDF-TrFE)/PANI and P(VDF-TrFE)/gold also exhibit similar polarization hysteresis loops. Moreover, in many cases, the all-polymer systems show a larger electric field induced strain response than that of films with gold electrodes under identical measurement conditions. The experimental results suggest that as far as the P(VDF-TrFE) film itself is concerned, the conductive PANI electrodes function in a very similar manner as that of the gold electrodes.

This work was supported by ONR through Grant N00014-98-1-0254.

REFERENCES

1. Q. M. Zhang, V. Bharti, , and X. Zhao, Science, 280, 2101(1998)
2. J .M. Herbert, Ferroelectric Transducers and Sensors; Gordon and Breach Science Publishers: New York, 1982
3. T. Ueno, H. Arntz, S. Flesch, and J. Bargon, J. Macrol. Sci., A25(12), 1557(1988)
4. Y. Tezuka, K. Aoki, and K. Shinozaki, Synth. Met., 30, 369(1989)
5 M. Angelopoulos, G. E. Asturias, S. P. Ermer, A. Ray, E. M. Sherr, A. G. MacDiarmid, M. Akhtar, Z. Kiss, J. Epstein.Mol. Cryst. Liq. Cryst., 160, 51(1988)
6 T. Kunugi, T. Kawasumi, T. Ito, J. Appl. Polym. Sci., 40, 2101(1990)
7.Y. Cao, P. Smith, A. J. Heeger, Synth. Met., 48, 91(1992)
8. J. K. Sinna, Rev. Sci. Instrum., 42, 696(1965)
9. J. Su, P. Moses, Q. M. Zhang, Rev. Sci. Instrum., 69, 2480(1998)
10. F. Zuo, M.Angelopoulos, A. G. MacDiarmid, A. Epstein, Phys. Rev., B36, 3475(1987)

COMPARATIVE STUDY OF SOLVENT EFFECT ON THE ELECTROCHEMICAL DEPOSITION OF POLYTHIOPHENE AND POLYBITHIOPHENE

T.K.S. Wong*, X. Hu**
*Photonics Laboratory, School of Electrical and Electronic Engineering,
Nanyang Technological University, Singapore 639798, ekswong@ntu.edu.sg
**Polymer Laboratory, School of Applied Science, Nanyang Technological University,
Singapore 639798, asxhu@ntu.edu.sg

ABSTRACT

Thin films of polythiophene and polybithiophene have been deposited by electrochemical polymerization onto indium tin oxide substrates using two different aprotic organic solvents: acetonitrile and benzonitrile. It was observed that when benzonitrile is used as supporting solvent, the initial deposit is always discontinuous and uniform thin layers could not be obtained. When acetonitrile is the solvent, the deposited layer is continuous. However, both the optical and mechanical properties of the polymer film are inferior. The difference is especially pronounced for thiophene monomers. Characterization results suggest a greater degree of polymerization in benzonitrile solvent.

INTRODUCTION

Electrochemical polymerization (ECP) is one of the main deposition techniques for conjugated polymers [1]. The advantages of ECP include simple apparatus, precise control of film thickness and reversible doping/de-doping of the deposited polymer. Despite these advantages, electrochemical polymerization is seldom used for fabricating the active layer of conjugated polymer electronic devices. The deposition methods usually used for conjugated polymer device fabrication are solution spin-coating, evaporation and organic vapor phase deposition (OVPE) [2]. One reason for this is that the electrochemical polymerization process involves a number of experimental parameters, such as the monomer type, electrolyte, solvent, electrode surface, temperature and applied electrical bias [3], which are mutually interrelated. As a result, the properties of electrochemically deposited polymer films are often difficult to control and optimize.

Amongst the various experimental variables, the solvent is known to have a significant effect on the ECP process. This is because the radical cation intermediates generated during ECP are sensitive to the nucleophilicities of the solvent medium and the electrolyte. Thus far, the effect of the protic solvents on ECP has been the most extensively studied. It is known that such solvents generally lead to oligomer formation and poor quality films [4]. However, for polypyrrole, the best films are deposited from acetonitrile containing a small amount (ca.1 wt%) of water. The role of water in deposition of polypyrrole remains controversial to date [5]. For the aprotic solvents, e.g. acetonitrile (AN), benzonitrile (BN) and propylene carbonate, the effect of the solvent on the polymerization is not well studied. In an earlier review on the polythiophenes (Pth) [6], it was pointed out that Pth films deposited from AN tend to be powdery whereas those deposited from BN tend to be more compact and robust. However, detailed experimental data was not provided. In this paper, we present a more systematic comparison of the role of the aprotic solvent on the polymerisation of thiophene and bithiophene. It will be demonstrated that the two selected solvents, AN and BN influence the initial stages of film formation on the substrate in significantly different ways. The

191

characterization techniques employed include optical microscopy, tapping mode atomic force microscopy (TMAFM), spectrophotometry and tensile testing. After presenting the experimental data, we compare the solvent properties of AN and BN using data taken from the available literature.

EXPERIMENT

The Pth and PBth samples used in this study were all deposited in a home made three electrode, single compartment electrochemical cell. Four monomer/solvent mixtures, namely thiophene/acetonitrile, thiophene/benzonitrile, bithiophene/acetonitrile and bithiophene/benzonitrile were prepared. Samples prepared from these were designated PTh/AN, PTh/BN, PBth/AN and PBth/BN respectively. The purities of the thiophene and bithiophene monomers were 99% and 96% respectively. The AN and BN solvents were of anhydrous or HPLC grade and were used without further purification. The concentration of the monomer was 0.1M in all solutions. The supporting electrolyte was 0.1M tetraethylammonium-tetrafluoroborate TEA-TBF$_4$ for all solutions. The TEATBF$_4$ was heated in a vacuum oven at 100°C for 6 hours prior to dissolution. Polymerisation was performed at constant potential using a EC&G 273A potentiostat. The applied potential was 1.4V (vs. Ag) for bithiophene and 2.4V (vs. Ag) for thiophene based solutions. The working electrode was indium tin oxide (ITO) glass pieces with sheet resistance of 150Ω/sq. The counter electrode was a platinum sheet. Prior to deposition, the cell was evacuated by rotary pump and purged with nitrogen gas for 30 minutes. The deposition was performed under a nitrogen blanket.

For each monomer/solvent combination, samples of different film thickness were prepared by varying the deposition time. The gross morphology of the films was observed with an optical microscope fitted with a CCD camera and image acquisition software. All optical images were acquired at the same magnification and feature dimensions were deduced using a patterned silicon calibration sample. Nanoscale morphology was studied by using a scanning probe microscope. Film thickness was measured by a Dektak3 surface profilometer. Usually five scans were made per sample and the average was taken. Optical absorption spectra were measured by a HP-8453 spectrophotometer. The mechanical properties of the films were measured by an Instron tensile tester.

RESULTS

Fig. 1 shows the optical micrograph of a PTh layer deposited using AN as the solvent. Deposition occurred as soon as voltage is applied. The Pth film is continuous and the surface is wrinkled. This morphology is very different from that observed when PTh is deposited using BN as the solvent (fig. 2). For BN, the initial deposit is not continuous and consists of distinct islands separated by either a very thin layer or bare substrate. The island density in fig. 2 is about $9 \times 10^3 cm^{-2}$ and becomes higher near the edges of the sample. The diameter of the nuclei ranged from 30μm-113μm and from profilometry, it is found that the height at the centre of each island in fig. 2 is about 280nm. As a result of these islands, the PTh film at first appeared mottled. As the deposition proceeds, the nuclei merged into stripes oriented parallel to the direction of current flow (fig. 3) and eventually formed a continuous layer.

When Bth was used as the monomer, a similar but less obvious pattern of morphology evolution occurred. With AN as solvent, the PBth layer is continuous (fig. 4). When the solvent is BN, the initial deposit was discontinuous. Since the contrast of the optical image is relatively poor, the optical image of the PBth layer is similar to fig. 4.

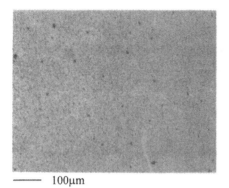

——— 100μm

Fig. 1 Optical micrograph of Pth film on ITO after 30s of growth in AN solvent.

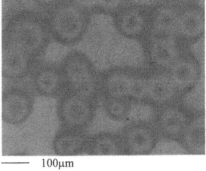

——— 100μm

Fig. 2 Optical micrograph of Pth film on ITO glass at an initial stage of growth from BN.

——— 100μm

Fig. 3 Optical micrograph of a thicker Pth/BN film showing merging of initial deposits.

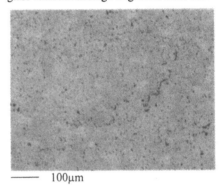

——— 100μm

Fig. 4 Optical micrograph of PBth on ITO deposited from BN solvent.

The nanoscale structure of the deposited layers was studied using TMAFM. Fig. 5 shows a 1μm×1μm TMAFM image of the Pth/AN sample shown in fig. 1. The surface consists of small nodules with a diameter of ~120nm. The morphology resembles cauliflower growth with pores being seen in between some of the nodules. Fig. 6 shows a 1.5μm×1.5μm TMAFM image of the Pth/BN sample shown in fig. 2. The scan was performed with the AFM tip manoeuvred above one of the circular nuclei. When the scan was performed in areas between the nuclei, the image consists of a smooth surface without the nodules. This is similar to the surface morphology observed when a bare ITO substrate not subjected previously to electrochemical deposition is imaged.

Fig. 7 shows a TMAFM image of a PBth film after 30s of growth in AN. The surface is similar to that in fig. 5 and consists of nodules about 50nm in diameter. These nodules are smaller than those in fig. 5 and appeared to be more ordered. As a result, the surface is less porous. Fig. 8 shows the TMAFM image of a PBth film after 10s of growth in BN. The film consists of bundles with the longitudinal axis aligned in a common direction. The bundles are not observed in all parts of the sample surface. In some sites, only the bare ITO is seen.

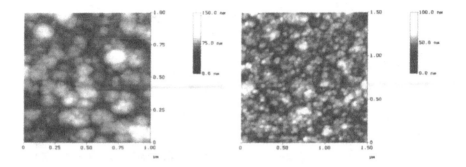

Fig. 5 TMAFM image of Pth film
after 60s of growth from AN.

Fig. 6 TMAFM image of Pth/BN film
acquired from an island deposit on ITO.

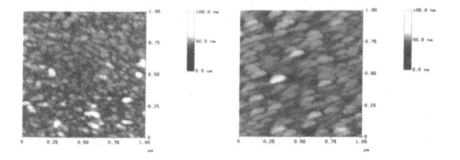

Fig. 7 TMAFM image of PBth film
on ITO after 30s of growth in AN.

Fig. 8 TMAFM image of PBth film
on ITO after 10s of growth in BN.

A total of 32 TMAFM images were acquired from the deposited samples and these were
analyzed for the average surface roughness, root mean square (RMS) surface roughness and
the average nodule diameter. The data from image analysis shows a similar pattern of nodular
growth for all the films studied.

Fig. 9 shows the optical absorption spectrum for the four types of samples deposited in this
study. The absorption coefficients are calculated by dividing the measured absorbance by the
film thickness using Beer's law. All four types of samples showed onset of optical absorption
at wavelengths below 700nm. By finding the intersection of the extrapolated rising section of
the absorption curve with the baseline, the onsets of absorption were determined to be
618nm, 621nm, 626nm and 615nm for the Pth/AN, Pth/BN. PBth/AN and PBth/BN samples
respectively. These correspond to photon energies around 2eV. The significant feature is that
when BN is used as the solvent, the peak absorption coefficient is higher for both thiophene
and bithiophene. The difference is especially pronounced for the PTh films. The peak
absorption coefficient of the Pth/BN films is about nine times that of the Pth/AN films.

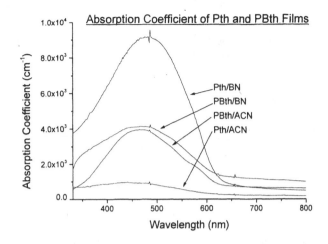

Fig. 9 Optical absorption spectra of Pth and Pbth films

For mechanical testing, thicker films were deposited as only free-standing films could be used in the tensile tester. Films deposited from BN were stretched at a rate of 0.5mm/min. The Young's modulus and tensile strength were measured to be 560MPa and 136kgcm^{-2} respectively. The corresponding values for an Al foil were 4.8GPa and 708 kgcm^{-2} respectively. Attempts to measure films polymerized in AN failed because these films were powdery and could not be detached from the substrate.

CONCLUSIONS

The TMAFM data for the Pth and PBth layers show that there is little difference in the basic film growth mechanism in the two solvents considered in this study. For both AN and BN solvents, the polymer layers are formed by initial nucleation followed by three dimensional growth. The significantly different morphologies seen in the optical images are due to a smaller nucleation density on the ITO substrate when BN is the supporting solvent. This is likely to be the result of a greater degree of polymerization in this solvent before precipitation onto the substrate occurs. This hypothesis is supported by the optical absorption data, which shows that the absorption coefficient of the PTh film is greater when BN is used as the solvent. This is because the absorption coefficient is related to the density of electron states in the $\pi-\pi*$ band and a higher value for the absorption coefficient suggests a more extensive delocalization of the π orbitals when BN is used as solvent. The higher mechanical strength of Pth/BN films may also be attributed to a higher degree of polymerization.

In the absence of in-situ data, it is instructive to compare the properties of acetonitrile and benzonitrile using tabulated values obtained from the literature [7,8]:

Table I: Comparison of Solvent Properties of Acetonitirle and Benzonitrile

Solvent Property	Parameter	Acetonitrile	Benzonitrile
Molecular Dipole	Dipole moment	11.7×10^{-30}Cm	10.7×10^{-30}Cm
Polarity	Dimroth-Reichardt, E_T^N	0.46	0.333
	Kamlet-Taft, π^*	0.66	0.88
Electron pair donicity	Gutmann's donor number, DN	14.1kcal/mol	11.9kcal/mol
	Kamlet-Taft, β	0.4	0.37
Hydrogen bonding ability	Acceptor number	18.9	15.5
	Kosower's, Z	71.3kcal/mol	65kcal/mol
	Kamlet-Taft, α	0.19	0.00

In the above table, polarity refers collectively to all those properties of the solvent that give rise to interaction forces between the solvent and solute molecules. Electron pair donicity describes the ability of the solvent to donate a pair of non-bonding electrons to another molecule to form a coordinate bond. Hydrogen bonding ability concerns the formation of bonds between a hydrogen atom on a solvent molecule and an electronegative atom on another molecule. It can be seen that for all parameters listed in Table I (except π^*), the values of corresponding solvent parameters for AN are higher than BN. This suggests AN has a stronger tendency to solvate solute molecules including possibly the radical cation intermediates generated during ECP. This may result in a lower degree of polymerization. With BN, the weaker solvation properties might have enabled more extensive polymerization and thus fewer nuclei were formed initially. However, further experimentation is required to verify this hypothesis and this should involve in-situ studies. Experimental techniques that may be useful include electrochemical scanning tunneling microscopy (ECSTM) and cyclic voltammetry (CV) of a series of acetonitrile/benzonitrile solvent mixtures.

REFERENCES

1. A. Diaz, J. Bargon, in *Handbook of Conducting Polymers*, edited by T.A. Skotheim (Marcel Dekker, New York, 1981), p.81.
2. T. Yamamoto, H. Wakayama, T. Fukuda, T. Kanbara, J. Phys. Chem. **96**, 8677 (1992); M. Baldo, M. Deutsch, P. Burrows, H. Gossenberger, M. Gerstenberg, V. Ban, S. Forrest, Adv. Mat. **10**, 1505 (1998).
3. G. Schopf, G. Koßmehl, in *Polythiophenes-Electrically Conductive Polymers*, (Springer-Verlag, Heidelberg, 1997), p.102.
4. F. Beck, Ulrich Barsch, Makromol. Chem. **194**, 2725 (1993).
5. M. Zhou, J. Heinze, J. Phys. Chem. B **103**, 8443 (1999).
6. J. Roncali, Chem. Rev. **92**, 711 (1992).
7. G.W.C. Kaye, T.H. Laby, *Table of Physical Constants*, (Longman, London, 1985), p. 191.
8. Y. Marcus, *The Properties of Solvents*, (Wiley, New York, 1999), p.131.

EQCM AND QUARTZ CRYSTAL IMPEDANCE MEASUREMENTS FOR THE CHARACTERIZATION OF THIOPHENE-BASED CONDUCTING POLYMERS

C. ARBIZZANI*, F. SOAVI*, M. MASTRAGOSTINO**
*University of Bologna, Dept. of Chemistry "G. Ciamician", Bologna, Italy
**University of Palermo, Dept. of Physical Chemistry, Palermo, Italy

ABSTRACT

EQCM was extensively used to investigate ion-transport phenomena during doping-undoping processes of electronically conducting polymers. Several early studies assumed that the polymer films were rigidly coupled to the quartz crystal so as to relate the mass change to the quartz crystal resonant frequency change *via* the Sauerbrey equation. However, the rigidity of electronically conducting polymer films is doubtful and it has to be demonstrated. Quartz crystal impedance analysis near the resonance is of paramount importance to get insight into the viscoelastic properties of the film [1] and to avoid misleading in the interpretation of EQCM data.

This contribution presents and discusses quartz crystal impedance measurements, performed with a Frequency Response Analyzer in the 9MHz region, and EQCM data collected for a dithienothiophene based polymer.

INTRODUCTION

EQCM represents a useful tool to explore the roles of anions, cations and neutral species in the electrochemical processes occurring in conducting polymer films. In particular the technique gives information about mass transport phenomena during electrode processes.

An alternating electrical potential applied across the piezoelectrical quartz crystal causes vibrational motion of the crystal with the vibrational amplitude parallel to the crystal surface. The vibrational motion of the quartz crystal results in a transverse acoustic wave that propagates perpendicularly to the crystal faces, and the standing wave condition is established when the acoustic wavelength is twice the crystal thickness. The deposition of a mass layer on one quartz crystal surface can be described as a change in thickness of the material in which the acoustic wave propagates and results in a change in the resonant frequency of the quartz crystal. If a change in thickness of the mass layer can be considered as a change in thickness of the quartz crystal, as in the case of thin films rigidly coupled to the quartz crystal, the Sauerbrey equation relates the mass layer change, Δm, to the quartz crystal resonant frequency change, Δf, as follows:

$$\Delta f = - 2f_0^2 \, \Delta m \,/[A \, (\mu_q \rho_q)^{1/2}] \tag{1}$$

where f_0 is the resonant frequency of the quartz crystal prior to the mass change and A, μ_q and ρ_q are the piezoelectrically active area, the shear modulus, and the density of the quartz, respectively.

In interpreting EQCM data for polymer layers, it becomes vital to determine whether or not the films are rigidly coupled to the quartz crystal. As a diagnostic of the rigidity of the polymer film, we performed quartz crystal impedance measurements with a Frequency Response Analyzer (FRA), and we carried out admittance analyses of the polymer coated quartz crystal referring to the models proposed for viscoelastic films. The FRA is generally used to characterize materials by electrochemical impedance studies in the region from 1mHz to

100kHz. We used the FRA equipment for a completely different purpose, and we investigated the quartz crystal impedance in the resonant frequency region, i.e. 9MHz, to verify the rigid film condition.

In this paper quartz crystal impedance measurements performed for different thickness of polydithieno[3,4-b:3',4'-d]thiophene (pDTT1) [2,3] films are presented and discussed. EQCM data collected during pDTT1 p-doping are also reported.

EXPERIMENT

The quartz crystals were 9 MHz AT-cut (SEIKO), coated with Pt electrodes of active area 0.196 cm^2. The crystal was clamped in the electrochemical cell by means of O-rings and with only one face in contact with the electrolyte.

EQCM data were collected with a SEIKO EG&G QCA917 quartz crystal analyzer. Quartz crystal impedance measurements were carried out with a PC interfaced Solartron 1255 Frequency Response Analyzer (FRA) with 10 mV AC voltage, in the 9MHz region. The polymer pDTT1 was electrosynthesized in galvanostatic conditions at 1 mAm^{-2} at room temperature in acetonitrile (ACN) – 0.1 M tetraethylammonium tetrafluoroborate (Et$_4$NBF$_4$) – 0.015 M DTT1. Polymer p-doping and undoping was performed by cyclic voltammetry (CV) with a EG&G PAR M270A potentiostat/galvanostat in propylene carbonate (PC) – 0.2 M Et$_4$NBF$_4$. An Ag electrode (-200 mV *vs.* saturated calomel electrode) was used as quasi reference electrode. All chemicals were reagent-grade products, dried and purified before use.

RESULTS AND DISCUSSION

For each frequency swept the FRA gives the real and imaginary components of the impedance of the system under study (Z_x). The impedance of the quartz crystal perturbed by a mass layer and a contacting liquid can be approximated by the modified Butterworth-Van Dyke (BVD) equivalent circuit [4] reported in Figure 1. A static capacitance C_0 arises between the electrodes located on opposite sides of the quartz crystal. Electromechanical coupling brings about a motional contribution Z_m, in parallel with the static capacitance. L_q, C_q and R_q represent respectively the inertial component related to the displaced mass during oscillation, the energy stored during oscillation, and the energy dissipation due to internal friction of the bare quartz crystal in air. The contacting liquid introduces a mechanical impedance, corresponding to Z_l, related to its density and viscosity [5]. Likewise, to take into account the mechanical contribution of a polymer film, the inertial mass component L_f and the element R_f, which represents energy losses due to the viscous nature of the film, are introduced in the motional branch of the equivalent circuit [1]. Contribution from elasticity of the film is ignored. To take into account parasitic impedances arising from the contacts and the instrumentation, an additional element Z_p is added to the equivalent circuit. The static capacitance and the parasitic impedance dominate the admittance away from resonance, while the motional contribution, Z_m, dominates near resonance. The elements in the motional arm were derived from the spectrum of the real part, G_y, of the total admittance corrected for parasitic impedances, referring to equation 2:

$$G_y = \frac{(R_q+R_l+R_f)}{(R_q+R_l+R_f)^2+[\omega\,(L_q+L_l+L_f)+1/(\omega C_q)]^2} \tag{2}$$

For a rigid film a mass change results in the shift of the admittance spectrum, i.e. G_y vs. ω, to lower frequencies without broadening: L_f changes and $R_f=0$ and the Sauerbrey equation holds. If the mass change is accompanied by a change of film viscosity and density, the admittance peak shifts and broadens, i.e. R_f and L_f both change, the film is not rigid and the Sauerbrey equation can not be used to relate frequency shift to mass change during deposition.

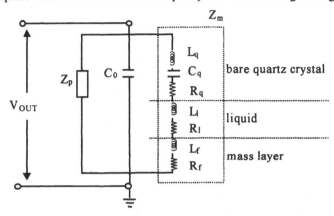

Figure 1. Modified Butterworth Van Dyke equivalent circuit.

Figure 2 reports the admittance spectra corrected for parasitic impedances and collected for p-doped pDTT1 films of different thickness. The evolution of the admittance plots demonstrates the deposition of non rigid films: upon deposition, the admittance plots shift to lower frequency and show diminution and broadening of the peak. These spectra were fitted to the modified BVD equivalent circuit and the results are shown in figure 3. Figure 3 displays the inductive and resistive components of the impedance representing the inertial mass and energy loss associated with the depositing films, respectively. The plots in figure 3 (a) show that the sensitivity of QCM significantly lowers beyond 30 mCcm^{-2} of electrosynthesis charge, i.e. for polymer film thickness beyond about 0.3 µm. This is not unexpected because it was already estimated for polybithiophene that the decay length for the acoustic wave in the polymer film is of this order of magnitude [6]. Hence beyond 30 mCcm^{-2} the measurement becomes less sensitive to the polymer. In conclusion, from the data in figure 3, upon electrosynthesis the Sauerbrey equation can not be used to relate the observed quartz crystal resonant frequency change to the mass change.

To investigate ion transport phenomena upon pDTT1 p-doping with EQCM, we first evaluated if the p-doping process determines viscosity changes in the nature of the films and we carried out impedance measurements near the resonance frequency of the quartz crystal coated by the p-doped and undoped polymer in PC- 0.2 M Et$_4$NBF$_4$. Even if pDTT1 was electrosynthesized with an electrosynthesis charge of 15 mCcm^{-2} so as to keep the system in the region of good sensitivity of the EQCM, PC led to a dumping of the admittance peak that lowered the sensitivity of our technique for the determination of R_f and L_f via the BVD model. In table I the fitted values of the total resistance for the doped and undoped polymer in PC – 0.2 M Et$_4$NBF$_4$ differ by an amount within the fitting error, while the frequency shift is about 200Hz, i.e. of the same order of that measured with EQCM (see Figure 4 and Table II). We thus considered the viscous nature of the polymer film constant upon p-doping-undoping and we applied the Sauerbrey equation for a study of ion transport phenomena in pDTT1.

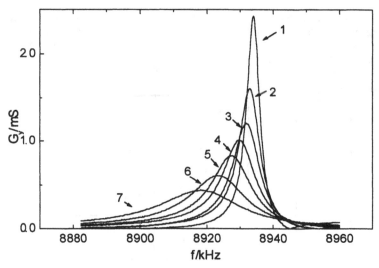

Figure 2. Admittance spectra corrected for parasitic impedances for the bare quartz crystal in air (1) and in the electrosynthesis solution coated with the p-doped pDTT1 electrosynthesized with 0 mC cm^{-2} (2), 5 mC cm^{-2} (3), 15 mC cm^{-2} (4), 30 mC cm^{-2} (5), 60 mC cm^{-2} (6), 120 mC cm^{-2} (7).

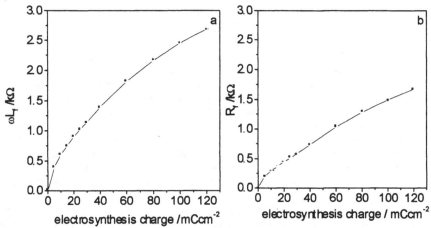

Figure 3. Inductive (a) and resistive (b) components of the impedance representing the inertial mass and energy loss associated with the depositing pDTT1 films vs. electrosynthesis charge.

Table I. Fitted values of the total resistance ($R_q+R_l+R_f$) and the resonant frequency (f_0) for the quartz crystal coated by the doped and undoped pDTT1 in PC – 0.2 M Et$_4$NBF$_4$.

pDTT1	($R_q+R_l+R_f$)/Ω	f_0 /Hz
p-doped	1664±20	8909450±18
undoped	1650±20	8909635±17

Figure 4 displays EQCM measurements during CVs in the p-doping domain at 50 mV/s in PC - 0.2M Et₄NBF₄ for pDTT1. Table II summarizes the results from CVs and reports the mass per electron (m.p.e.) value from the EQCM measurements, calculated from the Sauerbrey equation. The m.p.e. value is consistent with the entrance of the bare tetrafluoborate anion (BF₄⁻) in the polymer matrix upon p-doping.

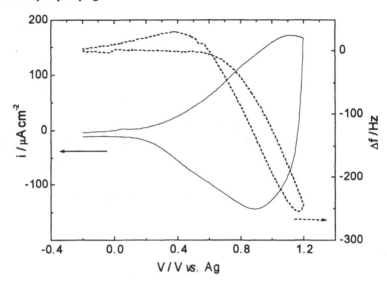

Figure 4. EQCM measurements during the CVs at 50 mV/s of pDTT1 (electrosynthesis charge: 15mC cm²) in PC – 0.2 M Et₄NBF₄.

Table II. Undoping charge (Q_u), coulombic efficiency (η), frequency shift (Δf) upon p-doping and mass per electron (m.p.e.) value for pDTT1 from the EQCM measurements during the CVs at 50 mV/s in PC – 0.2 M Et₄NBF₄.

Polymer	Q_u /mCcm⁻²	η (%)	Δf /Hz	m.p.e./gmol⁻¹
pDTT1	-1.76	95	-254	97

CONCLUSIONS

The FRA is a useful tool to investigate the viscoleastic properties of polymer films and to verify the reliability of EQCM data from the Sauerebrey equation. Quartz crystal impedance measurements near resonance demonstrated that pDTT1 is not a rigid film, and EQCM can not be used to evaluate the mass changes upon its electrosynthesis. Despite this, the admittance spectra of the p-doped and undoped polymer electrosynthesized with 15 mC cm⁻² displayed the same values of the real part of the admittance (G_y) at the resonance frequency. Thus we considered constant the viscous nature of pDTT1 upon p-doping-undoping and we carried out EQCM measurements for a study of ion transport phenomena. The m.p.e. value during p-doping-undoping indicated the entrance of the bare BF₄⁻ anion in the polymer matrix.

ACKNOWLEDGMENTS

We would like to acknowledge Progetto Finalizzato Materiali Speciali per Tecnologie Avanzate II - "Materiali polimerici e carboniosi per accumulatori e supercapacitori" - Contract n° 97.00912.PF34, for financial support, Dr. Marinella Catellani (Istituto di Chimica delle Macromolecole, CNR, Milano) for providing the DTT1 monomer and Professor Gabriele Cazzoli (Dept. of Chemistry "G. Ciamician", Bologna) for the useful discussion about the quartz crystal impedance measurements.

REFERENCES

1. D.A. Buttry and M.D. Ward, *Chem. Rev.* **92**, 1355 (1992).
2 C. Arbizzani, M. Catellani, M. Mastragostino and M.G. Cerroni, *J. Electroanal. Chem.* **423**, 23 (1997).
3 M. Mastragostino, C. Arbizzani, M.G. Cerroni and R. Paraventi, *Electrochemical Capacitors II*, Ed. F.M. Delnick, D. Ingersoll, X. Andrieu and K. Naoi, The Electrochemical Society Proceeding Series, Pennington, **PV 96-25**, 109 (1997).
4. H.L. Bandey, M. Gonsalves, A.R. Hillman, A. Glidle and S. Bruckenstein, *J. Electroanal. Chem.* **410**, 219 (1996).
5. S.J. Martin, V.E. Granstaff and G.C. Frye, *Anal. Chem.* **63**, 2272 (1991)
6. H.L. Bandey, A.R. Hillman, M.J. Brown and S.J. Martin, *Faraday Discuss.* **107**, 105 (1997).

Polythiophene Grafted on Polyethylene Film

N. Chanunpanich,[1,4] A. Ulman,[1,4]* Y. M. Strzhemechny,[2,4] S. A. Schwarz,[2,4] J. Dornicik,[5] M. Rafailovich,[5] J. Sokolov,[5] A. Janke,[3] H. G. Braun,[3] and T. Kratzmüller[3]

[1] Department of Chemical Engineering, Chemistry, and Materials Science, Polytechnic University, Brooklyn, New York 11201
[2] Department of Physics, Queens College of CUNY, Flushing New York 11367
[3] Institute for Polymer Research, Hohe Str. 6, D-01069 Dresden, Germany
[4] The NSF MRSEC for Polymers at Engineered Interfaces.
[5] Department of Materials Sciences and Engineering, State University of New York at Stony Brook, NY 11794-2275.

ABSTRACT

We have successfully grafted polythiophene on polyethylene (PE) film with a three reactions step: gas phase bromination on PE, yielding PE-Br; substitution reaction of PE-Br with 2-thiophene thiolate anion, following by chemical oxidative polymerization. The polymerization was carried out in a suspension solution of anhydrous $FeCl_3$ in $CHCl_3$, yielding a reddish PE-PT film after dedoping with ethanol. ATR-FTIR shows that the polythiophene (PT) was grafted on PE in the 2,5-position; on the other hand, PT homopolymer shows a small amount of 2,4 coupling. XPS reveals higher intensity of the S2p, including neutral and positive sulfur. SEM image reveals the island of PT on the PE film. AFM analysis found the thickness of the island is in the range of 120-145 nm. The conductivity of these thin films is in the range of 10^{-6} S/cm.

INTRODUCTION

During the past 10 years, there has been growing interest in electrically conducting polymers due to their potential applications. Among these polymers, polythiophene (PT) has attracted interest because of its high magnetic and optical properties, electrical conductivity, and environmental stability (both to oxygen and to moisture). PT can be prepared both by electrochemical and by oxidative polymerization of thiophene. The electrochemical polymerization of PT is carried out in an organic solvent such as acetonitrile, nitromethane, nitrobenzene, or propylene carbonate [1].

Polyethylene (PE) is one of the most interesting commodity polymers. However, it exhibits low surface energy, and different methods were developed to increase surface energy. Among those, hydrophilic polymers were successfully grafted on PE surface by direct plasma polymerization [2], pretreatment of PE surface with corona or Ar plasma prior to polymerization in the monomer solution [3], pre-swell organic initiator on the PE surface prior to polymerization in the hydrophilic monomer solution [4]. In addition, other monomers such as styrene, glycidyl methacrylate (GMA), 2-hydroxyethylmethacrylate (HEMA), can be grafted on PE surface by photo-irradiation [5], which the later products can be used in medical application.

Recently we reported that bromination of PE films improves their wettability [6]. Furthermore, the bromopolyethylene films (PE-Br) could be altered to amino terminated PE film (PE-S-Ph-NH$_2$) using nucleophilic substitution reaction. In addition, we also reported substitution reaction with other aromatic thiols [7]. Among others, we have successfully grafted thiophene thiol on PE films (PE-S-T). In principle, these films can be used as substrates in oxidative polymerization with thiophene, yielding polythiophene grafted on PE film. Here we report on the oxidative polymerization at the surface of PE-S-T films in a suspension of anhydrous ferric chloride ($FeCl_3$) in dry chloroform ($CHCl_3$) and thiophene.

EXPERIMENT

Chemicals. Low-density polyethylene (LDPE) was purchased from Acros Organic. 2-mercaptothiophene was purchased from Maybridge Chemical. Bromine, thiophene, anhydrous ferric chloride, and sodium metal were obtained from Fluka. All solvents were analytical grade and purchased from Fisher Scientific.

PE film preparation and synthesis. The procedures of film preparation and bromination have been reported elsewhere [6]. PE-S-T was carried out in 0.025M sodium 2-mecaptothiophene anion. The oxidative polymerization of PE-S-T was accomplished in a suspension of anhydrous $FeCl_3$ in dry $CHCl_3$.

Characterization. The PE-S-T and PE-PT and PT powder were investigated by the attenuated total reflection (ATR) FTIR (Nicolet 760 IR spectrophotometer, with a ZnSe crystal setup); UV-vis spectroscopy (DMS100 spectrometer); X-ray photoelectron spectroscopy, XPS (Kratos ES300 X-ray photoelectron spectrometer with a non-monochromatic Mg Kα x-ray source. The core level S2p, O1s, and C1s spectra were monitored at an electron takeoff angle of 15°. Deconvolution of complex peaks was performed using Origin nonlinear curve fit with the full width at half-maximum (fwhm) of Gaussian line shape of 1.3-1.4 eV. Quantitative measurements were made by correction of integrated peak intensities with the Scofield correction factors [8]; Atomic force microscopy, AFM (Digital Instruments); and Scanning electron microscope, SEM (Zeiss Gemini DSM 982).

RESULTS AND DISCUSSION

Grafting of thiophene onto PE film is easily done by a two step reaction, bromination PE film, following by substitution reaction of PE-Br with 2-mercapto thiophene in presence of 50% of 0.04M sodium ethoxide and DMF. Bromination of PE film in gas phase has been report recently [6]. Since 2-mercaptothiophene and its anion do not dissolve well in ethanol, DMF was added to increase solubility. Sodium ethanolate was added a little less in stoichiometric concentration to produce the nucleophilic thiolate, and to avoid the competition of dehydrobromination of PE-Br films.

The nucleophilic substitution reaction of PE-Br with 2-mercaptothiophene anion was carried out at 33°C for 20 h., yielding a light reddish PE-S-T film. The ATR-FTIR (Figure 1) spectrum of PE-S-T shows a characteristic of thiophene moieties at 1434 cm⁻¹ (νC=C); 1134 cm⁻¹ (vibration of the thiophene ring); 1018 and 806 cm⁻¹ (δ aro.C-H in plane and out of plane, respectively). S2p high resolution XPS of PE-S-T (Figure 2a) reveals two peak between 164.0±0.1 and165.3±0.2 eV, assigned to neutral sulfur and between 169.0±0.2, and

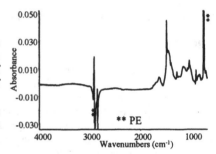

Figure 1 ATR-FTIR spectrum of PE-S-T after subtraction with PE film. The spectrum shows the appearance peak at 1434 cm⁻¹ (ν C=C); 1134 cm⁻¹ (stretching vibration of thiophene ring) 1018 cm⁻¹ (δ C-H in-plane); 806 cm⁻¹(δ C-Hb out-of-plane vibration).

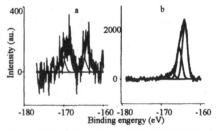

Figure 2. S2p high resolution XPS of (a) PE-S-T: and (b) PE-PT.

170.1±0.2 eV, assigned to sulfoxide compound, which was oxidized by air and/or during XPS running [9,10]. That oxidation occurs only when the film is exposure to the air is supported by the fact that if immediately following PE-S-T film formation oxidative polymerization is carried out to yield PE-PT, only the 164-168 eV appears in XPS (Figure 2 b).

Polythiophene can be synthesized by electrochemical, [11,12] and chemical polymerization. However, an inexpensive method of producing polythiophene is the oxidative polymerization of thiophene in a suspension solution of anhydrous $FeCl_3$ in dry chloroform ($CHCl_3$) [13] Niemi et al. [14] concluded that the polymerization started from solid $FeCl_3$. Hence, the active sites in the polymerization are the Fe^{3+} ions at the surface of the crystal. These surface Fe^{3+} ions have one unshared chloride and one free orbital; which is the source of their Lewis acidity. The soluble part of $FeCl_3$ is inert because it exists in a dimeric form without free orbitals.

The oxidative polymerization of PE-S-T with $FeCl_3$ and thiophene in $CHCl_3$ was carried out at 20°C for 5 min, yielding a reddish film (PE-PT) and a red powder of PT homopolymer after washing with ethanol. It has been established that chemical polymerization of thiophene can occurred via the 2,5, and 2,4 positions. However, the FTIR spectrum of PT powder (Figure 3) clearly reveals that the reaction preferentially occurs at the 2-5-positions. The spectrum (transmission, KBr pellet) shows the appearance of 3064 cm^{-1} (v C-H$_\beta$); the characteristic peaks of 2,5-disubstituted PT appear at 1489 cm^{-1}, 1451 cm^{-1}, and 1432 cm^{-1} (v C=C); 1066 cm^{-1} and 1037 cm^{-1} (δ aro. C-H in-plane); and 788 cm^{-1} (δ aro. C-H out-of-plane), which are agree to the literature [15,16]. The ATR-FTIR spectrum (Figure 4) of PE-PT shows the appearance of 3064 (v C-H$_\beta$ stretching); 1490 (v C=C); 1340 and 1200 cm^{-1} (stretching vibration of the thiophene ring); 1106-1020 cm^{-1} (δ aro. C-H in plane); 789 cm^{-1} (δ C-H out-of-plane of 2,5-disubstituted PT). The S2p band in the high resolution XPS of PE-PT is resolved into four peaks (Figure 2b) centered at 164.1, 165.4, 166.5 and 167.5 eV. The first two peaks are assigned to neutral S2p$_{3/2}$, and S2p$_{1/2}$, while the other two are associated with the formation of oxidized, positively charge sulfur $S^{\delta+}$ [17,18]

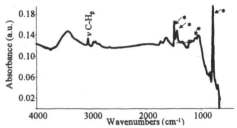

Figure 3. IR spectrum of PT. The spectrum shows the characteristic peak of thiophene at 3064 cm^{-1} (v C-H$_\beta$); 1489 cm^{-1}, 1451 cm^{-1}, and 1432 cm^{-1} (v C=C); 1066 cm^{-1} and 1037 cm^{-1} (δ aro. C-H in-plane); and 788 cm^{-1} (δ aro. C-H out-of-plane)

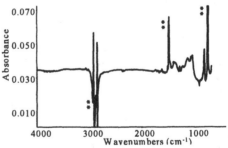

Figure 4. ATR-FTIR spectrum of PE-PT. The spectrum shows the appearance of 3064 cm^{-1} (v C-H$_\beta$); 1490 cm^{-1} (v C=C); 1340 and 1200 cm^{-1} (vibration of the thiophene ring); 1106-1020 cm^{-1} (δ aro. C-H in plane); 789 cm^{-1} (δ C-H out-of-plane of 2,5-disubstituted PT).

We turn now to the question "is the PT sorbed at or grafted on the PE films?" To evident the grafting polythiophene, a control experiment was carried out where an untreated PE film was added to the suspension of $FeCl_3$ in $CHCl_3$ and thiophene for 5 min at 20°C, yielding PE-PT (blank). Table I shows surface composition of PE-S-T, PE-PT, PE-PT (blank), and PT powder, calculated from XPS data. As expect, the composition of C and S in PT powder is close to theory $(C_{100}S_{20})$. The oxygen moiety confirmed that carbonyl moiety was form in the polythiophene chain. The sulfur intensity on PE-PT (blank) is a very small, with the mole ratio of $C_{100}S_{1.2}$. This is the evident that the large intensity of sulfur in PE-PT comes from grafted polythiophene that is formed by oxidative polymerization. Therefore, PT sorption on PE films can be ignored.

PE-PT film was further investigated by SEM, recorded at 20 000 magnification. SEM image of untreated PE film (Figure 5a) shows contrast between the soft part (amorphous) and hard part (crystalline) in the PE film. Figure 5b reveals islands of PT on PE film, showing that some parts of the film were not modified. The AFM analysis results are shown in Figure 6. That the curve line go down to zero is in agreement with the both SEM and AFM observations, i.e. that some part of the PE film was not modified. The thickness of PT grafted on PE film is in the range of 120-145 nm, suggesting that the surface has been grafted with polymers and not with oligomers. For 1200 - 1500 Å polythiophene chains the degree of polymerization should be ≥300. This conjugation length should be evident in the absorption spectrum.

Although UV-vis spectroscopy is not very sensitive to surface modifications, a shoulder at 410 nm was observed on PE-S-T (Figure 7). This shoulder is a characteristic of π-π^* transition of thiophene. The red shift to 516.9 nm for PE-PT indicates extensive π-electron delocalization as is expected from the polymer chain length. After doping with FeCl$_3$ in acetonitrile, two shoulders appear, one at 486.7 and the other at 360 nm. The π-π^* band shift from 516.9 nm to 486 nm suggest a shorter conjugation length, as a result of changing from monomeric cation radicals to dimeric cation radicals [19]. The shoulder at 360 nm can be assigned to uncharged thiophene units.

Having characterized the PE-PT films, it appears that there may not be connectivity of polythiophene chains at the PE surface to allow electrical conductivity. Therefore, we expected that these films will not exhibit high conductivity values, and only if improvement in grafting protocols is made, it will be possible to use them for conducting applications.

PE-PT is doped with 0.1% FeCl$_3$ in acetonitrile. As a result, the color turned from red to dark green. On the other hand, the film is easily dedoped by exposure

Table I Surface composition of PE-S-T, PE-PT, PE-PT(blank), and PT, calculated from XPS.

Samples	Surface stoichiometric
PE-S-T	$C_{100}S_{1.0}O_{4.8}$
PE-PT	$C_{100}S_{8.0}O_{5.9}$
PE-PT(Blank)	$C_{100}S_{1.2}O_{4.8}$
PT powder	$C_{100}S_{20.8}O_{16.9}$

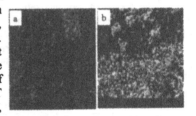

Figure 5. SEM of (a) PE-S-T and (b) PE-PT

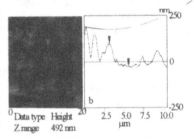

Figure 6. AFM analysis of PE-PT

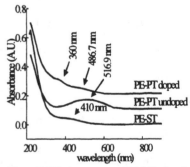

Figure 7. UV-vis spectrum of PE-S-T, doped and undoped PE-PT

to air or to ethanol. The color of the doped PE-PT film turn to purple after exposure to air. The conductivity of PE-PT film was calculated from equation (1)

$$\sigma = \frac{1}{\rho} = \frac{\ln 2}{\pi t}\left(\frac{I}{V}\right),$$ (1)

where σ is the conductivity (S/cm), ρ is resistivity (ohm-cm), t is thickness of PT film, I is current (1×10^{-9} A), and V is voltage drop (volt). The conductivity of doped PE-PT film was found to be very low ($\sigma = 1.4 \times 10^{-6}$ S/cm), confirming the expectation based on SEM and AFM images. However, other factors that should be considered are first the small thickness of the film that may result in mechanical breakdown upon connection to electrodes. Second, the dedoped during measurement by air. Finally, the full oxidation of the film to the bipolaron level, which does not conduct.

CONCLUSIONS

We have successfully grafted polythiophene on PE film with a three reactions step: gas phase bromination on PE, yielding PE-Br; substitution reaction of PE-Br with 2-thiophene thiolate anion, yielding PE-S-T, following by chemical oxidative polymerization. The polymerization was carried out in a suspension solution of anh. $FeCl_3$ in $CHCl_3$, yielding a reddish PE-PT film after dedoping with ethanol. ATR-FTIR shows that the PT was grafted on PE in the 2,5-position, on the other hand, PT homopolymer shows a small amount of 2,4 coupling. XPS reveals higher intensity of the S2p, including neutral and a small amount of positive sulfur. SEM image reveals the island of PT on the PE film. AFM analysis found the thickness of the island is in the range of 120-145 nm. The conductivity of this thin film is in the range of 10^{-6} S/cm.

Acknowledgements
Support for this project by the NSF through the MRSEC for Polymers at Engineered Interfaces is appreciated. We thank Prof. Kalle Levon and Dr. Lyubov Chigirinskaya for assisting in the conductivity measurements.

REFERENCES AND FOOTNOTES

1. (a) A.F.Diaz, J.A.Logan, J. Electroanal. Chem., **111**, 111 (1980); K. Imanishi, M. Sotoh, Y. Yasuda, R. Tsushima, and S. Aoki, ibid., **242**, 203 (1988); M. Sata, S. Tanaka, and K. Kaerigama, J. Chem. Soc., Chem. Commun., (1985), 713.

2. T-M. Ko. S.L. Cooper, J. Appl. Polym. Sci. **47**, 1601 (1993).

3. (a) J. Zhang, K. Kato, Y. Uyama, and Y. Ikada, J. polym. Sci.: part A. **33**, 2629 (1995); H. Iwata, A. Kishida, M. Suzuki, Y. Hata, and Y. Ikada, ibid., **26**, 3309 (1988); N. Inagaki, S. Tasaka, and Y. Goto, J. Appl. Polym. Sci. **66**, 77 (1997).

4. K. Kildal, K. Olafsen, and A. Stori, J. Appl. Polym. Sci. **44**, *1893* (1992).

5. Y. Li, J.M. Desimone, C-D. Poon, and E.T. Samulski, J. Appl. Polym. Sci. **66**, 883 (1997); G. Bai, X. Hu, and Q. Yan, Polym. Bull. **36**, 503 (1996).

6. N. Chanunpnainch, A. Ulman, Y.M. Strzhemechny, S.A. Schwarz, A. Janke, H.G. Braun, and T. Kratzmüller, Langmuir **15**, 2089 (1999).

7. A coming up report "Surface Modification of Polyethylene Films via Bromination: Reactions of Brominated Polyethylene with Aromatic Thiolate Compounds.

8. Sensitivity factor of C, S, O are 1.0, 2.63, and 1.18 respectively.

9. J. Lukkari, K. Kleemola, M. Meretoja, T. Ollonqvist, and J. Kankare, Langmuir **14**, 1705 (1998).

10. K. Bandyopadhyay, M. Sastry, V. Paul, and K. Vijayamohanan, Langmuir **13**, 866 (1997).

11. S. Hotta, S.D.D.V. Rughooputh, A.J. Heeger, and F. Wudl, Macromolecules **20**, 212 (1987).

12. B.L. Funt, S.V. Lowen, Synth. Met. **11**, 129 (1985).

13. T.J.J.M. Kock, B. de Ruiter, Synth. Met. *79*, 215 (1997); M. Lanzi, P.C. Bizzarri, and C.D. Casa, ibid., **89**, 181 (1997); J. Lowe, S. Holdcroft, Macromolecules **28**, 4608 (1995); P. Buvat, P. Hourquebie, ibid., **30**, 2685 (1997).

14. V.M. Niemi, P. Knuuttila, J.-E. Österholm, and J. Korvola, Polymer **33**, 1559 (1992).

15. O. Inganäs, B. Liedberg, W. Chang-ru, and H. Wynberg, Synth Met. **11**, 239 (1985).

16. S. Otta, W. Shimotsuma, and M. Taketani, Synth Met. **10**, 85 (1984/85).

17 S.C. Ng, P. Fu, W-L. Yu, H.S.O. Chan, and D.L. Tan, *Synth. Met.*, **87**, 119 (1997).

18 S.C. Ng, S.S.O Chan, P. Miao, and K.L. Tan, *Synth. Met.*, **90**, 25 (1997).

19. M.G. Hill, K.R. Mann, L.L. Miller, and J.–F. Penneau, J. Am. Chem. Soc. **114**, 2728 (1992).

A conductive composite film by permeation method

JinWei Wang, M. P. Srinivasan*

Department of Chemical & Environmental Engineering, National University of Singapore, 10 Kent Ridge Crescent, Singapore 119260, **Singapore**

ABSTRACT

A new method of making conductive composite films by permeation of the conducting guest species into the host is reported. A layer of poly(3-n-dodecyl thiophene) (P3ddt) is embedded at the surface of polyimide by permeation of the monomer or polymer (in solution in tetrahydrofuran or chloroform) into a solution of polyamic acid in n-methyl pyrrolidinone or dimethyl acetamide. The resulting composites were imidised and polymerized (if necessary). Chemical imidisation yielded composite films that retained the conducting polymer even when the composite was subjected to solvent extraction. The films were conductive upon doping with iodine and recovered conductivity when they were exposed to iodine vapor subsequent to thermal de-doping. Thermogravimetry showed that the amount of thiophene incorporated into the polyimide was higher for permeation of the polymer than that of the monomer; however, the amount of p3ddt incorporated by the latter method was still higher than the amount that could be incorporated by blending polyamic acid with p3ddt. The levels of conductivity and speed of recovery for doped films were also higher for the permeated films. Results of scanning electron microscopy suggested that the higher mobility afforded by contact in the liquid state have contributed greater entanglement between the constituents leading to higher thermal and solvent resistance of the conducting constituent. The permeation method could be adopted to form composite films in solvent systems that are not completely miscible.

INTRODUCTION

Conducting polymers have been intensively studied in the past two decades. New conductive polymers such as polythiophene, polypyrrole, poly(p-phenylene), poly(p-phenylene sulphide), polyaniline and their derivatives have been developed . Among these polymers, the pure, unsubstituted polythiophene is an intractable material. By introducing a flexible side chain at the 3-position of the thiophene ring, the polymer is made fusible, melt processable, and soluble in common organic solvents [1]. The processability of the PATs implies that it is possible to obtain a wide variety of polymer blends or composites with thermosetting or thermoplastic polymers. In addition to providing mechanical and thermal reinforcement, these new polymeric materials may yield further interesting properties. Conductive polymer composites which combine the electrical conductivity of polythiophene, polypyrrole or its derivatives with good mechanical properties of insulating polymers such as polystyrene [2,3], ethylene vinyl acetate copolymer [4,5], polyimide [6,7,8], and rubber [9] have been reported.

Among insulating matrix polymers, polyimide has been popular because of its well-known thermal and mechanical stability. Polypyrrole-polyimide composite films were prepared by electrochemical and chemical oxidation of pyrrole on polyimide substrates [7]. A surface absorption method was also reported [8]. These methods use the imidised polymer as the substrate, and therefore will have to contend with limited diffusion of the conducting species or its precursor into the rigid host matrix. Furthermore, these techniques specifically deal with absorbing the monomer and subsequently polymerising it on the polyimide surface. The nature of the interaction between the guest and the host, the extent of polymerisation, and the stability of the composites to thermal and solvent environments are not apparent from the above methods.

We report a method for incorporating a conducting constituent at the surface of an insulating matrix by forming a composite of poly (3-dodecylthiophene) and polyimide. The significant difference in this method relative to earlier methods is that the constituents of the composite are contacted with each other in the solution states. The resulting films show more uptake of p3ddt and retain the conducting component even after extraction with solvent. The effect of the processing conditions on the conductivity, the thermal properties and the morphology of the composite were investigated.

EXPERIMENTAL

PAA : Polyamic acid (PAA) precursor was prepared by dissolving diaminodiphenyl ether (DDE) in N,N'-dimethylacetamide (DMAc) or N-methyl-2-pyrrolidinone (NMP) (all from TCI, Japan). Subsequently solid pyromellitic dianhydride (PMDA) was added in stoichiometric amounts to this solution and the solution was stirred for two hours at 25°C under a stream of argon or nitrogen to obtain a lemon-yellow, viscous solution [10].

P3ddt : Neutral poly (3-n-dodecylthiophene) was prepared by the method of Sugimoto et al [11]. 0.1 N 3-dodecylthiophene in chloroform was polymerised by oxidation in 0.4 N FeCl$_3$ solution in chloroform at 0°C, room temperature or higher under nitrogen atmosphere. The resulting mixture was then poured into methanol to precipitate the polymer. This polymer was washed several times and then refluxed in methanol in order to remove the oxidant residue and oligomers. The purified polymer was ready for use after vacuum drying at 30°C for 1 hr.

Films from polymer : 10% PAA in NMP was taken in a beaker to form a thin layer of solution. A 1%-5% p3ddt solution in THF was poured slowly along the side of the beaker so that the latter formed a separate layer over the former. This 2-layer solution was stored in a dessiccator or nitrogen-purged glove box for 4 to 8 hr. Subsequently, the solution was dried in a vacuum oven at 30°C for 1 hr and heated at 80°C for 1 hr. The obtained films were smooth but were coloured differently on either side and for different compositions.

Films from monomer: Neat 3-dodecylthiophene was spread on the surface of the PAA solution and stored for 2 to 4 hr. The solution was then dried in a vacuum oven at 80 °C for 1 hr. The film was subsequently immersed in 0.4 N FeCl$_3$ in chloroform for 2 hrs at 0 °C to polymerize the monomer and also rendered it conductive. The film was then washed with chloroform and methanol and dried at 80 °C for 1 hr.

The dried films were then subjected to chemical imidisation, solvent extraction and characterised in terms of surface resistivity, spectroscopy and thermogravimetry.

Chemical Imidisation

Chemical imidisation was performed by immersing the films in a solution of acetic anhydride, pyridine and benzene (in the ratio 1:1:10 by volume) for 24 hr followed by drying at 80°C for 2 hr [12]. Subsequently, the film was heated to 200°C for 2 hr to ensure removal of residual solvent.

Film Treatment

Doping: The films were placed in a vial filled with iodine vapour at room temperature for periods ranging from 2 hr to 72 hr.

Solvent extraction: The permeated films were refluxed in chloroform for 4 hr to check the stability of the film with solvent.

Characterization

Resistivity: Surfave Resistivity was measured using an RS 392-185 Surface Resistivity Probe.

Thermogravimetry: Thermogravimetric analyses were performed on a Setaram Labsys [TM] instrument using an open aluminum crucible and a flow of nitrogen. The temperature was ramped from 50°C to 1000°C at the rate of 10°C per minute.

Scanning Electron Microscopy (SEM): The cross-sectional morphology of the films was examined by means of a JSM-T220A scanning electron microscope. The specimens were coated with gold with a JFC-1100 ion sputterer at 10kV and 10mA for 5 minutes before examination.

RESULTS and DISCUSSION

Surface Resistivity

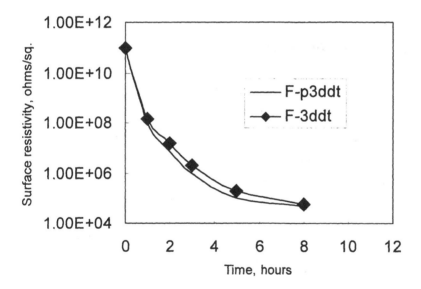

Figure 1. Surface resistivity versus doping time
F-3ddt: film from from 3ddt after polymerisation and chemical imidisation;
F-p3ddt: film from pre-formed P3ddt as the conducting constituent.

Figure 1 shows the resistivity of the films with doping time from the above two processes. The surface of the unimidised composite films is conducting after doping with iodine. However, since chemical imidisation is followed by heating at 200°C to ensure completion of imidisation and solvent removal, the surface loses conductivity during the heating step. Therefore, Figure 1 represents the recovery of conductivity of p3ddt subsequent to thermal dedoping. As seen from Figure 1, the conductive properties of the composites are similar for films from both processes. The samples showed decrease of resistivity to less than 10^5 Ω / sq within 8 hours of exposure to iodine. Previous studies [13,14] and ours (to be published) have shown that p3ddt is dedoped when heated to temperature beyond 150°C, and can be redoped by exposure to iodine. The rate of recovery of conductivity decreases with increasing temperature. Therefore, it is important to investigate the effect of processing conditions, especially thermal treatment, on the conductive properties. As a result, the process conditions to make the films and the effect of these process conditions on the conductivity of the final films have to be seriously considered for making conductive composite films.

Thermogravimetry

Thermogravimetric analyses were performed in order to investigate the thermal properties of the materials and to study the effect of the processing conditions on performance of the films obtained by the permeation method. Figure 2 shows the TG results of films from monomer

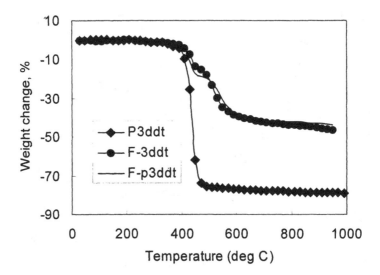

Figure 2. Thermaogravimetry of pure P3ddt and composites.
P3ddt: pure poly 3-dodecyl thiophene;
F-3ddt: film from from 3ddt after polymerisation and chemical imidisation;
F-p3ddt: film from pre-formed P3ddt as the conducting constituent.

(F-3ddt) and polymer (F-p3ddt) as spreading layers after chemical imidisation and solvent extraction. Both the curves show two stages of weight loss; the first stage between approximately 400°C and 470°C is due to loss of the P3ddt and the second stage is the decomposition of polyimide. However, the film made from the pre-formed polymer shows greater uptake of the conductive constituent. Compared with the pristine P3ddt with one single stage decomposition from around 350°C to 470°C, the P3ddt in the composite films show increase of initial decomposition temperature to around 400°C. This difference suggests that the presence of p3ddt in a thermally insulting matrix such as polyimide may increase thermal stability of the conducting layer [7,13].

Scanning electron microscopy

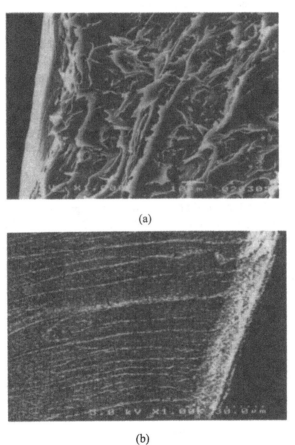

(a)

(b)

Figure 3. SEM cross sectional micrographs
(a) film from pre-formed P3ddt as the conducting constituent;
(b) film from 3ddt after polymerisation and chemical imidisation.

Figure 3 a and b show the cross-sectional morphologies of the two films after extraction with chloroform. The persistence of the p3ddt layer near the surface indicates the robust structure of the conducting surface. The entanglement of the conducting polymer with the insulating and rigid polyimide chain may account for the stability of the conductive layer. However, the polyimide matrix in these two films shows different morphologies; Figure 3b shows line normal to the surface, representing penetration of p3ddt into the polyimide substrate. This may be the consequence of greater penetration of the 3ddt monomers(before polymerization) compared to the diffusion of the pre-polymer(Figure 3a). Further, this would explain the lower mechanical strength of the composite in Figure 3b, since the intrusion of the conductive layer into the polyimide matrix may have compromised the tenacity of the latter. On the other hand, since imidisation was preceded by polymerization of 3ddt, the catalyst (FeCl$_3$), solvent (CHCl$_3$) or the reaction may have affected the mechanical properties of polyimide.

CONCLUSIONS

Conductive composite films were successfully prepared from p3ddt with polyimide by permeation mothod. This method is conducive to chemical imidisation method for the polyimide. It is also suitable for constituents that are soluble in solvent that are partially miscible. P3ddt in the films from these methods showed higher decomposition temperatures compared the pure polymer, indicating that the polyimide matrix provided some extent of protection. The thermal and solvent stability suggested that p3ddt was entangled in the rigid polyimide chains.

REFERENCES:

1. R.Sugimoto, S Takeda., H. B. Gu and K.Yoshino, Chem. Express 11, 635-638 (1986).
2. S. Hotta, S. D. D. V. Rughooputh and A. J. Heeger, Synth. Met. 22, 79-87 (1987).
3. E. Ruckenstein and J. S. Park, J. Appl. Polym. Sci. 42, 925 (1991).
4. J. E. Osterholm, J. Laakso and P. Nyholm, Synth. Met. 28, C435-C444 (1989).
5. K. S. Ho, K. Levon, J. Mao and W. Y. Zheng, Synth. Met. 55-57, 3591-3596 (1993).
6. L. H. Dao, X. F. Zhong, A. Menikh, R. Paynter and F, Martim, Annu. Tech. Conf. Soc., Plast. Eng. 49, 783 (1991).
7. B. Tieke and W. Gabriel, Polymer 31, 20 (1990).
8. M. B. Meador, D. H. Green, J. V. Auping, J. R. Gaier, L. A. Ferrara, D. S. Paradopoulos, J. W. Smith, D. J. Keller, J. Appl. Polym. Sci. 63, 7, 821-834 (1997).
9. Y. Sun, E. Ruckenstein, Synth. Met., 74, 145-150 (1995).
10. Sroog, C. E.; Engrey, A. L.; Abramo, S. V.; Berr, C. E.; Edwards, W. M.; and Olivier, K. L., Journal of Polymer Science: Part A, 3, 1373-1390 (1965).
11. E. Sugimoto, S. Takeda, H. B. Gu, K. Yoshino, Chem. Express 1, 11, 635-638 (1986).
12. A. L. Endrey, U. S. Patent 3,179,633, 5pp (1965).
13. J. W. Wang, M. P. Srinivasan, Synthetic Metals 105, 1-7 (1999).
14. S. A. Chen, and J. M. Ni, Polymer Bulletin 26, 673-680 (1991).

ELECTROCHEMICAL INVESTIGATION OF
2, 2'-DIAMINOBENZYLOXYDISULFIDE

Y.Z. Su, K.C. Gong
Polymer Structure & Modification Res. Lab., South China University of Technology,
Guangzhou, China 510641, pskcgong@scut.edu.cn

ABSTRACT

A new conducting polymer, poly(2, 2'-diaminobenzyloxydisulfide), has been proposed as a positive material suitable for secondary lithium batteries. With the aim of better understanding the process of polymerization and depolymerization of 2,2'-diaminobenzyloxydisulfide(DABO). The redox behavior, kinetic reversibility and adsorption of DABO have been investigated at platinum electrode in acetonitrile/tetrahydrofuran solution by using linear sweep voltammetry, the cyclic sweep voltammetry and the rotating disk electrode technique. These experiments clearly showed the reaction is chemically reversible but kinetically slow at ambient temperature and charge transfer is the rate-determining step, but chemical dimerizaton is at equilibrium. The results are common to many organic disulfides. Furthermore, the striking observation from cyclic voltammograms is the smaller separation of the anodic and cathodic peak owing to the specific structure of DABO, compared with other organic disulfides. This results indicates the redox reaction of DABO is higher kinetically reversible and poly (DABO) positive material is expected to deliver higher power output or energy efficiency.

INTRODUCTION

The disulfide bonds, so common in peptides and proteins, play an important role in remaining biological activity and stable conformation. The biologically important interconversion and electrochemical properties of cysteine and cystine had been studied [1,2]. The formation and cleavage processes of the disulfide bond, in molecular conversion between cysteine and systine by oxidation of the thiol group and reduction of the disulfide bond, has been considered as charging and discharging process of energy during the last few decades [3-7]. Disulfide compounds have been proposed as the new organic/polymeric energy storage materials based on the reversible polymerization-depolymerization process (2SH☐S-S). Theoretical energy content of these materials far exceeds that of conventional battery materials as well as those of other candidate materials such as intercalation compounds and conducting polymers.

However, disulfide compounds are not electronically conductive, and usually exhibit a low redox reaction rate at room temperature. It is necessary to find an appropriate redox electrocatalyst to improve electrode kinetics. An effective way for accelerating the reaction rate of disulfide compounds has been found by mixing them with conducting polymer. Oyama and his coworkers [8,9] has studied the effects of polyaniline and copper (II) salts on 2,5-dimercapto-1,3,4-thiadiazole for rechargeable lithium batteries. The authors have investigated the electrochemical activity of thiokol/polyaniline and thiokol/polypyrrole composites [10-12]. Recently, Naoi et al introduced a new energy storage material: poly (2,2'-dithiodianiline), which have higher electrochemical activity and higher electrical conductivity by virtue of its intermolecular electrocatalytic effect [13,14].

Here the authors present a novel conducting polymer compound: poly (2,2'-diaminobenzyloxydisulfide) containing -O-S-S-O- bonds confined between the chains of polyaniline [15]. With the aim of better understanding the processes of polymerization and depolymerization of 2, 2'-diaminobenzyloxydisulfide. This contribution examines the redox behavior, kinetic reversibility and adsorption of DABO at platinum electrode in acetonitrile/tetrahydrofuran solution by using linear sweep voltammetry, cyclic sweep voltammetry and rotating disk electrode technique.

EXPERIMENTAL

Materials

DABO was prepared as described in a previous publication [15]. Acetonitrile (AN) and tetrahydrofuran (THF) were purified by distillation in used manner. All other chemicals were of reagent grade and used without further purification.

Electrochemical measurements

Linear and Cyclic sweep voltammetric measurements were made with a standard three-electrode, two-compartment electrochemical cell employing a HDV-7C potentiostat/galvanostat in conjunction with a DCD-3 functional generator. The voltammograms were recorded on a 3086 x-y recorder. In rotating disk electrode (RDE) experiments, an ATA-1A electrode rotator and speed controller were used to modulate the rotation speeds of platinum disk electrode (2mm, diameter). Before each experiment, the working electrode was polished to a mirror finish with diamond past (10^{-3}mm) and rinsed with acetone.

The auxiliary electrode was a glass-carbon. The reference electrode was Ag/AgCl (sat. KCl) electrode. The working electrode was a platinum electrode.

Unless other stated, electrolyte solutions consisted of 0.1M $LiClO_4$ in AN/THF (2/1, v/v) and electrode potentials were measured vs. an Ag/AgCl reference electrode in all experiment. The electrolyte solutions were purged with dry argon for about 10 min prior to measurements and gas flow was maintained upon solution throughout the experiments.

RESULTS AND DISCUSSION

Redox behavior of DABO

Most organodisulfide does almost not dissolve in water and only dissolve in a few of organic solvents. The mixture solution consists of AN and THF (AN/THF=2/1, v/v) owing to DABO monomer's low solubility in AN but relative higher solubility in THF. And taking the relative dielectric constants of AN (298K, 37.5) and THF (298K, 7.5), the AN/THF volume ratio should not be lower than 2:1. The AN/THF volume ratio in all electrolyte solutions is 2:1.

The linear sweep voltammograms for reduction of DABO to the thiolate anions at different sweep rates are shown in Fig. 1(a). Owing to difficulties in synthesis of the dithiol form of DABO, the linear sweep voltammograms for the oxidation of the thiolate anions to DABO were obtained after electrolysis of DABO for about half an hour at –0.35V, as shown in Fig. 1(b).

The analysis of i_p and E_p in both cathodic and anodic direction at different scan rates indicates that the relationships between E_p and $\ln(i_p)$ (shown in Fig.2), between $\ln(v)$ and E_p and between i_p and $v^{1/2}$ are approximately linear. These results confirm that no deposition of product of redox reaction on the electrode and the redox couples are kinetically hindered.

At different sweep rates, the plot of $\ln(i_p)$ vs. E_p gives slopes of $-\alpha_c F/RT$ for cathodic and $\alpha_a F/RT$ for anodic processes. Similarly, the plot of E_p vs. $\ln(v)$ gives the slopes of $-RT/2\alpha_c F$ for the cathodic processes and $RT/2\alpha_a F$ for the anodic processes. The transfer coefficients calculated from the obtained data are $\alpha_c = 0.45$ and $\alpha_a = 0.54$, obtained from $d[\ln(i_p)]/d[E_p]$, and $\alpha_c = 0.47$ and $\alpha_a = 0.52$, obtained from $d[E_p]/d[\ln(v)]$, and are in nearly agreement with each other. Furthermore, the summation of the observed cathodic and anodic transfer coefficients being close to one indicates that the number of electrons transferred in the rate-determining step is equal to one.

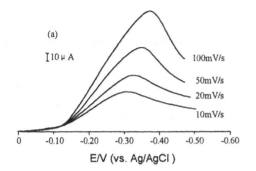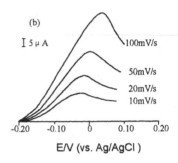

Fig.1. Linear sweep voltammograms on platinum disk electrode in AN/THF containing 0.1M LiClO$_4$ at different sweep rates: (a) reduction of DABO (0.026M) to the thiolate anions; (b) oxidation of the thiolate anions to DABO after electrolysis of DABO (0.034M) for 30m at –0.35V.

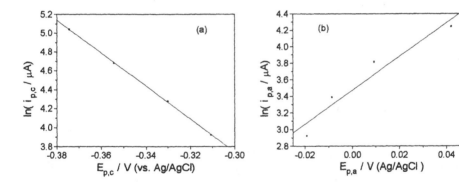

Fig 2 Plot of logarithm of peak current ($\ln i_p$) vs. peak potential (E_p): (a) reduction process; (b) oxidation process.

Electrode kinetics of DABO

According to the previous studies [16-18], the overall, stoichiometric reaction for organo disulfide /thiolate redox couples can be described as:

$$RSSR + 2e = 2RS^-$$ (1)

If one suppose that adsorption is negligible, the relationship between current, i, and overpotential at the electrode surface, η, can be described by:

$$i=i_0[(C_{RS^-,s}/C_{RS^-,b})^{\gamma_{RS^-}} \exp(\alpha_a F \eta /RT) - (C_{RSSR,s}/C_{RSSR,b})^{\gamma_{RSSR}} \exp(\alpha_c F \eta /RT)] \quad (2)$$

where i_0 is exchange current, γ_i, $C_{i,s}$ and $C_{i,b}$ are reaction order, concentration at the electrode surface and concentration in bulk solution for oxidation and reduction.

Under the condition that the electrode reaction is completely controlled by kinetics, Eq(2) can be reduced to the Butler-Volmer equation:

$$i=i_0[\exp(\alpha_a F \eta /RT) - \exp(\alpha_c F \eta /RT)] \quad (3)$$

The relation between current at an RDE and the convective-diffusive limiting current, i_l, can by described as:

$$i/i_k=[1-i/i_l]^{\gamma_i} \quad (4)$$

where i_k is the pure kinetic current, the anodic or cathodic reaction order for species i can also be expressed as:

$$\gamma_i = d[\log i]/d[\log(1-i/i_l)] \quad (5)$$

The plot of $\log(i)$ vs. $\log(1-i/i_l)$ at constant potentials and different rotation rates are shown in Fig. 3. The slopes of the plots at different potentials represent the cathodic order, γ_{RSSR}, or the anodic order, γ_{RS^-}. The value of γ_{RSSR} is close to 0.5 and that of γ_{RS^-} near 1. This indicates that the electrode reaction takes the EC route with charge transfer as the rate determining step and chemical reaction at equilibrium, i. e.

$$2RS^- \rightarrow 2RS^· + 2e^- \quad (6)$$

$$2RS^· \rightarrow RSSR \quad (7)$$

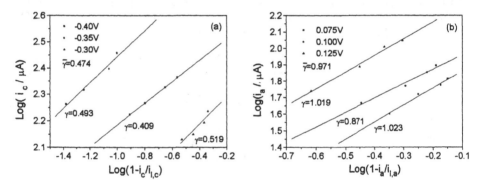

Fig.3. Plot of $\log(i)$ vs. $\log(1-i/i_p)$ at different rotation rates for constant potentials at a platinum disk electrode in AN/THF containing 0.1M LiClO$_4$: (a) oxidation of DABO (0.024M); (b) oxidation of the thiolate anions after electrolysis of DABO (0.034M) for 30m at −0.35V.

Redox behavior of poly(DABO)

Kinetic reversiblity is an important parameter for energy storage material; the higher the kinetic reversibility of the redox reaction, the higher the power output at a given energy efficiency of the system or the higher the energy efficiency at a given power output. The kinetic reversibility of the redox couples was assessed according to the separation of the anodic and cathodic peak potential in cyclic voltammograms.

Cyclic voltammograms of DABO on a platinum wire (diameter 0.5mm, length 5cm) electrode in AN/THF containing 0.1M LiClO$_4$ are shown in Fig. 4(a). In the first sweep of the potential from 0V to a positive potential, no oxidative peaks were observed. However just after one reduction peak observed as swept from a positive to a negative potential, ane oxidation peak appeared as the potential was swept from a negative to a positive potential at the very start. The monomer DABO showed reduction and oxidation peaks at −0.01 and −0.33V vs. Ag/AgCl, respectively, sweep rate 20mV/s. Fig. 4(b) is the cyclic voltammograms of poly(DABO) on a platinum wire (diameter 0.5mm, length 5cm) electrode in N-methyl-2-pyrrolidon (NMP) solution containing 0.1M LiClO$_4$. The redox behavior of poly(DABO) was similar to that of its monomer. Both reduction and oxidation peaks were observed at −0.11 and −0.25V vs. Ag/AgCl, respectively, sweep rate 20mV/s. The peak separations between the reduction and oxidation potentials of poly(DABO) (ca. 0.14V) were narrower than those of DABO (ca. 0.32V) at 20mV/s.

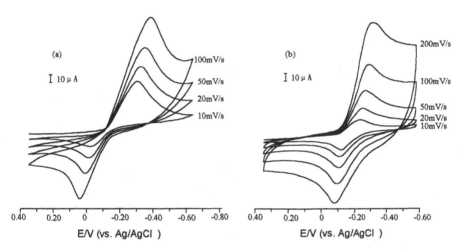

Fig. 4 (a) Cyclic voltammograms of DABO (0.022M) in AN/THF containing 0.1M LiClO$_4$.
(b) Cyclic voltammogarams of poly(DABO) (0.016M) in NMP solution containing 0.1M LiClO$_4$.

The cyclic voltammograms demonstrated that the oxidation potendial of poly(DABO) was lower than that of its monomer DABO and the reduction potential of it was more positive than that of its monomer. That is, the cleavage (reduction process) of S-S bonds and the recombination (oxidation process) of poly(DABO) was easier and more reversible than those of DABO monomer. The cleavage and recombination efficiency of S-S bonds can be greatly improved maybe due to the heavy electron donating characteristics of O-S-S-O bonds confined between the chains of polyaniline and the good catalysis of conducting polyaniline.

CONCLUSIONS

The electrochemical behaviour of DABO has investigated by linear, cyclic sweep voltammetry and the rotating disk electrode technique. The reaction orders for the disulfide-thiolate redox couple are one in the anodic direction and half an one in the cathodic direction. The transfer coefficients are estimated and the the reaction pathway takes the EC route. The reaction is chemically reversible but kinetically slow at ambient temperature, and change transfer is the rate determining step but chemical dimerzation is at equilibrium. These results are common to many other organic disulfides.

Furthermore, form a comparison of the cyclic voltammograms between DABO monomer and poly(DABO), it becomes clear that the cleavage and recombination of S-S bonds in poly(DABO) are easier and more reversible than those of DABO monomer owing to the specific structure of poly(DABO).

ACKNOWLEDGMENTS

This work was financially supported by the National Science Foundation of China (#59836232).

REFERENCES

1. I. R. Miller, J. Teva, J. Electroanal. Chem., **36**, p. 157(1972).
2. A. Fava, G. Reichenbach, J. Am. Chem. Soc., **89**, p. 6696(1967).
3. S. J. Visco, L. C. De Jonghe, J. Electrochem. Soc., **135**, p. 2905(1988).
4. S. J. Visco, L. C. De Jonghe and M. B. Amand, J. Electrochem. Soc., **136**, p. 661(1991).
5. M. Liu, S. J. Visco, L. C. De Jonghe, J. Electrochem. Soc., **138**, p. 1891(1991).
6. K. Naoi, M. Menda, H. Ooike and N. Oyama, Proc. **31**st Bat. Symp. Jpn., p. 31(1990).
7. K. Naoi, M. Menda, H. Ooike, et al., J. Electroanal. Chem., **318**, p. 395(1991)
8. N. Oyama, J. M. Pope and T. Sotomura, J. Electrochem. Soc., **144**, p. L47(1997).
9. Q. Chi, T. Tatsuma, M. Ozaki, et al., J. Electrochem. Soc., **145**, p. 2369(1998).
10. K. C. Gong, W. S. Ma, Mat. Res. Soc. Symp. Proc., **411**, p. 351(1996).
11. W. S. Ma, Z. B. Jia, K. C. Gong, China Synthetic Rubber Industry (China), **20**, p. 91(1997).
12. K. C. Gong, W. S. Ma, Mat. Res. Soc. Symp. Proc., **461**, p. 87(1997).
13. K. Naoi, K. Kawase, M. Mori, et al. J. Electrochem. Soc., **144**, p. L173(1997).
14. K. Naoi, K. Kawase, M. Mori, et al., Mat. Res. Soc. Symp. Proc., **496**, p. 309(1998).
15. Y. Z. Su, W. S. Ma, K. C. Gong, Mat. Res. Soc. Symp. Proc., **575**, p.   (1999).
16. M. Liu, S. J. Visco, L. C. De Jonghe, J. Electrochem. Soc., **136**, p. 2570(1989).
17. M. Liu, S. J. Visco, L. C. De Jonghe, J. Electrochem. Soc., **137**, p. 750(1990).
18. S. Picar, E. Genies, J. Electroanal. Chem., **408**, p. 53(1996).

ON THE NATURE OF HETEROGENEITY IN VACUUM DEPOSITED POLYANILINE FILMS

V.F. IVANOV, A.A. NEKRASOV, K.V. CHEBERYAKO, O.L. GRIBKOVA,
A.V. VANNIKOV
A.N. Frumkin Institute of Electrochemistry RAS, Leninskii prospect 31, Moscow 117071,
Russia. E-mail: vanlab@glasnet.ru.

ABSTRACT

Heterogeneity of polyaniline and other conductive polymers in intermediate oxidation states is fundamental problem of physics and chemistry of these ones. Usually only advanced methods may be used for immediate detection of the heterogeneity in the molecular scale range. For the first time we have observed the process of the heterogeneous net-like structure formation in macroscopic scale under the oxidation of the evaporated polyaniline films by aqueous HNO_3. and other oxidative agents. Formation of heterogeneous structure is explained in terms of nonequilibrium thermodynamic and chemical kinetics.

INTRODUCTION

Heterogeneity relates to the principal features of conducting polymers. In this regard polyaniline (PAn) is one of the most important examples. Our present view on PAn heterogeneity come from the investigations of ESR [1,2,3,4], NMR [5,6] and electrical conductivity [7,8,9]. The STM [10], AFM [11] and microscopic [12] data are also directly related to the subject under consideration. In compliance with this view, coexistence of crystalline "metallic islands" dispersed in "non-conducting" amorphous phase is proposed for PAn in an intermediate oxidation states.

Electrochemical synthesis or spin casting may be used to produce thin films of PAn on various substrates, but neither of the methods is compatible with modern technology necessary to construct working circuits. In this respect vacuum thermal evaporation is of particular interest. Earlier studies [13,14] revealed that thin films can be produced also by this method, and many questions arose concerning the nature of the layers obtained. Further investigations have substantiated the complexity of this problem [15,16,17,18,19,20,21,22]. For example, when either emeraldine salt or base were used as a starting material XPS studies reveal almost fully reduced state of PAn in the deposited films. It is of value that the thickness of these films ≥ 0.1 μm. According to our data [22] formation of leucoemeraldine-like form during evaporation is due to the hydrolysis of quinone-imine fragments by strongly bonded water resulted in formation of appropriate amine- and oxygen-containing functional groups. Molecular weight and other properties of the thermally deposited PAn depend strictly on the temperature range of evaporation. The most PAn-like material (MW ~ 1500 D) was obtained in the temperature range 275-325°C. Immediately after evaporation the deposited films (in contrast to leucoemeraldine) may be easily dissolved in many common organic solvents like ethanol, benzene, acetone etc. Optical absorption spectra also indicate some changes in the electronic structure, as they are not identical to those of leucoemeraldine [16]. In addition, the films are characterized by the absence of ESR signal and poor conductivity ($\sim 10^{-7}$ S/cm) [17]. According to our data, regeneration of PAn-like properties may be achieved by the cyclic acid-base treatment of the as-deposited films on air [16]. After this treatment electronic spectra, ESR data and cyclic voltammetric curves of the vacuum evaporated PAn are the same as these ones for common PAns. Furthermore, as a result of such a treatment the vacuum evaporated PAn also loses the solubility in common organic solvents. The latter fact was attributed to the formation of three-dimensional intermolecular aggregates of donor-acceptor type by their nature. Nevertheless, the electrical conductivity of these films ($\sim 10^{-4}$ S/cm [17]) was much less than that for common PAns (10^2 S/cm and higher). It should be particularly emphasized that the conductivity of very thin (≤ 100 Å) acid doped vacuum deposited PAn films evidently may be very high concidering plasmons exist in the far-IR [21]. So, it is reasonable to assume that the local structure in microscopic scale for vacuum deposited films, in general, is identical with this one for common PAn, whereas character of self-organization on higher structural levels is not.

EXPERIMENTAL

Emeraldine salt powder was prepared by chemical oxidation by ammonium persulfate [15]. Then the salt was treated by concentrated ammonia solution to obtain emeraldine base. PAn films were deposited on polished glass substrates in the course of vacuum thermal evaporation of the emeraldine base at 275-325 °C. The fractions having lower evaporation temperature were preliminary sublimated on a shutter. Residual pressure in the vacuum system during the evaporation was less than 1.0×10^{-3} Pa.

Spectral studies were carried out using a Beckman DU-7HS single beam spectrophotometer. X-ray diffraction measurements were performed on diffractometer DRON 6. Optical microscopic studies were carried out using a MBI-6 ("LOMO") microscope.

RESULTS AND DISCUSSION

According to our X-ray diffraction data there is no any significant amount of crystalline phase in the bulk of the as prepared vacuum thermally deposited PAn film. Thickness of the as-deposited film was about 0.2 μm. As mentioned above, the as-deposited PAn differs significantly from common PAn and regeneration of a more PAn-like structure may be accomplished in the course of cyclic acid-base treatment on air. With the purpose to restore original structure of the polymer in present study the films were subjected to a post-deposition oxidative treatment in 2 M aqueous HNO_3. In Fig. 1a the changes of the electronic absorption spectra during the oxidation are presented.

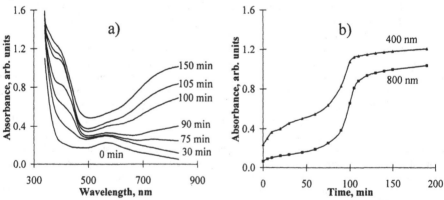

Fig 1. Spectra (a) and kinetics of absorbance change at characteristic wavelengths (b) in the course of oxidation of vacuum deposited PAn in 2 M aqueous HNO_3.

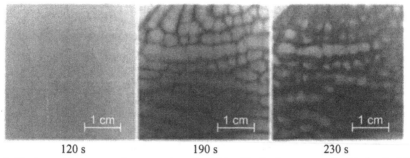

120 s   190 s   230 s

Fig. 2. Photographic images of vacuum deposited PAn film in the course of oxidation in 3.3 M aqueous HNO_3. The treatment time is indicated under the images.

We observe growth of the absorption within spectral ranges near 400 nm (usually assigned to cation-radicals absorption [23]) and near 800 nm (localized polarons absorption [24]). The growth of the band near 800 nm in this case is far more intense and the band itself is more clearly pronounced than that observed by us earlier [16], when using oxidation by oxygen of air between the acid-base treatment cycles. This means that this oxidizing procedure is more efficient to produce PAn-like structure in the vacuum deposited films and, as a result, more conducting areas are apparently formed within low-conductive polymer matrix. The kinetics of absorption at some characteristic wavelengths are presented in Fig. 1b. The form of kinetic curves is typical for autocatalytic process. It is seen that there is an induction period in growth of absorption at the two characteristic wavelengths followed by a stage of rapid growth, which further reaches a plateau. It should also be noted that in these conditions, judging from the coloring, the oxidation proceeds rather slowly and uniformly on the whole area of the sample. However, for more concentrated solutions (more than 3 M HNO_3) a remarkable phenomenon is observed. It consists in the fact that after some induction period a rapid formation of net-like structure, which can be visually observed in macroscopic scale, occurs on the whole area of the film (Fig. 2). Some details of this process were described by us earlier [25]. Green color of the net-like structure elements indicates that they are composed of emeraldine form of PAn. During further course of the process growth of these structure elements is primarily observed, which gradually results in coloring of the whole area of the film. At final stages of the process evaporated PAn layer acquires blue color of pernigraniline. As a characteristic feature of the process one should mention existence of the following two stages: formation of nuclei and their growth. Such a situation is typical for the second type phase transition. Taking into account tremendous differences in properties of semi-oxidized and reduced forms of PAn such a comparison seems to be justified. In case of further increase of the acid concentration formation of the macroscopic structure proceeds more rapidly. At concentrations more than 5 M of HNO_3 the visual observation becomes practically impossible. In these conditions almost the whole area of the film is rapidly transferred into ultimately oxidized state. While using other oxidants, for example, aqueous solutions of ammonium persulfate, the macroscopic structure formation can be observed only at concentrations less than 10^{-3} M. In this case the structure has essentially different character and comprises blue colored rather compact areas resembling spots separated by weakly colored spacers. At higher concentrations of the oxidant, similarly to the process in nitric acid, there is practically no formation of the macroscopic structure. Rapid and uniform coloring of the whole area of the film is observed instead.

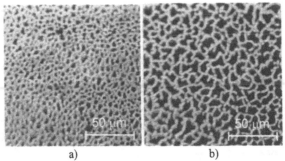

a)   b)

Fig. 3. Microphotographs of vacuum deposited PAn film after 190 s (see Fig. 2) of oxidation in
3.3 M aqueous HNO_3: a) weakly oxidized area; b) heavily oxidized area.

The above described formation of the net-like structure may be considered (since it is composed of conducting emeraldine form of PAn) as formation of a conducting net providing charge carriers transport along the whole area of the film similarly to the situation, which takes place in composite structures like PAn(CSA)-PMMA [26]. However, as mentioned above, this is not the case because conductivity of the oxidized sample exceeds no 10^{-4} S/cm [17]. The reason of this apparently consists in microscopic structure of the oxidized areas. Naturally, microphotographs of the oxidized areas at different stages of their formation, which are presented in Fig. 3, show that these areas are actually inhomogeneous and comprise green colored patterns composed of conducting emeraldine separated by grayish-yellow colored spacers composed of non-conducting leukoemeraldine. Thus, rather low conductivity of the evaporated layer observed during macroscopic measurements is due to its heterogeneous structure, which is forming under essentially non-equilibrium conditions in autocatalytic process. In this respect the phenomena observed, apparently, are close in their nature to dissipative structures formed in Belousov-Zhabotinskii reactions. It is known that necessary conditions for the dissipative structures formation also include non-equilibrium character of the system and existence of autocatalytic stages during the process [27]. However, characteristic feature of the system considered consists in the fact that, as distinct from homogeneous single-phase systems, which are usually considered in connection with Belousov-Zhabotinskii reactions, our system is initially double-phase one. This, apparently, significantly complicate a task of creating a mathematical model describing adequately all stages of the process.

CONCLUSIONS

Thus, it was found that oxidation of PAn films deposited by vacuum thermal evaporation in aqueous solutions of oxidants occurs non-uniformly. This leads to formation heterogeneous structure, whose characteristic sizes are determined by the nature and concentration of oxidant. The phenomena found by us may evidently serve as the base to understand the mechanisms of processes causing formation of heterogeneous structure of conducting polymers.

ACKNOWLEDGEMENTS

The study was supported by the International Science and Technology Center (project 872) and Russian Foundation for Basic Research (grants No. 99-03-32077 and No. 96-15-97320).

REFERENCES

1. F. Lux, G. Hinrichsen H.-K. Roth, V.I. Krinichnyi, Macromol. Chem. Macromol. Symp. **72**, p. 143 (1993).
2. F. Lux, G. Hinrichsen, V.I. Krinichyi, I.B. Nazarova, S.D. Cheremisov, M.-M. Pohl, Synth. Met. **53**, p. 347 (1993).
3. M. Lapkowski, E.M. Genies, J. Electroanal. Chem. **279**, p. 157 (1990).
4. S.M. Long, K.R. Cromack, A.J. Epstein, Y. Sun, A.G. MacDiarmid, Synth. Met. **55**, p. 648 (1993).
5. P.K. Kahol, B.J. McCormic, J. Phys.: Condens. Matter **3**, p. 7963 (1991).
6. H.H.S. Javadi, K.R. Cromack, A.G. MacDiarmid, A.J. Epstein, Phys. Rev. B. **39**, p. 3579 (1989).
7. A.J. Epstein, J.M. Ginder, F. Zuo, H.S. Woo, D. Tanner, A.F. Richter, M. Angelopoulos, W.S. Huang, A.G. MacDiarmid, Synth. Met. **21**, p. 63 (1987).
8. Z. Wang, A. Ray, A.G. MacDiarmid, A.J. Epstein, Phys. Rev. B. **43**, p. 4373 (1991).
9. R.S. Kohlman, J. Joo, Y.Z. Wang, J.P. Pouget, H. Kaneko, T. Ishiguro, A.J. Epstein, Phys. Rev. Lett. **74**, p. 773, (1995).
10. D.A. Bonnell, M. Angelopoulos, Synth. Met. **33**, p. 301 (1989).
11. M. E. Vela, G. Andreasen, R.C. Salvarezza , A. J. Arvia, J. Chem. Soc., Faraday Trans. **92**, p. 4093 (1996).
12. K. Aoki, Y. Teragashi, Abstracts of the Joint International Meeting of the Electrochemical Society and International Society of Electrochemistry, August 31 - September 5, 1997, Paris, France, p. 1447 (1997).
13. A. Angelopoulos, G.E. Asturias, S.P. Ermer, A. Ray, E.M. Scherr, A.G. MacDiarmid, M. Akhtar, Z. Kiss and A.J. Epstein, Mol. Cryst. Liq. Cryst. **160**, p. 151 (1988).
14. K. Uvdal, M. Lögdlund, P. Dannetun, L. Bertilsson, S. Stafström, W.R Salaneck, A.G. MacDiarmid, A. Ray, E.M. Scherr, T. Hjertberg and A.J. Epstein, Synth. Met. **29**, p. E451 (1989).
15. V.F. Ivanov, O.L. Gribkova, A.A. Nekrasov, A.V. Vannikov, J. Electroanal. Chem. **372**, p.57 (1994).
16. A.A. Nekrasov, V.F. Ivanov, O.L. Gribkova, A.V. Vannikov, Synth. Met. **65**, p.71 (1994).
17. V.F. Ivanov, A.A. Nekrasov, O.L. Gribkova, A.V. Vannikov, Electrochim. Acta, **41**, p.1811 (1996).
18. V.F. Ivanov, A.A. Nekrasov, O.L. Gribkova, A.V. Vannikov, Synth. Met. **83**, p. 249 (1997).
19. A.A. Nekrasov, V.F. Ivanov, O.L. Gribkova, A.V. Vannikov, Electrochim. Acta **44**, p. 2317 (1999).
20. D.M. Cornelison, T.R. Dillingham, E. Bullock, N.T. Benally, S.W. Townsend, Surf. Sci. **343**, p. 87 (1995).
21. R.V. Plank, N.J. DiNardo, J.M. Vohs, Synth Met. **89**, p. 1 (1997).
22. V.F. Ivanov, I.V. Gontar', A.A. Nekrasov, O.L. Gribkova, A.V. Vannikov, Russ. J. Phys. Chem. **71**, p. 125 (1997).
23. D.E. Stilwell and S.-M. Park, J. Electrochem. Soc. **136** p. 427 (1989).
24. Y. Min, Y. Xia, A.G. MacDiarmid and A.J. Epstein, Internat. Conf. Sci. Tech. Synth. Met., July 24-29, 1994, Seoul, Korea. Abstracts. p. 287 (1994).
25. V.F. Ivanov, O.L. Gribkova, A.A. Nekrasov and A.V. Vannikov, Mendeleev Commn. p. 4 (1998).
26. R. Menon, C.O. Yoon, D. Moses, A.J. Heeger, Handbook of Conducting Polymers, Second Edition, edited by T.A. Scotheim, R.L. Elsenbaumer, J.R. Reynolds, Marcel Dekker, New

York, 1998, p.27-84.

27. R.J. Field, Oscillations and Traveling Waves in Chemical Systems, edited by R.J. Field and M. Buger, John Wiley and Sons, New York, 1985, pp. 75-116.

Polymer Gels and Muscles

POLYMER ELECTROLYTE ACTUATOR DRIVEN BY LOW VOLTAGE

K. Oguro*, K. Asaka*, N. Fujiwara*, K. Onishi** and S. Sewa**
*Osaka National Research Institute, Midorigaoka 1-8-31, Ikeda, Osaka 563-8577, JAPAN, oguro@onri.go.jp
**Japan Chemical Innovation Institute, Osaka, JAPAN

ABSTRACT

Composites of perfluorinated polymer electrolyte membrane and gold electrodes bend in response to low-voltage electric stimuli and work as soft actuators like muscles. The composites were prepared by chemical plating. Charge on the electrode induces electric double layer and electro-osmotic drag of water by cation from anode to cathode through narrow channels in the perfluorinated ion-exchange resin. The electro-osmotic flow of water swells the polymer near the cathode rather than anode, and the membrane bends to the anode. The actuator comprises polymer electrolyte, electrodes, counter ion, solvent, lead wires, etc. Each component affects the performance of the actuator. Surface area of electrode and species of counter ion have drastic effect on voltage-displacement response. The response may depend on water channel structure of the polymer electrolyte. Modification of these factors improved the performance and resulted in the deflection over 360 degrees at a film actuator of 10 mm length. A tubular actuator was demonstrated as a multidirectional actuator. These actuators are applicable to artificial muscle, micro robots, or micro medical equipment inside body.

INTRODUCTION

Electrochemomechanical polymers have attracted a great interest as soft actuators in recent years. Large deformation of polyelectrolyte gels under electric fields has been demonstrated [1, 2]. A problem is the high voltage applied on them that induced water electrolysis and gas evolution. A polymer electrolyte membrane plated with platinum has deformed under low voltage in water or saline solution [3, 4]. An effective membrane was the perfluorinated sulfonic acid membrane which was commonly used for proton exchange membrane fuel cells (PEM-FC) or solid polymer electrolyte water electrolysis. The platinum plating method on the membrane was developed originally for the water electrolysis [5, 6]. The first actuator of the polymer electrolyte-metal composite was made of perfluorosulfonic acid and platinum [3]. Platinum is a good catalyst for water electrolysis and hydrogen evolution which is not favorable *in vivo* use. The acidic membrane needs a noble metal for the electrodes. Gold was the most promising electrode for the composite. A new chemical plating process of gold on the membrane was developed by an ion-exchange method with cationic gold complexes and interfacial reduction on the surface of the membrane [7]. Perfluorocarboxylic acid membrane was expected to give high response because of its higher ion-exchange capacity than perfluorosulfonic acid.

EXPERIMENT

In general, the composite was prepared by the chemical plating method which consists of ion-exchange reaction of Na^+ in the cation exchange membrane with a cationic platinum or gold complex, followed by a reduction process in an aqueous solution of anionic reducing agent. The interfacial reduction form platinum or gold electrode layer with large surface area, strong adhesion and high electric surface conductivity.

Perfluorosulfonic acid membrane (Nafion® 117 (Du Pont, 0.9 meq/g-resin, 0.18 mm thickness), perfluorocarboxylic acid membranes, Flemion®(Asahi Glass Co., Ltd., 1.44 meq/g-resin, 0.14 mm thickness and 1.8 meq/g-resin, 0.17 mm thickness and tube of 0.3-0.8 mm external diameter) were plated with the complex of $[Au(phen)Cl_2]Cl$ (phen = 1, 10-phenanthoroline) and the reducing agent of Na_2SO_3. Sodium salt of perfluorocarboxylic acid has the following chemical structure

$$\left[\begin{array}{l} \text{CF}_2\text{CF} - (\text{CF}_2\text{CF}_2)_n \overline{} \\ | \\ (\text{OCF}_2\text{CF})_m - \text{O}(\text{CF}_2)_3\text{COO}^-\text{Na}^+ \\ | \\ \text{CF}_3 \end{array} \right]$$

where m is 0 or 1.

Fine glass powder was blown to the dry membrane with compressed air to roughen surface of the membrane. After washing and cleaning the surface, the membrane was boiled in water to swell. The swollen membrane was immersed in the aqueous ca. 10^{-2}M solution of the gold complex to exchange ions for 12 hrs at room temperature. The impregnated cationic gold complex was reduced with the aqueous ca. 10^{-3}M solution of Na_2SO_3 at 40-70°C. The cationic gold complex has the structure:

The ion-exchange and reducing processes were repeated up to 8 times to make thick gold layers on the membrane. Sodium cation in the composite was exchanged with various alkali metal ions or alkylammonium ions by immersing the composite in aqueous chloride salt solutions for 12 hours at room temperature.

Tubes of perfluorocarboxylic acid resin were plated only on outside by the same process as the membranes. The plated electrode was cut into four parts as shown in Figure 7 with an excimer laser abrasion process.

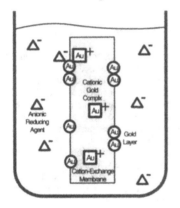

Figure 1: A schematic of interfacial reduction process to plate gold electrode.

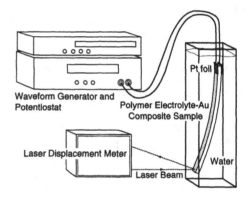

Figure 2: A schematic of the set-up for evaluating bending response.

Figure 2 is a schematic of the set-up for evaluating bending response. The plated polymer electrolyte was cut into a ribbon of 1 mm width and 8 mm length. The ribbon was supported vertically in water by clumping in platinum foil to make electric contact with the gold electrode. Electric signals were supplied from a Yokogawa AG1200 waveform generator and a Hokuto Denko HA-501G potentiostat/galvanostat. Displacement of the free end at a distance of 5 mm from the fixed point was measured by a Keyence LC-2220/2100 laser displacement meter directed through the glass sample cell at room temperature.

Repeated cycles of ion exchange and reduction resulted in dendritic growth of. To evaluate the surface area capacitance of electric double layer was measured on the samples plated multi times. The capacitance increased with the number of the cycles almost in proportion. The surface area of the interface between electrode and polymer electrolyte is proportional to the capacitance.

Typical waveforms of current and displacement response of the composites driven by step voltage are shown in Figures 3. The sample is sodium perfluorosulfonate (Nafion® 117) with Pt electrodes. The complex waveform indicates multiple mechanism for the response. The first action is very fast and the peak appears within 10 ms after the step signal. We call this response as "quick action". The quick action decays in a few seconds. After ca. 5 seconds the second peak toward the cathode appears in this case. We call this response as "slow action". The slow action also decays in a minute.

The direction of the quick action is fixed to the anode, but that of the slow action depends on cations, materials and other environments. Quick action of an anion-exchange membrane is small but opposite to that of cation-exchange membranes.

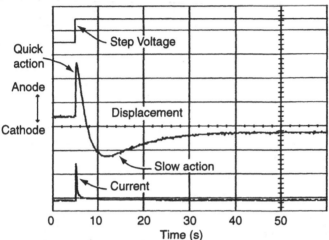

Figure 3: Waveforms of displacement and current under step voltage (-1V to +1V): Strip of sodium perfluorosulfonate Pt composite bends quickly to the anode in 10 ms, a slow action to the cathode in 5 s and gradual decay of the peak follows.

A membrane without electrode does not move, if a current flow through the surrounding electrolyte solution. Mechanical stress on the composite does not induce an electric potential on the electrodes.

Surface area of the electrode is almost proportional to the response and discussed elsewhere [8]. Cation species have the secondary dominant effect on the response. Large effects were observed on bulky organic cations. Figure 4 shows response waveforms for perfluorocarboxylic acid membrane (1.8 meq/g) plated with gold electrode for 5 cycles. The composite containing sodium ion moves fast but the displacement remains small (Figure 4a). The same composite containing tetra-n-butylammonium ion moves slower but larger (Figure 4b). Ionic current flowed quickly and decayed exponentially as a simple electric circuit of a series connection of resistance and capacitance. The ammonium ion shows longer time constant of current decay and higher resistance of the composite.

The strip of the sample containing sodium cation (Figure 4a) bends quickly to the anode and the displacement decays gradually and partly. The composite with alkylammonium cation bent slower than that with sodium cation but gave larger deflection than that with sodium cation.

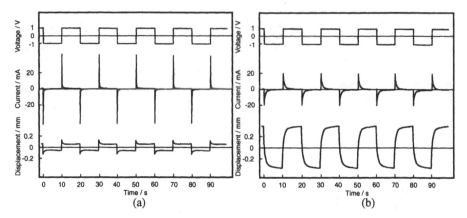

Figure 4: Waveforms of voltage, current and displacement of perfluorocarboxylic acid membrane; (a) containing sodium ion, (b) containing tetra-n-butylammonium ion.

The high response of the material gave some demonstration of small devices. Figure 5 shows the micro gripper which was made from a sheet of the composite. It has eight fingers of 15 mm long and catches a ball within a second in water.

Figure 5: Micro gripper

Figure 6: A schematic of tubular actuator with four electrodes.

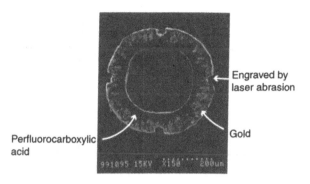

Engraved by
laser abrasion

Gold

Perfluorocarboxylic
acid

991895 15KV X160 200um

Figure 7: Scanning electron microphotograph of
cross-section of a tubular actuator (400 μm in diameter).

The tubular actuator has four lateral electrodes to bend all directions with combined electric signals. It moves slower than strips but finer tube can move faster. Practical usage of the tubular actuator is the tip of active catheter for intravascular neurosurgery.

DISCUSSION

The difference between sodium ion and alkylammonium ion is remarkable and systematic. Alkali ions and alkali-earth ions gave series of response and water transfer characters. In the series of alkali metals, lithium ion responds with the largest displacement and transfers most water through the membrane. As shown in Figure 8 electro-osmotic drag of water from anode to cathode is the most feasible mechanism of the quick action of Figure 3. The fast decay of the quick action can be assignable to water diffusion in the membrane [9].

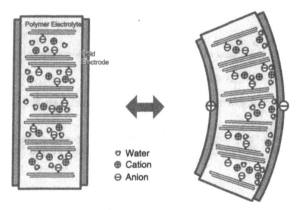

Figure 8: Schematics for bending model of water dragging by cation.

Perfluorosulfonic acid membrane is well known to have hydrophilic spherical clusters [10]. The clusters were estimated to be 4 nm in diameter and connecting with narrow ionic channels to make ion-water network in the membrane. Water transport by alkali-metal ions is explained by dragging of hydrated molecules around the ion. The bulky cation in the narrow channel plays as a piston in a cylinder to push large amount of water from anode to cathode as shown in Figure 9. The cluster structure of perfluorocarboxylic acid is not known but it assumed to be a similar structure. The electro-osmotic flow of water should flow through the narrow hydrophilic channels in the membrane and it must depend strongly on the cluster structure.

The mobility of the bulky cation in the channel seems to be low and causes the high resistance against the ion current and the slow response in "quick action".

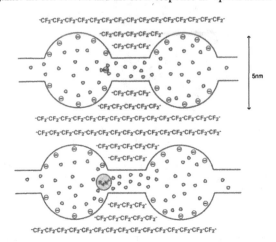

Figure 9: A schematic of water transfer effect of sodium ion and tetraalkylammonium ion in the hydrophilic domain of perfluoro ion-exchange membrane.

That electrochemomechanical property of polymer electrolyte metal composite depends on the channel structure of perfluorinated ion-exchange resin.

CONCLUSION
Polymer electrolyte actuator is;
safe, durable, easy to make, easy to miniaturize, driven under low voltage, and quick.

Mechanism of "slow action" is still open.

New polymer is expected, that may have ionic channels of appropriate size and structure.

ACKNOWLEDGEMENT
This work was partly supported by New Energy and Industrial Technology Development Organization (NEDO). Perfluorocarboxlic acid membranes were offered from Asahi Glass Co., Ltd.

REFERENCES

1. T. Tanaka, I. Nishio, S.-T. Sun, S. Ueno-Nishio, *Science*, 218, 467 (1982)

2. Y.Osada, H.Okuzaki, H.Hori, *Nature*, 355, 242, (1992)

3. K. Oguro, Y. Kawami and H. Takenaka, *Bulletine of Government Industrial Research Institute Osaka*, 43, p.21 (1992).

4. K. Oguro, Y. Kawami and H. Takenaka, "Actuator element", U.S.Patent No. 5268082, (1993).

5. H. Takenaka, E. Torikai, Y. Kawami and N. Wakabayashi, *Int. J. Hydrogen Energy*, 7, 397 (1982).

6. E. Torikai and H. Takenaka, Jpn. Patent No.57-134586 (1982).

7. N. Fujiwara, Y. Nishimura, K. Oguro and E. Torikai, 190th Meeting of The Electrochemical Society, San Antonio, Tex., October, 1996.

8. S. K. Oguro, K. Asaka, N. Fujiwara, K. Onishi and Sewa, *Proc. SPIE*, 3669, 64 (1999).

9. K. Asaka, K. Oguro, Y. Nishimura, M. Mizuhata and H. Takenaka, *Polymer J.*, 27, 436 (1995)

10. T. D. Gierke, G. E. Munn, and F. C. Wilson, *J. Polym. Sci., Polym. Phys. Ed.*, 19, 1687 (1981).

POLYMER-GEL PHASE-TRANSITION AS THE MECHANISM OF MUSCLE CONTRACTION

GERALD H. POLLACK
Dept. of Bioengineering, University of Washington Box 357962, Seattle, WA 98102-7962

ABSTRACT

The thesis offered here is that the muscle-contraction mechanism is similar to the mechanism of contraction in many artificial muscles. Artificial muscles typically contract by a phase-transition. Muscle is thought to contract by a sliding-filament mechanism in which one set of filaments is driven past another by the action of cyclically rotating cross-bridges—much like the mechanism of rowing. However, the evidence is equally consistent with a mechanism in which the filaments themselves contract, much like the condensation of polymers during a phase-transition. Muscle contains three principal polymer types organized neatly within a framework. All three can shorten. The contributions of each filament may be designed to confer maximum strength, speed and versatility on this biological machine. The principles of natural contraction may be useful in establishing optimal design principles for artificial muscles.

FUNCTIONAL ELEMENTS

The essence of the textbook structure of muscle is shown in Fig. 1. The fundamental molecular unit is the sarcomere, bounded on either end by the Z-line. Muscle shortening is the summated shortenings of the many sarcomeres in series and in parallel. Sarcomeres in series build velocity, whereas sarcomeres in parallel build force.

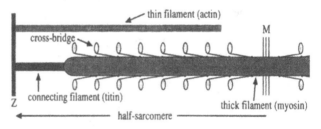

Figure 1. Textbook view of muscle-unit structure. The sarcomere is symmetrical about the middle (M) region. It consists of three longitudinally oriented filaments: thick, thin and connecting. Cross bridges arrayed along the thick filament are thought to interact with the thin filament to mediate contraction.

The sarcomere is built of three principal proteins—or polymers. The central polymer is the thick filament, which is built of multiple repeats of the protein myosin. The thick filament connects at either end to the connecting filament, which in turn connects to the Z-line. In vertebrate muscle the connecting filament is built largely of the protein titin. Titin is built of tandem repeats of discrete domains, the principal one being the immunoglobulin-like domain, which repeats many times between thick filament and Z-line. The thin filament is also a polymer. Along with regulatory proteins of various kind, it consists of multiple repeats of the protein actin. Thus, all three longitudinally oriented structures of the sarcomere are biological polymers.

These three polymers are cross-linked along their length: actin and connecting filaments are cross-linked through the Z-line; thick filaments cross-linked to parallel thick filaments through M-lines; and while cross-bridges (Fig 1) are thought to make transient cross-links with thin filaments, structural evidence obtained by freeze-substitution and thin sectioning (Baatsen et al., 1988) indicates that the cross-bridges form permanent cross-links between thick filaments, much like the M-line connections. There is also evidence that thin and connecting filaments are cross-linked along their length by lateral struts (Trombitas et al., 1988). Evidence for such interconnecting links is considered in greater detail in a book (Pollack, 1990), which also considers the functional role of these cross-links. Thus, one way or another, the extensive cross-links found in polymer gels also exist between muscle polymers.

Like the polymer gel, the cross-linked network of the sarcomere is invested with solvent—an aqueous salt solution. The solvent remains trapped within the cross-linked network; it does not leak out. This can be seen in "skinned" specimens that have been removed from the cell: such specimens do not easily lose solvent. Again, this phenomenon is similar to that of the gel, the solvent presumably held by strong hydrophilic interactions between protein surfaces and water, which retains the water within the gel's polymeric framework. The network also has the "feel" of a gel; i.e, it is resilient to the touch. Thus, muscle has all the essential features of a polymer gel except that its polymers are not randomly dispersed, but well organized into an organized framework.

BASIC ACTION

If muscle contraction were to involve a phase transition, the anticipated dimensional change would be largely axial. This follows because of the polymers' orientation: polymers are aligned along the axis of the muscle, and shortening of these polymers would produce shortening along this axis. On the other hand, polymeric fragments of actin and myosin can be used to construct random gels, in which case contraction should be isotropic. Such gels were many years ago considered as essential models of the contractile process from which much muscle biochemistry was gleaned. Figure 2 shows a classical experiment in which an acto-myosin gel is exposed to increasing concentrations of ATP (the fuel for contraction). Nothing happens until the concentration just crosses a threshold; then the gel shrinks massively, and at the same time the water is forced out. This type of "critical" behavior, based on the crossing of a

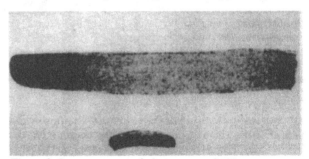

Figure 2: In response to a slight increase of ATP concentration, the actomyosin gel undergoes massive volume change. From Szent-Gyorgyi (1951).

threshold, is typical of polymer-gel phase-transitions.

Critical behavior in muscle contraction is also seen when muscle polymers are naturally oriented as in real muscle. In Fig. 3 the concentration of the physiological trigger, calcium, is progressively

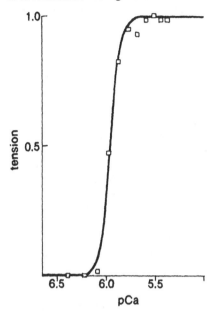

increased. The contractile specimen, devoid of membrane, is immersed in a physiological solution in which the level of free calcium is progressively increased. Once the concentration crosses threshold, full tension is observed within a narrow calcium concentration window. Similar behavior is observed when calcium is replaced by other divalent cations such as barium or strontium.

Similarly critical behavior is seen when the organic solvent is progressively replaced by an aqueous solvent; see Fig. 4. Again, the contraction is largely all-or-none, within a narrow window of organic/aqueous solvent ratio. Such behavior is a classical signature of a polymer gel phase-transition. Thus, razor-edge behavior is preserved in the naturally oriented polymeric system just as it is in the random muscle gel.

Figure 3: *Effect of increase of calcium concentration on isometric tension development in single rabbit muscle cells. From Pollack (1990).*

If muscle contracts like an ordinary polymer gel—how is contraction mediated? The evidence is most consistent with a mechanism by which the polymers themselves contract. Ample evidence in the literature implies that all three polymers shorten.

Before outlining this evidence I should briefly reiterate that this view is out of accord with the generally held view. The current view is that the cross-bridges attach transiently to the thin filament and swing in oar-like fashion to propel thin filaments to slide over thick (Fig. 1); this shortens the sarcomere. Although this is the model of the day, the paradigm is contradicted by numerous observations (for review, see Iwazumi, 1970; Pollack, 1983; Schutt and Lindberg, 1992; Oplatka, 1996, 1997).

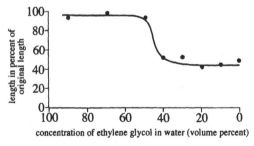

Figure 4: *Effect of solvent variation on contraction in rabbit muscle. From Pollack (1990).*

These conflicts include: evidence that the filaments themselves shorten; evidence that the cross-bridges do not swing; evidence of large active forces with little or no overlap of thick and thin filaments; evidence of synchronous dynamics over large domains, which is out of accord with the stochastic nature of this theory; evidence that force is not proportional to the number of active cross-bridges; etc.

The newer paradigm suggested here is detailed in a book (Pollack, 1990). Some newer elements are considered in a review article (Pollack, 1996). In this paradigm all three filaments shorten. This shortens the sarcomere. I will briefly sketch some of the evidence that leads to this kind of mechanistic view.

CONNECTING FILAMENTS

Connecting filaments are elastic-like polymers. When unactivated muscles are forcibly stretched, connecting filaments bear the tension, and are strained. If the muscle is then released, connecting filaments recoil and the tension is dissipated. Connecting filaments are thereby responsible for the muscle's so-called resting tension. The resting tension ensures that the muscle, or sarcomere, returns to its natural length, and also keeps the thick filament centered in the sarcomere. All of this discussion refers to muscles that are not stimulated to contract actively; thus, the force range under consideration is relatively modest.

Behavior of the connecting filament has been elegantly revealed in experiments on isolated molecules of titin (Rief et al., 1997; Tskhovrebova et al., 1997). In the former study ramp length changes imposed on titin molecules brought about sawtooth-like tension changes. This implied a one-by-one unfolding of the filament's tandem immunoglobulin-like domains—each unfolding event giving rise to one "tooth." The length change occurs as the compact "beta-barrel" domain structure of the immunoglobulin domain gives way to an extended random structure (Fig. 5). Many such unfoldings lengthen the filament.

Figure 5. Unfolding of immunoglobulin domain results in discrete length change. Unfolding of successive domains yields large-scale length change of connecting filament.

In a conceptually similar experiment (Tskhovrebova et al., 1997) the titin molecule was stretched and held. During the holding phase, the stress-relaxation decline of tension occurred in stepwise fashion, not smoothly. Again, the implication is progression of discrete unfolding events, each one dropping the tension by an increment.

Stepwise length changes are also observed under similar conditions in intact sarcomeres (Yang et al., 1998). In these experiments the single myofibril is interrogated. The single myofibril is the smallest functional muscle unit that retains the natural structure of muscle (Fig. 6).

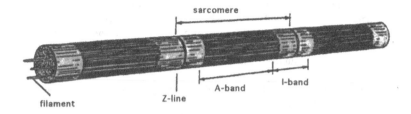

Figure 6. *Structure of an intact myofibril. A-band corresponds to thick filament. Connecting filaments lie in I-band.*

When the unactivated myofibril is stretched or released by a linear ramp, the sarcomere-length change is stepwise. An example is shown in Figure 7.

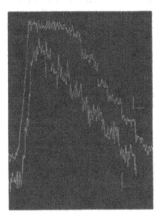

Figure 7. *Time course of sarcomere-length change during trapezoidal length change imposed on single isolated myofibril. Calibration bars: 20nm; 1sec.*

The observation of myofibril sarcomere steps confirms that discrete behavior persists at levels of organization considerably higher than that of the single molecule—for the myofibril sarcomere contains hundreds of filaments in parallel. In fact, discrete behavior remains detectable all the way up to the cellular level (Granzier et al., 1985), reinforcing the highly cooperative nature of the step.

The character of the step implies that it qualifies as a phase transition. First, the onset and termination are abrupt; it appears to have two states. Second the step is not an isolated event: the fact that it occurs over like domains of many different molecules at the same time implies a quasi-all-or-none behavior, which again is characteristic of a phase-transition. The simplest interpretation is that the phase-transition progresses collectively along parallel connecting filaments, by one domain at a time.

THICK FILAMENTS

The thick filament is built of successive repeats of the molecule myosin. The myosin molecule consists of a long alpha-helical coiled-coil rod-like tail, which lies in the thick filament backbone, and two globular heads or cross-bridges, which radiate from the filament at nominally right angles. The rods overlap in a staggered pattern. Hence, the filament surface shows a helical feature, and the helix is mirror-symmetrical about the filament's midpoint.

Although it is generally accepted that thick filaments do not change length during contraction, the evidence is mixed. Classical studies imply a constancy of filament length under most conditions. But some thirty-plus papers published since those classical studies report substantial shortening during contraction (for review, c.f. Pollack, 1983). These studies include many species, invertebrate and vertebrate muscles, heart and skeletal muscle. They employ many techniques ranging from electron microscopy to various types of optical microscopy. And they also include isolated thick filaments extracted from the muscle and exposed to physiological activating agents. The evidence for thick filament shortening is therefore appreciable.

A way in which thick filaments might shorten is if the myosin rods were able to slide past one another. The filament could then "telescope" to a shorter length. The driving force for such telescopic action would lie in the filament itself, presumably in the myosin rods that comprise the backbone. In this context it has been shown in many experiments that the alpha-helical coiled-coil rod is able to shorten by a helix-coil transition (for review c.f. Pollack, 1990). The helix-coil transition is a phase-transition in which the molecular structure undergoes radical change. It is a force-producing process: because the equilibrium length of the random coil is near zero, the coil will always want to shorten from the extended length. The retractive force depends on the degree of extension, much like a spring. Thus, the transition shortens the myosin molecule in much the same way as a wool sweater is shortened by excess heat. Here, however, the transition is reversible.

A model illustrating how localized shortening in a myosin

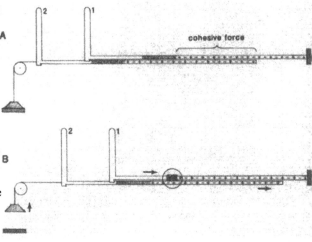

Figure 8: *Mechanism by which local shortening in one myosin rod can propel the adjacent rod to slide. Only two of the many molecule are shown.*

rod could generate rod sliding is shown in Figure 8. Adjacent molecules are held together by cohesive forces derived from molecular surface charges, which alternate semi-regularly along the surface. The unstable zone of one molecule near the midpoint of the filament undergoes helix-coil transition. (The transition occurs symmetrically on either side of the filament's midpoint.) The transition results in local shortening, which draws the remainder of the filament toward the mid-zone. Then the same shortening event occurs in the next molecule along the filament, and the next, etc. In such a way the filament shortens step by step. And because molecules are cross-linked to respective molecules on the next parallel filament, the transition is cooperative over the muscle cross-section.

This mechanism is in good agreement with structural evidence from X-ray diffraction (Huxley and Brown, 1967; Yagi and Matsubara, 1984). This X-ray pattern shows that during active sarcomere shortening the spacing between molecules along the thick filament does not change; only the intensity changes. This is precisely what is anticipated. The X-ray pattern is dominated by contributions from those molecules along the thick filament that have not yet shortened, for they remain regularly arrayed; others have shortened by variable amounts and do not therefore contribute significantly. As the number of transitioned molecules increases, the intensity contribution therefore diminishes. The model is therefore in good agreement with ultrastructural and X-ray diffraction evidence.

If thick filaments shorten during contraction, the likelihood, then, is that shortening occurs one molecule at a time in each half-filament. The filament will then shorten in steps. If the thin filament is bound to the cross-bridges during these filament-shortening steps, the sarcomere will likewise shorten in steps. In such a way the myosin phase-transition contributes to the shortening of the sarcomere.

THIN FILAMENTS

The thin filament may bring about stepwise length changes as well. Unlike the thick and connecting filaments, whose shortening can directly shorten the sarcomere, thin filaments are differently situated (Fig. 1). Because of their arrangement they would need to facilitate contraction in a different way. One potential mechanism is an inchworm-like process—similar to that reported in PAMPS gels (Osada and Gong, 1993). In this mechanism the long narrow gel undergoes repeated cycles of curling and straightening. The ends of the gel are hooked to a ratchet in such a way that each cycle results in a step advance. Repeated cycling advances the gel by an appreciable magnitude.

The same principle could apply in the actin filament. If the filament were to pass through cycles of curling and straightening, this action could be harnessed to propel the filament toward the center of the sarcomere. More realistically, such curling would arise locally from a phase-change, which would then propagate along the filament, to produce a worm-like reptation. Each cycle of propagation would result in a step of translation. The extent of translation would be a function of the number of waves. A schematic illustration of such an inchworm process is shown in Figure 9.

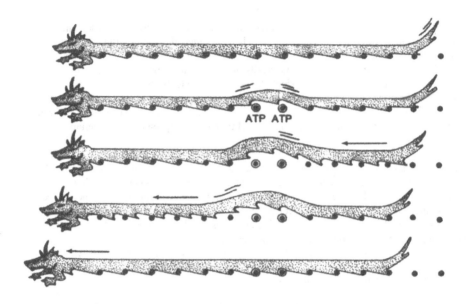

Figure 9. *Reptation model explaining translation of thin filament over thick filament. Black dots represent myosin cross-bridges. In this model a phase-transition propagates from tail to head, advancing the thin filament by one notch.*

The wave-like motion that would be anticipated in the actin filament is broadly observed (for review, *c.f.* Pollack, 1996). The evidence draws from as early as four decades ago when undulations were directly observed to propagate along actin-filament bundles responsible for active streaming. It also follows from modern studies of single actin filaments. Such motion could be generated by a local molecular structural change, which is observed morphologically (Menteret et al., 1991) and in actin crystals (Schutt and Lindberg, 1992). That structural changes can propagate along the filament is confirmed by the observation that the binding of a ligand to one end of the actin filament affects the physical and mechanical properties of the entire filament (Prochniewicz et al., 1996). Indeed, direct microscopic visualization of translating actin filaments shows a translation pattern quite strikingly characteristic of snake-like motion (deBeer et al., 1998).

Evidence for inchworm-like behavior is inferred as well from dynamics of the intact sarcomere. During active sliding of thin filaments past thick, the pattern of shortening is stepwise. The size of the step is an integer multiple of the linear spacing between actin monomers (Blyakhman et al., 1999). A histogram demonstrating this result is shown in Figure 10.

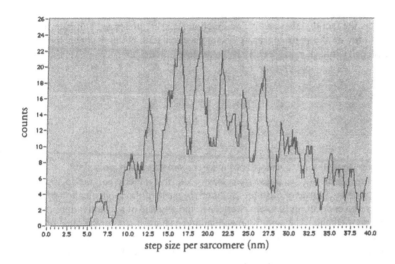

Figure 10. Histogram of step size measured during active contraction of single isolated myofibrils. Spacing between major peaks is 2.7 nm.

The histogram of Figure 10 shows multiple peaks, indicating steps of discrete size. Although inter-peak spacing is not precisely uniform, there is a strong tendency for peaks to repeat at regular intervals. The best-fit spacing is an integer multiple of 2.71 nm. This value is equal to the linear advance of actin monomers along the thin filament. Thus, the step is a function of the repeat of actin.

Such "quantal" behavior is inevitable if an actin monomer is bound to the (immobile) cross bridge between translation steps; the filament must then translate by n x the actin advance (Fig. 11). That actin and myosin bind to one another is well recognized, and it is presumed that such binding is responsible for the sustenance of active tension. Successive bindings give rise to the inevitable quantum step. The size of the quantum is consistent with the inchworm-like mechanism.

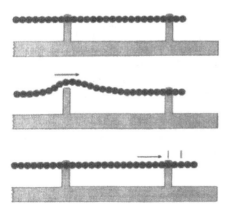

Figure 11. Inchworm advance. The model predicts that the size of the step advance will be an integer multiple of the actin monomer spacing.

245

Of particular interest is the fact that these active contraction steps are preserved at higher levels of organization. Not only do the steps appear in the single sarcomere, but they appear as well at the level of the single cell (Pollack et al., 1977). If the process is considered as a phase transition, the transition is, again, highly cooperative.

CONCLUSION

There is ample evidence that all three polymers of the sarcomere shorten in a discrete, cooperative manner—much the way as phase transitions occur in synthetic polymeric systems. The biological transition may, however, be a lot "smarter" because of the sophistication of its responsiveness. Thus, the connecting filament shortens in the absence of any activation; it behaves as a discrete elastic band. The actin filament undergoes transition after the level of activation crosses threshold. Because the actin filament, but not myosin, is found in relatively simple motile cells, the actin mechanism may be primitive. Like other primitive mechanisms, it is limited in its capacity and unable to function under too high a load—just as a weight hung on the caterpillar's tail can inhibit upward climbing, even though the caterpillar may still cling. Beyond this critical load, the operative agent is the thick filament, which can shorten under the highest of loads. Indeed, the helix-coil transition has been demonstrated to produce a force that accounts quantitatively for maximum force muscle can produce (Harrington, 1979). Thus, the muscle polymer gel capitalizes on the advantages of all proteins. This is perhaps why muscle, a highly evolved organelle, is as versatile as it is.

One upshot of these considerations is that the mechanism in muscle contraction is not as different from the mechanisms used to construct artificial muscles as had been thought. Both employ polymer-gel phase-transitions. Like the polymer gel, the muscle can be triggered by any variety of stimuli ranging from a change of pH, salt, temperature, solvent type, and even electrical current. It may therefore be possible to construct artificial muscles of greater speed and efficacy by looking closely at how the task of biological contraction is accomplished in nature. This remains for the future.

REFERENCES

1) P. H. W. W. Baatsen, K. Trombitas, and G. H. Pollack, "Thick filaments of striated muscle are laterally interconnected", *J. Ultras. & Mol. Str. Res. 97*, pp. 39-49, 1988.
2) K. Trombitas, P. H. W. W. Baatsen, and G. H. Pollack, "I-bands of striated muscle contain lateral struts", *J. Ultras. & Mol. Str. Res. 100*, pp. 13-30, 1988.
3) G. H. Pollack, Muscle and Molecules: Uncovering the Principles of Biological Motion, 1990.
4) T. Iwazumi, "A new field theory of muscle contraction", Ph. D. Dissertation, Univ. of Pennsylvania, 1970.
5) G. H. Pollack, "The cross-bridge theory." *Physiol. Reviews 63*, pp. 1049-113, 1983.
6) C. E. Schutt, and U. Lindberg, "Actin as a generator of tension during muscle contraction", *Proc. Nat'l. Acad. Sci. 89*, pp. 319-23, 1992.
7) A. Oplatka, "The rise, decline, and fall of the swinging crossbridge dogma", *Chemtracts Bioch. Mol. Biol. 6*, pp. 18-60, 1996.

8) A. Oplatka, "Critical review of the swinging cross-bridge theory and of the cardinal active role of water in muscle contraction", *Crit. Rev. Biochem. Mol. Biol. 32*, pp. 307-60, 1997.

9) M. Rief, M. Gautel, F. Oesterhelt, J. M. Fernander, and H. E. Gaub, "Reversible unfolding of individual titin immunoglobulin domains by AFM", *Science 276*, pp. 1109-12, 1997.

10) L. Tskhovrebova, J. Trinick, J. A. Sleep, and R. M. Simmons, "Elasticity and unfolding of single molucules of the giant muscle protein titin", *Nature 387*, pp. 308-12, 1997.

11) P. Yang, T. Tameyasu, and G. H. Pollack, "Stepwise dynamics of connecting filaments measured in single myofibril sarcomeres", *Biophys. J. 74*, pp. 1473-83, 1998.

12) H. L. M. Granzier, and G. H. Pollack, "Stepwise shortening in unstimulated frog skeletal muscle fibres", *J. Physiol. 362*, pp. 173-88, 1985.

13) H. E. Huxley and W. Brown, "The low angle X-ray diagram of vertebrate striated muscle and its behaviour during contraction and rigor", *J. Mol. Biol. 39*, pp. 383-434, 1967.

14) N. Yagi and I. Matsubara, "Cross-bridge movements during slow length change of active muscle", *Biophys. J. 45*, pp. 611-4, 1984.

15) Y. Osada, J. Gong, *Prog. Polym. Sci. 18*, pp. 187, 1993.

16) G. H. Pollack, "Phase transitions and the molecular mechanism of contraction." *Biophys. Chem. 59*, pp. 315-28, 1996.

17) J. F. Menetret, W. Hoffman, R. R. Schroder, R. Gapp, R. S. Goody, "Time-resolved cryo-electron microscopic study of the dissociation of actomyosin induced by photolysis of photolabile nucleotides." *J. Mol. Biol. 219*, pp. 139-44, 1991.

18) E. Prochniewicz, Q. Zhang, P. A. Janmey, D. D. Thomas, "Cooperativity in F-actin: Binding of gelsolin at the barbed end affects structure and dynamics of the whole filament." *J. Mol. Biol. 260*, pp. 756-66, 1996.

19) E. L. deBeer, A. M. A T. A. Sontrop, M. S. Z. Kellermayer, G. H. Pollack, "Actin-filament motion in the in vitro motility assay has a periodic component."*Cell Motil. Cytoskel. 38*, pp. 341-50, 1997.

20) F. Blyakhman, T. Shklyar, G. H. Pollack, "Quantal length changes in single contracting sarcomeres" *J. Mus Res. Cell Motility, 20*, 529-538, 1999.

21) G. H. Pollack, T .Iwazumi, H. E. D. J. ter Keurs, and E. F. Shibata, "Sarcomere shortening in striated muscle occurs in stepwise fashion", *Nature 268*, pp. 757-9, 1977.

22) W. F. Harrington, "On the origin of the contractile force in skeletal muscle."*Proc. Nat'l. Acad. Sci. 76*, pp. 5066-70, 1979.

BIOCHEMICAL SYNTHESIS AND UNUSUAL CONFORMATIONAL SWITCHING OF A MOLECULAR COMPLEX OF POLYANILINE AND DNA

Ramaswamy Nagarajan, Sukant K. Tripathy, Jayant Kumar
Lynne A. Samuelson[*], Ferdinando F. Bruno[*]
Department of Chemistry and Physics, Center for Advanced Materials, University of Massachusetts Lowell, Lowell MA 01854
[*]Natick Soldier Center, U.S. Army Soldier & Biological Chemical Command, Natick, MA 01760

ABSTRACT

Polyaniline (Pani) is one of the most interesting conducting polymers because of its promising electrical properties and unique redox tunability. Recently a new template assisted, enzymatic synthetic approach has been developed which yields a water-soluble and conducting complex of polyaniline and the template used. The mild reaction conditions in this approach allow for the utilization of delicate biological systems as template materials, such as DNA. Here we report the extension of this enzymatic approach to the synthesis of Pani on a DNA matrix. DNA serves as a suitable template for complexation and formation of conducting polyaniline. The redox behavior of Pani induces reversible conformational change in the secondary structure of DNA. Circular dichroism results suggest that the secondary structure of the DNA may be reversibly switched through manipulation of the redox state of the polyaniline.

INTRODUCTION

Polyaniline (Pani) has received significant interest due to its ease of synthesis, stability and interesting electrical properties. However the commercial application of this polymer has been limited due to the difficulty in obtaining the polymer in a processable form. Current investigations have been directed towards improving the processability, electrical properties, stability and environmental compatibility of Pani. Several electrochemical and chemical methods have been reported for the synthesis of Pani [1,2].

More recently chemical/electrochemical synthesis of Pani in the presence of an anionic polyelectrolyte has been reported [3,4]. The polyelectrolyte helps align the monomer molecules prior to the polymerization, provides the counter-ion for charge compensation in the polyaniline formed after polymerization and maintains water solubility. The low pH conditions involved in this approach however limits the choice of polyelectrolyte template to polymers that are stable in strongly acidic conditions. A biochemical approach towards the synthesis of polyaniline in near neutral pH conditions involves the use of an enzyme, horseradish peroxidase (HRP)[5]. Enzyme catalyzed polymerization of several phenols and anilines using HRP has been reported [6,7]. However in the absence of a polyelectrolyte matrix the polymer formed is of low molecular weight, highly branched and quickly precipitates out of solution. When appropriate template is present a linear, water soluble conducting form of polyaniline is obtained A detailed study of the enzymatic synthesis of Pani in the presence of an anionic polyelectrolyte, poly (sodium 4-styrene sulfonate) (SPS) has also been reported [8]. This polyelectrolyte assisted enzyme catalyzed approach provides the benefits of a benign synthetic route and opens new possibilities for the use of biological polyelectrolytes such as DNA as templates.

249

The electrostatic interaction between the phosphate groups (from DNA) and the protonated aniline was used in this study to align the aniline molecules prior to polymerization and promotes a para-directed coupling of the aniline molecules. The formation of Pani on DNA was found to induce changes in the secondary structure of DNA. The Pani also exhibits a template-guided macro-asymmetry. Conformational changes in DNA in the DNA-Pani complex could be accomplished by the reversible doping and dedoping of Pani.

EXPERIMENT

Horseradish peroxidase (HRP, EC 1.11.1.7) type II, (150-200 units /mg) solid was purchased from Sigma Chemical Co. (St. Louis, MO). Calf Thymus DNA was purchased from Worthington Biochemical Corporation (Freehold, N.J). Aniline monomer (purity 99.5%) and hydrogen peroxide (30% by weight) were purchased from Aldrich Chemicals. Inc, Milwaukee, WI, and were used as received. All other chemicals were of reagent grade or better.

Synthesis of DNA-Pani complex

A 1.0 mM calf Thymus DNA solution was prepared by dissolving the required amount of DNA in 10ml of sterilized 10 mM sodium citrate buffer maintained at pH 4. The solution was stored in the refrigerator for 48 hours prior to reaction. The concentration of DNA was determined by the UV absorbance at 258 nm. To this solution, 0.9μl (1mM) aniline and a catalytic amount of HRP (0.15 mg) were added. The polymerization was carried out by the drop-wise addition of hydrogen peroxide (0.098M), over a period of 240 seconds. In order to prevent the precipitation of the complex, the total amount of hydrogen peroxide was limited to $1/5^{th}$ of the stoichiometric amount, calculated with respect to the aniline concentration.

Characterization of Pani complexes

UV-Vis spectra and Circular dichroism (CD) spectra were obtained from the DNA-Pani solutions in a quartz cuvette with a path length of 1 mm using a HP diode array detector photometer (type HP8452A) and Jasco CD spectrometer J-720, respectively.

RESULTS

The use of poly (alkylene phosphates) as dopants mixed with Pani in N-Methylpyrrolidnone solution has been reported earlier [9]. However the Pani used in these cases was chemically polymerized before the introduction of the dopant. More recently Pani has been synthesized enzymatically in the presence of poly (vinyl phosphonic acid) (PVP) [10]. The Pani formed in this process was self doped and remained complexed with the PVP. The phosphate groups in PVP provide the counter-ion for charge compensation leading to a self-doped PVP-Pani complex. The PVP-Pani complex was redox active and could be reversibly doped and dedoped with acid and base respectively. It was believed that the phosphate groups in DNA would behave in a similar manner, and thus PVP was used as the proof-of- concept for extension of this enzymatic approach to DNA.

DNA-Pani

When aniline monomer is added to a DNA solution at pH 4.0, the aniline molecules become protonated and the electrostatic interaction between the protonated aniline and the phosphate groups in the DNA causes the anilinium ions to closely associate with the DNA. The interaction between aniline and other polyanionic matrices has been investigated and reported elsewhere [11]. The association of the protonated monomer on the DNA matrix facilitates a predominantly para-directed coupling and deters parasitic branching during the polymerization. The high proton concentration around the phosphate groups also provides a unique local environment that permits polymerization of aniline at a higher pH than the conventional chemical synthetic methods. Polymerization is accomplished using HRP and hydrogen peroxide. The UV-Vis spectra of the DNA-Pani complex recorded at 5 and 80 minutes after initiation of the polymerization are shown in Fig I. The UV-Vis spectrum obtained after 5 minutes indicates polaron absorption bands at 420 nm and 750 nm. The bipolaron band at 750 nm is attributed to the formation of pernigraniline in the initial stages of the reaction. The 750nm band diminishes after 80 minutes, while the 420 nm and 310- 320-nm bands increase in intensity indicating the formation of the emeraldine salt.

Fig I. UV spectra of DNA-Pani   **Fig II. CD spectra of DNA-Pani**

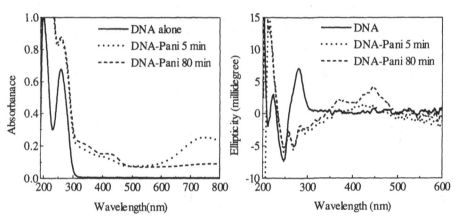

Conformational changes due to Pani formation

The CD spectra of DNA-Pani complex at 5 and 80 minutes are presented in Fig II. Prior to the formation of Pani, the DNA is present as the 'B' polymorph [12]. The CD spectrum of DNA-Pani complex obtained after 5 minutes indicates that the secondary structure of DNA changes significantly during the formation of Pani. The positive peak at 220 nm increases in intensity, while the negative peak at 245 nm reduces in intensity. The positive shoulder at 270 nm changes to a new negative peak with fine structure. There is very little change in the CD in the visible region. After 80 minutes, the CD spectrum does not change significantly in the region between 190nm and 300nm. However two broad, positive peaks appear at 365 nm and 445 nm. It can be concluded that significant changes in the DNA conformation occur within the first few minutes of the polymerization.

Conformational studies on calf thymus DNA reported earlier [12,13] indicate that, changes in secondary structure of the DNA in solution is influenced by the nature of the solvent and concentration of salt. The addition of large amounts of salt to a DNA solution has been known to cause a polymorphic transition, to the over-wound polymorph (from the 'B' form to 'C' form). The change in CD spectrum of DNA caused by the formation of Pani is similar to what is observed by the presence of inorganic salts. The UV-Vis spectrum indicates that the polyaniline exists in the emeraldine salt form, wherein DNA provides the necessary template environment to form conducting polyaniline. The shielding of charge on the phosphate groups by the Pani induces the change in the secondary structure of DNA leading to the formation of the over-wound polymorph. The emergence of CD peaks in the visible region indicates that the helical polyelectrolyte template (DNA) induces macroscopic order in the polyaniline. Chirality/optical activity has been observed in chemically and electrochemically synthesized complexes and colloids of Pani and (1R)-(-)-10-camphorsulfonic acid [14,15,16]. The macro-asymmetry in Pani observed in these cases has been explained in terms of the Pani chain adopting a preferred one-sense helical screw, which is caused by electrostatic and H-bonding interactions between Pani and the anionic dopant [17].

Conformational switching during doping-dedoping of Pani

The conformation of the DNA in the DNA-Pani complex can be switched with almost complete reversibility by doping and dedoping the polyaniline. The pH of the DNA-Pani solution was increased by the addition of a few microliters of NaOH solution. The UV-Vis spectra and the CD spectra were recorded at pH 4 and pH 9 [Fig III, IV].

Fig III. UV-Vis spectra of DNA-Pani on dedoping and redoping

Fig IV. CD Spectra of DNA-Pani on dedoping and redoping

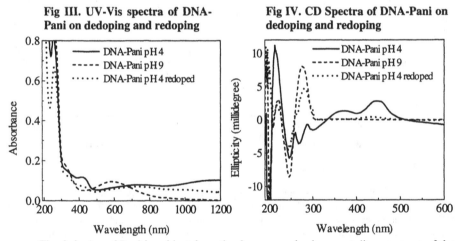

The dedoping of Pani is evident from the decrease and subsequent disappearance of the polaron bands at 404 and 755 nm in the UV-Vis spectrum. Simultaneously, the exciton transition of the quinoid rings at 564 nm and the π-π^* transition of the benzenoid rings at 320 nm emerge. The CD spectrum of DNA-Pani changes significantly during the dedoping process and at pH 9, the spectrum is very similar to the CD spectrum of DNA without the Pani. The neutralization of Pani minimizes the electrostatic interaction between the DNA and Pani resulting in the uncoiling of DNA, back to its native state [Fig II].

If the Pani is redoped by the addition of HCl, the charge on the Pani forces the DNA back to the over-wound form as indicated by the CD spectrum Fig [IV]. The recovery of the CD and UV spectra is partial and improves with time. The substantial loss in the CD peaks in the visible region may be explained in terms of significant loss of the macro-asymmetry in the Pani and the precipitation of the Pani molecules that were weakly complexed with DNA. This explanation can also be extended to the incomplete recovery of peaks in the DNA region of the CD spectra.

The dedoping and redoping process was repeated after 24 hours. The CD and UV spectra obtained after 24 hours after redoping shows improved recovery. The DNA-Pani solution was subsequently subjected to another dedoping and redoping cycle and the CD spectra were obtained indicate better conformational reversibility [Fig V]. The changes in CD bands at 285 nm, 275nm and 245nm were reported to be indicative of a polymorphic transition between the 'B', 'C' and "A" form of DNA. The average conformational recovery of the CD bands in this region for the second dedoping-redoping cycle is over 75%. Further, the significantly higher conformational recovery in the second cycle is presumably due to higher contribution from the complexed Pani to the doping and redoping process. In other words the Pani strands that are weakly complexed with DNA are lost during the first doping-dedoping cycle and the Pani that remains after the first cycle is substantially well complexed with the DNA. It is concluded that the observed changes in CD spectra are the consequence of conformational switching of DNA, between the loosely wound form containing 10.4 base pairs per turn (when the Pani is dedoped), and a tightly wound form containing less than 10 base pairs per turn (when Pani is doped).

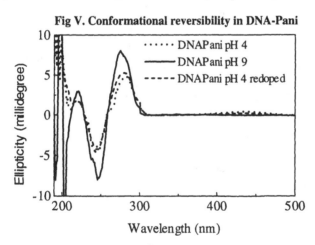

Fig V. Conformational reversibility in DNA-Pani

······ DNAPani pH 4
——— DNAPani pH 9
----- DNAPani pH 4 redoped

Ellipticity (millidegree)

Wavelength (nm)

CONCLUSION

An enzymatic route to the synthesis of a molecular complex of DNA and polyaniline is presented. The phosphate groups in the hydrophilic exterior of the DNA have been used to electrostatically attract the monomer molecules and promote a para-directed coupling. The DNA template provides the counterions necessary for charge compensation and maintains the Pani in the doped form. The helical template induces a macro-asymmetry in the Pani. The formation of Pani on DNA causes significant change in the secondary structure of DNA resulting in a more tightly polymorphic form of DNA. The conformation of DNA in the complex can be reversibly switched by doping and dedoping the Pani. These studies suggest exciting new possibilities in the use of this complex for the fabrication of nano-wires, biosensors, and diagnostic tools and for fundamental studies to "probe" the structure and function of DNA.

ACKNOWLEDGEMENTS

We thank Prof. David Kaplan and the Tufts Biotechnology center for assisting with the circular dichroism characterization and Dr. W. Liu for his technical assistance and helpful discussions.

REFERENCE

1. C.H. Yang and T. C. Wen *J. Electrochem. Soc.*, **141**, p. 2,624 (1994)
2. A.F. Daiz, J.A. Logan, J.Electroanal Chem, **111**, p. 111, (1980)
3. J -M. Liu, L. Sun, J. Hwang, H. S. C. Yang, *Mat. Res. Soc. Symp. Proc.*, **247**, p. 601 (1992)
4. L Sun, H. Liu, R. Clark, S.C. Yang, *Synth. Met* , **84**, p.67 (1997)
5. L.A. Samuelson, A. Anagnostopoulos, K.S Alva, J. Kumar, S.K. Tripathy, *Macromolecules.* **31**, p.4376-4378 (1998)
6. J.S. Dordick, *Enzyme Microb. Technol.*, **11**, p.194 (1989)
7. J. A. Akkara., K.J Senecal, D. L. Kaplan, *J. Polym. Sci., Polym. Chem.*, **29**, p.1591 (1991)
8. W. Liu, J. Kumar, S.K. Tripathy, K.J Seneacal, L. A. Samuelson, *J. Am.Chem. Soc.*, **121**, p.71-78 (1999)
9. Kulszewicz-Bajer, I.; Pretula, J.; Pron, A.; *J. chem. Soc. Chem. Commun.*, p. 641-642 (1994)
10. R.Nagarajan, S.K. Tripathy, L.A. Samuelson, F.F.Bruno, (unpublished)
11. W. Liu, A.L. Cholli, R. Nagarajan, S. K. Tripathy, F.F. Bruno, L.A. Samuelson, submitted to *J. Am.Chem. Soc* (1999)
12. C. A. Sprecher, W.A. Basse, W.C. Johnson, *Biopolymers*, **18** p. 1009-1018 (1979)
13. Bokma, J.T.; Johnson, W.C.; Blok, J.; *Biopolymers*, **26**, p. 893-909 (1987)
14. M. R. Majidi, L.A.P. Kane-Maguire, A. P. Leon, G.G. Wallace, *Polymer*, **36**, 18, p. 3597 (1995)
15. M. R. Majidi, L.A.P. Kane-Maguire, A. P. Leon, G.G. Wallace, *Polymer*, **35**, 14, p.3113 (1994)
16. J.N., Barisci, P.C. Innis, L.A.P. Kane-Maguire, I.D. Norris, G.G. Wallace, *Synth. Met.*, **84**, pp.181-183 (1997)
17. M. R. Majidi, S.A. Ashraf, L.A.P. Kane-Maguire, I.D. Norris, G.G. Wallace, *Synth, Met.*, **84**, p.115 (1997)

ENZYMATIC TEMPLATE SYNTHESIS OF POLYPHENOL

Ferdinando F. Bruno, Ramaswamy Nagarajan*, Jena S. Sidhartha, Ke Yang*, Jayant Kumar*, Sukant Tripathy* and Lynne Samuelson.
Materials Science Team, Natick Soldier Center, U.S. Army Soldier, Biological, Chemical Command, Natick, MA 01760; * Departments of Physics and Chemistry, Center for Advanced Materials, University of Massachusetts Lowell, Lowell, MA 01854.

Abstract

An alternative, biocatalytic approach for the synthesis of a new class of water soluble and processable polyphenols is presented. In this approach, the enzyme horseradish peroxidase (HRP) is used to polymerize phenol in the presence of an ionic template. The template serves as a surfactant that can both emulsify the phenol monomer and growing polyphenol chains and provide water solubility of the final polyphenol/template complex. This approach is a simple, one step synthesis where the reaction conditions are remarkably mild and environmentally compatible. The final product is a water soluble, high molecular weight complex of polyphenol and the template used. The approach is also very versatile as numerous templates may be used to build in specific functionalities to the final polyphenol complex. Polystyrene sulfonates (SPS), lignin sulfonate and dodecyl benzene sulfonates (micelles) are the templates investigated in this study. Thermal analysis and UV-Vis spectroscopy shows that these complexes have exceptional thermal stability and a high degree of backbone conjugation. Electrical conductivities on the order of 10^{-5} S/cm and third order nonlinear optical susceptibilities ($\chi^{(3)}$) of 10^{-12} esu are also observed. In the case of the SPS template, under certain conditions, a sol gel complex may be formed. This enzymatic approach offers interesting opportunities in the synthesis and functionalization of a new class of processable polyphenolic materials.

Introduction

Phenolic resins are of major importance for a wide variety of electronic and industrial applications including coatings, laminates, and bonding agents just to name a few [1, 2]. However, the toxic nature of the starting reagents (formaldehyde) and the difficult reaction conditions required for the synthesis of these polymers has seriously restricted their use in commercial markets [1, 2, 3]. To address these limitations, the use of enzymatic catalyst systems has been investigated. Horseradish peroxidase (HRP) was used under various reaction conditions (aqueous, mixed solvents, micelles) to catalyze the oxidative polymerization of phenol and aniline derivatives [4, 5, 6, 7]. Although these enzymatic approaches significantly improved the mildness of the reaction conditions, the polymers that formed were limited in their thermal, mechanical or/and electronic properties. Another approach involved enzymatic polymerization of phenol at the air-water interface of a Langmuir trough [8]. Although this approach improved these properties, large-scale production using this technique is not practical for commercial use.

In this study, an alternative enzymatic approach for the polymerization of phenol is presented that produces a water soluble, high molecular weight polymer, with improved thermal, mechanical and electronic properties. This approach was first demonstrated in our laboratories for the polymerization of aniline to form a water soluble and processable form of conducting polyaniline [9]. It involves the use of an ionic template that forms a local environment that serves to emulsify the monomer and growing polymer chains, minimize parasitic branching and complex with the polymer to maintain water solubility [10]. In this work, we extend this

Mat. Res. Soc. Symp. Proc. Vol. 600 © 2000 Materials Research Society

approach to the polymerization of phenol in the presence of several polyelectrolyte templates. The synthesis, characterization and properties of these new polyphenolic complexes will be discussed.

Experimental Section

Horseradish peroxidase (EC 1.11.1.7) was purchased from Sigma Chemical Co. (St. Louis, MO) as a salt free powder. The specific activity was of 240 purpurogallin units/mg solid. Phenol, polystyrene sulfonated (SPS, Mw 1.0M and 70K g/mol), dodecyl benzene sulfonate (DBSA), hydrogen peroxide (30% solution), phosphate buffer, and all the solvents (reagent grade or better) were purchased from Aldrich (Milwaukee, WI). Lignin sulfonate (Lignosol SFX-65) was obtained from Lignotech USA (Rothschild, WI). Enzymatic polymerization was carried out in a 10 mL of aqueous phosphate buffer (10 mM, pH 7.0), phenol (71.3 mM) and equimolar concentrations (with respect to the phenol) of SPS (based on the repeat unit) and hydrogen peroxide (no more than 20 mM added every 5 minutes). The HRP concentration was 0.1-0.15 mg/mL and was added prior to the hydrogen peroxide.

The reactions were carried out for 30 minutes at room temperature and the final products were dialyzed using Centricon concentrators (10,000 cut off, Amicon Inc., Beverly, MA). The samples were dried under vacuum at 50°C and stored until further analysis. The percent yield was typically 95% or higher. Control samples, using denatured enzyme, were prepared following the same procedure. Leaving in buffered water at 100°C for 30 minutes denatured the enzyme. The denatured HRP was then tested using purpurogallin, and was found to be inactive. Spectral characterization of the polymers and controls was performed with a Perkin-Elmer Lambda-9-UV-Vis-Near IR spectra-photometer (Norwalk, CT). FTIR spectra, of the samples deposited on ZnSe, were collected on a Perkin Elmer FTIR 1720X. Thermal gravimetric analysis (TGA) and differential scanning calorimetry (DSC) analysis were conducted using a TA instrument 2950 and a DSC instrument 2910 (New Castle, DE) respectively. The TGA and the DSC measurements were carried out under nitrogen at a rate of 10°C/min.

Static light scattering (SLS) measurements were performed on SPS, SPS-Phenol (control), SPS-Polyphenol polymer and DBSA-polyphenol using a Brookhaven instrument (Model SG-7B Rigaku Denki, Japan). A stock solution of SPS was prepared by dissolving an appropriate amount of polymer in 0.01 M of a filtered, pH 7.0 sodium phosphate buffered solution. The stock solution was sonicated for one hour at 25°C and then re-dialyzed for 24 hours. 0.1M of NaCl was added to the solution and filtered multiple times using a 0.45 μm filter. Similar procedures were followed for SPS-Phenol prior to and after polymerization. However, SPS-Polyphenol was dissolved in a 50:50 DMSO/water solvent mixture. A series of concentrations required for the measurements was prepared by diluting the stock solution with filtered buffer solution added directly into the scattering cell. Similar procedures were used for the DBSA/polyphenol system. SLS measurements were carried out using a homemade goniometer and a scattering cell chamber. The excitation source, a linearly polarized 35 mW He-Ne laser (Melles-Griot, Model No 05-LHP-927) operating at 632.8 nm, was housed on the goniometer fixed arm. The detection unit, which consisted of an Avalanche Photo Diode (RCA make, Model # SPCM-100), a discriminator and a pre-amplifier, was mounted on the goniometer horizontally rotating arm along with the collection optics. A standard cylindrical sample cell was placed centrally to a quartz vat containing the index matching liquid. This is a standard technique used to suppress the reflection from the cell walls. The angular distribution of scattered intensities, for different concentrations was obtained at an interval of 15°. The distribution ranged from 30° to 150°. Finally the data was analyzed using Zimm-plot software supplied by Brookhaven Instruments (Vers. 2.04). The increment in refractive index with respect to concentration was measured at the same wavelength as the excitation source for the SLS measurements using an interferometric refractometer (Wyatt technology, Model No. 7 Wyatt/optilab 903).

The electrical conductivity of all polyphenol samples was measured using a four-point probe. In each case the polyphenol complex was first dried in a vacuum oven at 60°C for 48 hours and stored in the desiccator until being compacted into a disc using a standard IR die in a hydraulic press. The discs were dried in the desiccator for several days before measuring the conductivity. Electro-absorption spectroscopy was carried out on a high molecular weight SPS-polyphenol film. The film was spin coated on an indium tin oxide (ITO) glass substrate and an aluminum electrode was evaporated on the top. A sinusoidal electric field (f = 500 Hz, 5V rms) was applied to the sample. A light beam, from a tungsten lamp passing via a monochromator, was incident normally on the sample. The electro-absorption signal, ΔI, which is defined as the change in the output intensity I, was detected by a lock-in amplifier (Stanford Research System SR830) set at twice the electrical modulation frequency (2f). The output intensity I (without the electric field) was measured by using a chopper. A microcomputer (AT-IBM) was used to synchronize the wavelength change of the monochromator and the data reading for the lock-in amplifier.

Results and Discussion

The enzymatic polymerization of phenol in the presence of the SPS template was determined using UV-Vis spectroscopy (Fig. 1). To monitor the progress of the reaction a series of different H_2O_2 concentrations was injected into the monomer solution. For all curves other than the phenol control, a large broad absorption tail is observed from 300 to 800 nm that is indicative of an extended degree of backbone conjugation. Similar absorption has been previously observed in the enzymatic polymerization of phenol using the LB technique [8]. These curves also show that as the amount of initiator (H_2O_2) is increased, the absorption intensity increases. It must be pointed out that large H_2O_2 injections were avoided to maintain high enzymatic activity. However, it was found that when the amount of H_2O_2 was as high as 100 or 200 µL, an SPS-polyphenol gel formed. These gels showed temperature stability up to 90°C where no phase transition, triggered by temperature, pH or salt concentration was observed. The gels could be dried but after drying they could not be redissolved in water. The dry polymer could only be redissolved in a DMSO/water (50/50 by volume) solvent mixture. Similar UV-Vis spectra were observed for the DBSA/polyphenol and Lignin/polyphenol however gel formation did not occur with these systems. Fourier transformer infrared (FTIR) spectroscopy of the lignin/polyphenol complex is given in Figure 2. Three important absorptions are observed in the spectra that support the formation of polyphenol in this enzymatic reaction. The first is a significant shift of approximately 70 cm^{-1} toward a lower frequency in the polymeric OH stretch region (3000-3500 cm^{-1}) with respect to the monomer.

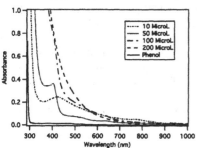

Figure 1. Phenol control and polyphenol synthesized with different concentration of H_2O_2.

This is most likely due to stronger hydrogen bonding in the polymer structure. The second notable peak is in the 1550-1650 cm^{-1} region and is assigned to conjugation of the double bond in the main chain.

The third peak of interest is in the 1130-1330 cm^{-1} region and is assigned to the presence of ether groups. This formation of ether linkages in the enzymatic polymerization of phenol has been previously observed [11]. Similar spectral features were found for the other systems.

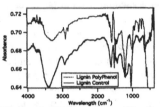

Figure 2. FTIR spectra of lignin and of lignin/polyphenol.

The molecular weight of the SPS/polyphenol complex was determined using SLS. As a control, the molecular weight of just the SPS template (10^6 Mw) was first determined. It was found that the molecular weight and the radius of gyration of the SPS were of 1.08 x 10^6 g/mol and 81 nm respectively, which was in good agreement with data provided by the supplier. When phenol monomer was added to the SPS (prior to polymerization) only a small increase in molecular weight (1.16 x 10^6 g/mol) and radius of gyration (87 nm) were observed. However, after polymerization, the molecular weight was found to increase to 1.29 x 10^7 g/mol and the radius of gyration showed a substantial decrease to 58 nm. These changes would be expected if an increase in the molecular interaction of the SPS with the phenol after polymerization. The Flory-Huggin's interaction parameter, which is a measure of goodness of the solvent, is 0.472, 0.466 and 0.485 for SPS, SPS-Phenol (before polymerization) and SPS-polyphenol (after enzymatic polymerization) respectively. This parameter value is below 0.5 for all three cases. This indicates that the solvent is good for all polymers [12]. Thus the molecular weight of the complex SPS-polyphenol is as high as 10M, but at this time we cannot differentiate the molecular weight of the polyphenol from that of the entire complex. The molecular weight of the DBSA/polyphenol micellar complex was found to 6.07 x 10^6 g/mol. The high molecular weight in this case cannot be attributed to the template and is thus believed to be due to a large degree of aggregation of the polyphenol within the micelles. These results show that the presence of an ionic template in these reactions leads to the formation of high molecular weight complexes.

The TGA in Figure 3 shows the thermal stability of the complex Lignin/polyphenol, e. g., after polymerization. It was found that a significant amount of complex (75-80%) remains after heating the polymer to 400°C. A first degradation is observed at 320°C, which is assigned to the beginning of lignin degradation and at 600°C over 65% residue is left. The residue at 600°C for the DBSA/phenol and SPS/Phenol systems was of 61% and 78% respectively. These results were similar to that found for insoluble polyphenols synthesized from organic media (5).

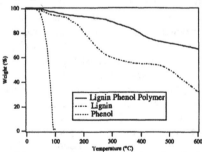

Figure 3. TGA of lignin, phenol and of lignin/polyphenol.

The electrical conductivity of the SPS/polyphenol was found to be 1.0 x 10^{-5} S/cm. The $\chi^{(3)}$ value for the SPS-polyphenol system was found to be 1.0x10^{-12} esu. Considering that a significant part of this complex is an electrically and optically inactive template, these are relatively high values and suggest substantial electronic conjugation of the polyphenol. These results are encouraging and warrant further investigations using other interesting templates that can contribute their own electrical and/or optical properties to the complex.

Conclusions

A new biological route for the direct synthesis of water soluble, high molecular weight polyphenol is described. This approach is advantageous in that it is simple, biochemically mild and requires minimal separation and purification. Varying the polyelectrolyte and concentration of enzyme initiator can be used to optimize the final polymer's properties. This process is general in that numerous phenol functional comonomers and polyelectrolytes may be employed to produce new phenolic resins. Furthermore, it is anticipated that suitable immobilization of the enzymes used with this approach could potentially lead to a facile, reusable cost effective and environmentally friendly approach for large scale production of processable phenolic resins.

Acknowledgments

FFB acknowledges funding under NAS grant and Dr. Mario Cazeca for electrical measuraments.

References

1. G. L. Brode, *Kirk-Othmer Encyclopedia of Chemical Technology*, (John Wiley & Sons: New York, 1982), Vol. 17, p. 384.
2. D. M. White, G. D., Cooper, *Kirk-Othmer Encyclopedia of Chemical Technology*, (John Wiley & Sons: New York, 1982), Vol. 18, p. 594.
3. J. J. Clary, J. E. Gibson, R. S. Waritz, *Formaldehyde: Toxicology, Epidemiology, Mechanisms*, (Decker, New York, 1983).
4. J. S. Dordick, M. A. Marletta, and A. M. Klibanov, *Biotechnol. Bioeng.*, **30**, 31, (1987).
5. J. A. Akkara, K. J. Senecal, D. L. Kaplan, *J. Polym. Sci. Part A: Polym. Chem.*, **29**, 1561, (1991).
6. S. Kobayashi, R. Ikeda, J. Sugihara *Macromolecules*, **29**, 8702, (1996) and ref. in the paper.
7. A. M. Rao, V. T. John, R. D. Gonzalez, J. A. Akkara, D. L. Kaplan, *Biotechnol. Bioeng.*, **41**, 531, (1993).
8. F. F. Bruno, J. A. Akkara, L. A. Samuelson, D. L. Kaplan, B. K. Mandal, K. A. Marx, J. Kumar, S. Tripathy, *Langmuir*, **11**, 889, (1995).
9.W. Liu, J. Kumar, S. Tripathy, K. Senecal, L. Samuelson, *J. Am. Chem. Soc.* **121**, 71, (1999).
10. W. Liu, A. L. Cholli, R. Nagarajan, S. K. Tripathy, F. F. Bruno, L. A. Samuelson, to be published in *J. Am. Chem. Soc.* (1999)
11. Higashimura, H.: Fujisawa, K.: Moro-oka Y.: Kubota M.: Shiga A.: Terahara A.: Uyama H.: Kobayashi S.: *J. Am. Chem. Soc.* **120**, 8529, (1998).
12. L. H. Sperling, *Physical Polymer Science*, (John Wiley & Sons: New York, 1986), p. 384.

THE CONSTITUTIVE RESPONSE OF ACTIVE POLYMER GELS

S.P. MARRA, K.T. RAMESH, A.S. DOUGLAS
Department of Mechanical Engineering
The Johns Hopkins University
Baltimore, Maryland 21218

ABSTRACT

Active polymer gels can achieve large, reversible deformations in response to environmental stimuli, such as the application of an electric field or a change in pH level. Consequently, great interest exists in using these gels as actuators and artificial muscles. The goal of this work is to characterize the mechanical properties of ionic polymer gels and to describe how these properties evolve as the gel actuates. Experimental results of uniaxial tests on poly(vinyl alcohol)-poly(acrylic acid) gels are presented for both acidic and basic environments. These materials are shown to be to be slightly viscoelastic and compressible and capable of large recoverable deformations. The gels also exhibit similar stress in response to mechanical deformation in both the acid and the base.

INTRODUCTION

An active polymer gel is composed of a crosslinked polymer network and interstitial fluid. Environmental changes, such as a change in the pH of the fluid or the application of an electric field, can result in a shrinking or swelling of the gel. Large volume changes (as large as 500 percent for some gels) accompanied by relatively low densities (near that of water) have made these gels of great interest in a variety of applications. Many devices have been developed which incorporate a polymer gel as an actuating device. Examples of these include finger-like gripping devices and camera lens dust-wipers.[1, 2] Polymer gels are also being developed for internal drug delivery systems and other medical applications.[3]

Much research has been conducted on active polymer gels, especially in the area of swelling kinetics.[4, 5] Some constitutive models have been developed to describe the mechanical properties of these materials. However these models are typically limited to small deformation theory and swelling from the dry state or only in response to concentration changes.[6]

The goal of this work is to characterize the mechanical properties of polymer gels undergoing large elastic deformations, and to describe the coupling of this behavior with the actuation (shrinking/swelling) response due to environmental changes. The specific polymer gel considered is poly(vinyl alcohol)-poly(acrylic acid) (PVA-PAA), and actuation is induced through changes in the pH of the interstitial fluid.

POLYMER GEL FABRICATION

The polymer gel used in this work is fabricated from poly(vinyl alcohol) (PVA) and poly(acrylic acid) (PAA). PVA-PAA gels have been studied previously by a number of researchers including Shin et al.,[7] Chiarelli et al.,[8] Genuini et al.,[9] Caldwell and Taylor,[10] and Chiarelli and De Rossi.[6]

PVA-PAA gel fabrication is conducted on-site and begins by separately dissolving 1 gram each of PVA and PAA in deionized water. Once both polymers are dissolved, the solutions are combined and mixed thoroughly. The combined solution is then poured into a clean glass pan where it remains until all of the water has evaporated. A thin (<100μm) film of PVA and

PAA is then peeled off the glass pan and placed in an oven at ~145°C for 30 minutes, during which time crosslinking occurs. Finally, the film is boiled in a 0.6M sodium hydroxide solution for 30 minutes, which improves the magnitude of actuation.

The resulting gel film shrinks in acidic solutions and swells in basic solutions. The gel film can also be stretched considerably while remaining elastic (such behavior is called finite-elasticity). Unfortunately, the film is very fragile and tears quite easily, so careful handling is required.

EXPERIMENTAL TECHNIQUES

The mechanical characterization of a finite-elastic material requires a combination of loading conditions. To achieve this, a biaxial testing system was developed which can deform a thin film specimen in two directions simultaneously.

The biaxial testing system consists of four stepper motors which drive four linear actuators. In a uniaxial test, two of these actuators are attached to a dog-bone shaped specimen through end-clamps (see Figure 1). For biaxial tests all four actuators are attached to the specimen. As the actuators move away from each other, the specimen is stretched. Submersible load cells are used to measure the forces in each direction produced in deforming the specimen. Actuator motion trajectories are prescribed using motion control software and the load cell data is acquired using LabView software.

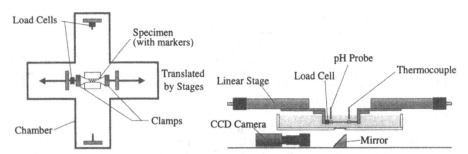

Figure 1 – Schematic of the biaxial testing system. The plan view on the left shows the cross-shaped environmental chamber and the specimen mounting for a uniaxial test (the linear actuators are not shown). The side view on the right shows a pair of linear actuators (only two are used for a uniaxial test) and the imaging system.

Prior to testing, ink markers are applied to the specimen. As the specimen is deformed, these markers move with the gel. A CCD camera and frame-grabber card are used to record the positions of the markers during a test and to save the images to a hard disk. After the test is complete, these images are processed to determine the actual deformation of the specimen. The deformations and load cell data are used to calculate the stretches (current length divided by original length) and stresses in the specimen.

The specimen is suspended inside a polyethylene chamber which may be filled with either acidic or basic solutions. During a test the environmental temperature is measured with a thermocouple and the pH of the testing fluid is measured with a pH meter.

RESULTS

A PVA-PAA gel specimen in 0.1M citric acid and in 0.5M sodium bicarbonate is presented in Figure 2. The ratio of volume in the base (swollen state) to volume in the acid is approximately 2.25.

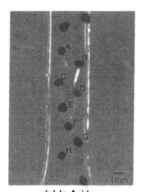 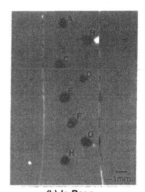

(a) In Acid   *(b) In Base*

Figure 2 – A PVA-PAA uniaxial test specimen (a) in 0.1M citric acid and (b) in 0.5M sodium bicarbonate. Dots of ink have been applied to the surface of the gel to track the deformation during a test. The ratio of volume in the base (swollen state) to volume in the acid is approximately 2.25.

Uniaxial sinusoidal extension tests were performed on separate PVA-PAA gel specimens in 0.1M citric acid and in 0.5M sodium bicarbonate. All specimens were the same size, with cross-sectional dimensions 5mm wide by approximately 100μm thick and a gauge length of 20mm.

The stretches in the extensional (λ_1) and lateral (λ_2) directions during a test in the acid are presented in Figure 3. These values were determined from the marker positions. The volume ratio (current volume divided by original volume), calculated from the stretches and assuming isotropy, is also presented. The slight increase in volume ratio during the stretching deformation indicates that the gel is not incompressible. The asymmetry in the volume ratio curve during each cycle is a result of a slight phase difference between the stretches in the extensional and lateral directions due to the viscoelasticity of the gel. This may be attributed to the viscoelasticity of the polymer network itself and/or to the relative movement of the fluid within the network.

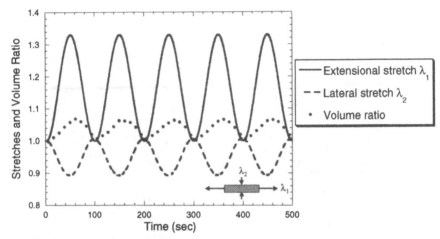

Figure 3 – Stretches and volume ratio versus time for a PVA-PAA gel specimen stretched uniaxially in 0.1M citric acid.

The stretches during a test in the base are presented in Figure 4. Note that these stretches are caused by the *mechanical* deformation only, and prior to testing the gel already had an effective stretch due to the swelling from the acid state (the *swelling* stretch was measured as approximately 1.31 in each direction).

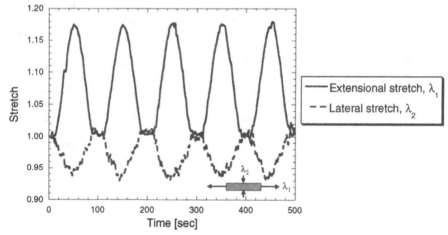

Figure 4 – Stretches (due only to the mechanical deformation) versus time for a PVA-PAA gel specimen stretched uniaxially in 0.5M sodium bicarbonate.

The Cauchy stress versus extensional stretch for separate PVA-PAA specimens in acid and in base are presented in Figure 5. Both specimens appear to respond similarly for the same amount of stretch. Both specimens clearly exhibit large elastic deformations (although the gel cannot be stretched as far in the base without rupturing). The relatively low stresses produced during these deformations is evidence that these gels are very compliant. The slight hysteresis in the stress for the gel in acid is a further indication of the viscoelasticity of the gel. Similar hysteresis may occur for the gel in the base, but the scatter in the data may prevent identification of the hysteresis.

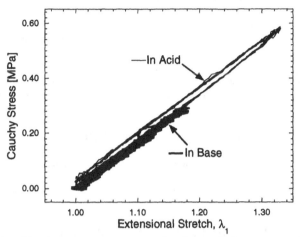

Figure 5 – Cauchy stress versus extensional stretch for separate PVA-PAA specimens in 0.1M citric acid and in 0.5M sodium bicarbonate (multiple cycles are presented).

DISCUSSION

For the same mechanical stretch, both the gel in the acid and the gel in the base give similar stress responses. So for the same cross-sectional area, both gels produce similar forces. However, the gel in the base is swollen; the polymer network has expanded and more fluid has entered the gel. Hence, for the same cross-sectional area, there is less polymer in the swollen gel. This indicates that the polymer network is actually stiffer in the base than in the acid.

A possible reason for the increase in stiffness of the network in the base is that in the swollen state, the polymer side groups are ionic and repel one another. This mutual repulsion reduces the number of possible configurations for the polymer segments, thus reducing the entropy of the network. Some uncoiling of the polymer segments may also occur during the swelling process, thus further reducing the entropy. From classic rubber elasticity theory[11] it can be shown that as the entropy decreases, the stiffness of the network increases.

ACKNOWLEDGEMENTS

Funding has been provided by the Army Research Laboratory (contract number DAAL01-96-2-0047 and through the Army Research Office (contract number DAAH04-95-2-0006). Funding has also been provided through the National Defense Science and Engineering Graduate Fellowship Program.

REFERENCES

1. T. Kurauchi, T. Shiga, Y. Hirose, A. Okada, in *Polymer Gels Fundamentals and Biomedical Applications*, edited by D. DeRossi, K. Kajiwara, Y. Osada, and A. Yamauchi (Plenum Press, New York: Plenum Press, 1991), pp. 237-246.

2. Y. Bar-Cohen, S. Leary, M. Shahinpoor, J.O. Harrison, J. Smith, SPIE Smart Structures and Materials Symposium, Electroactive Polymer Actuators and Devices, Newport Beach, CA, 1999.

3. A.S. Hoffman, in *Polymer Gels Fundamentals and Biomedical Applications*, edited by D. DeRossi, K. Kajiwara, Y. Osada, and A. Yamauchi (Plenum Press, New York: Plenum Press, 1991), pp. 289-297.

4. T. Tanaka and D.J. Fillmore, Journal of Chemical Physics, **70**, pp. 1214-1218, (1978).

5. A. Peters and S.J. Candau, Macromolecules, **21**, pp. 2278-2282, (1988).

6. P. Chiarelli and D. DeRossi, Progress in Colloid and Polymer Science, **78**, pp. 4-8, (1988).

7. H.S. Shin, S.Y. Kim, Y.M. Lee, Journal of Applied Polymer Science, 65, pp. 685-693, (1997).

8. P. Chiarelli, P.J. Basser, D. DeRossi, and S. Goldstein, Biorheology, **29**, pp. 383-398, (1992).

9. G. Genuini, G. Casalino, P. Chiarelli, and D. DeRossi, in *NATO ASI Series*, vol. F57, edited by G.E. Taylor (Springer-Verlag , Berlin, 1990) pp. 341-360.

10. D.G. Caldwell and P.M. Taylor, International Journal of Engineering Science, **28**, pp. 797-808, (1990).

11. L.R.G. Treloar, *The Physics of Rubber Elasticity*, 3rd ed. (Clarendon Press, Oxford, 1975).

ELECTROACTIVE NONIONIC POLYMER GEL
-SWIFT BENDING AND CRAWLING MOTION -

T. HIRAI*, J. ZHENG, M. WATANABE, H. SHIRAI, M. YAMAGUCHI
*Faculty of Textile Science and Technology, Shinshu University,
3-15-1 Tokida, Ueda-shi 386-8567, Japan, tohirai@giptc.shinshu-u.ac.jp

ABSTRACT

We have found a nonionic polymer gel swollen with nonionic dielectric solvent can be actuated by applying an electric field. The motion was not only far much faster than conventional polyelectrolyte gel materials, but also far much bigger in deformation. The motion was completed, for instance, within 60 ms and the deformation reached over 100%. The heat loss is negligible compared to that of polyelectrolyte gels. The deformation is not only bending, but also crawling. The principle was suggested to be charge-injected solvent dragging in the gel. The force was suggested to be proportional to the square of the electric field and proportional to the dielectric constant of the solvent. The principle was suggested to be promising and applicable to other conventional polymers.

INTRODUCTION

Electroactive polymer actuators have been attracted strong attention as one of the smart materials, particularly, as a possible candidate for an artificial muscle. One of them is polymer gel, which is swollen with solvent[1]. Polymer gels have been considered to possess high potential as an artificial muscle through their soft textures and their largely deforming property. As electroactive polymer gels, polyelectrolyte gels have been extensively investigated since they contain ionic species that are directly affected by the electric field[2]. These gels have been pointed out, however, have some difficulties in practical application, such as poor durability, electrochemical consumption on the electrodes. The electrochemical reactions on the electrodes accompany large electric current or heat generation.

On the other hand, nonionic polymer gels swollen with nonionic solvent have hardly been investigated as electroactive polymer materials. Poly(vinyl alcohol) gel swollen with dimethylsulfoxide have been demonstrated to be actuated very efficiently by applying a d.c. electric field[3-4]. The phenomenon was explained as an electrostrictive one, although the detailed mechanism was remained uncertain.

In this paper, we present the other features of the electroactive nonionic polymer gel materials. They are swift bending and crawling motions. The motion speed and the magnitude of the deformations are much larger than those reported on polyelectrolyte gel materials. The concept will be suggested to be applicable to the other conventional nonionic polymer materials.

EXPERIMENT

Materials

Poly(vinyl alcohol) (PVA) was purchased from Kuraray Co. The degree of polymerization and the degree of saponification are 1700 and 98% (in mol), respectively. The residual acetate groups in the PVA were completely saponified with 12N-sodium hydroxide aqueous solution. After the purification, the saponified PVA was used for the gel preparation.

Poly(vinyl chloride) (PVC) (degree of polymerization = 1100) was purified by reprecipitation using tetrahydrofurane as a good solvent and methanol as a poor solvent. Other chemicals were reagent grade, and used without further purification.

PVA gel preparation was carried out by endowing chemical crosslinks in physically crosslinked PVA gel. First, physically crosslinked PVA gel was prepared from a solution of PVA, DMSO, and water. The composition was PVA: DMSO: water=10:72:18 in weight. By keeping the solution at -20 °C, the solution was easily changed into a physically crosslinked gel. Thus prepared physical gel was immersed overnight in a glutaraldehyde solution, and then an aliquot 1N-HCl solution was added into the solution containing the physical gel and warmed at 30 °C. After 30 min reaction, the chemically crosslinked gel was thoroughly rinsed with water. The chemically crosslinked gel was immersed in methanol for deswelling and then in DMSO for re-swelling. By repeating this deswelling and re-swelling process three times, the chemically crosslinked DMSO gel was served for measurements.

Preparation of plasticized PVC was carried out by conventional method, using tetrahydrofurane as a solvent and dioctylphthalate as a plasticizer.

Measurements

Deformation induced by a dielectric field was recorded either by CCD camera for large deformation or by laser stain gauge for small deformation. During the deformation measurements, the electric current was also recorded at the same time.

Raman spectra measurement was carried out using a Renishow Raman microscope spectrometer. The laser beam used for excitation was He-Ne laser.

RESULTS AND DISCUSSION

PVA-DMSO gel

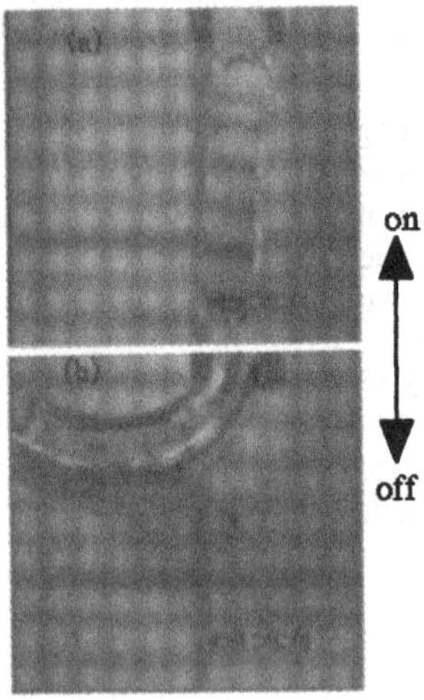

Bending Motion: PVA gel whose DMSO content is more than 98wt% was found to be actuated efficiently by applying dc field. The size of the gel was 10 mm in length, 5 mm in width, and 2 mm in thickness. In Figure 1, PVA gel whose surfaces were coated with gold electrodes was shown to bend at the field strength of 500 V/2mm within 60 ms attaining the bending angle of about 90°. The strain was held stable as far as the field was on. The electric current observed at the steady state was ca.30 µA.

This performance is much more efficient than that observed in Nafion membrane. Heat generation is less than 1/10 than those of the polyelectrolyte gels. Under the experimental condition, chemical consumption was not observed. The gel was found to be actuated in air, and seems to be stable at least for a couple of hours although the durability is not enough.

Dependence of the bending curvature on the applied electric field was shown in Figure 2. The curvature is proportional to the square of the applied electric field. The strain induced in the bending deformation was found to exceed 110% in outer surface, and 50% in inner surface, as the volume of the gel was kept constant during the deformation and the thinning deformation in the field direction was 8%. With increasing the field strength, the gel can bend

Figure 1. Bending motion of PVA-DMSO gel induced by dc electric field. 500 V/2mm

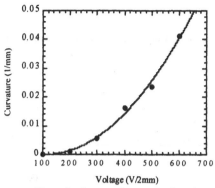

Figure 2. Curvature is proportional to the square of an electric field.

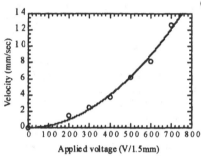

Figure 3(a). Dependence of flow velocity (mm/s) of DMSO on an applied field (V/mm).

nearly 180°. Nevertheless, large deformation breaks not only the thin gold electrodes but also the gel itself.

For the clarification of the bending deformation, we investigated the electric property of the solvent and the solution, and found that the electric field application to the dielectric solution induces (1) the flow and (2) sharp asymmetrical pressure gradient in the solution. Figure 3 illustrate the features. (1) Flow rate of DMSO was investigated using a pair of comb-like electrodes, the distance between which is 2 mm, and was found to be proportional to the square of an applied electric field as shown in Figure 3(a). (2) When the field was applied to a pair of electrode plates, the distance between which is 2 mm, DMSO was pulled up between them. (Figure 3(b)) The electrodes were vertically suspended, and the bottom tips were dipped in the liquid of DMSO. The height is proportional to the square of the applied field. The liquid surface is almost symmetric, although slight gradient seems to exist. When similar experiment was carried out on a PVA-DMSO solution (PVA content is 2 wt%), the feature is very different from that of DMSO solvent as shown in Figure 3(c). The solution was found to climb up onto the cathode. Similar process can be expected to happen in the PVA-DMSO gel, causing the extreme pressure gradient in the gel and the remarkable deformation of the gel shown above.

Solvent-pulling-up phenomenon can be explained by the following relationship.

$$h = \frac{1}{2\rho g}(\varepsilon - \varepsilon_0)E^2 \qquad (1)$$

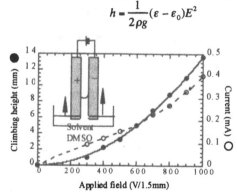

Figure 3(b). Dependence of DMSO level and an electric current on an applied field.

where h, ρ, ε, ε_0, and E are the height of liquid pulled up, density of the liquid, dielectric constant of the liquid, dielectric constant of air, and the electric field applied, respectively.

The curve shown in Figure 3(b) was calculated according to the above equation, suggesting that the equation (1) can quantitatively explain the phenomenon. While in the case of PVA-DMSO solution, the explanation is no more available. We explained the phenomenon in Figure 3(c) as follows: The injected charge is stabilized by being transferred from DMSO to PVA or entrapped by PVA, thus is slightly depressed the discharging rate. For quantitative explanation, further investigation is required.

In order to explain the pressure gradient electrically induced in the solution, we employed the idea of charge injected solvent dragging deduced on the dielectric solution[5].

$$\Delta p = \frac{9}{8}\varepsilon(\frac{V - V'}{d})^2\frac{x - x_0}{d} \qquad (2)$$

where Δp, V, x, and d are pressure generated, voltage applied, distance between the electrodes, respectively. V' and x_0 are fitting parameters to optimize complicated processes which might encounter on the vicinities of the electrodes. In equation (2), complicated processes such as turbulent flow, charge transfer among the chemical species in the solution and so on are not took into account. Although equation (2) can qualitatively explain the result shown in Figure 3(a), the equation does not explain the phenomena shown in Figure 3(c).

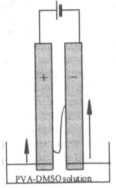

Figure 3(c). Electrically induced asymmetric climbing phenomenon of PVA-DMSO solution.

We applied equation (2) for PVA-DMSO gel, though the diffusion of DMSO in the gel under an electric field is not clarified yet. The pressure Δp estimated for pure DMSO is 3.6 kP and we assumed this value is applicable for the gel, since the gel contains 98 wt% of DMSO. In Figure 1, the bending angle can reach nearly 90°, the strain induced on outer surface was estimated to reach 113% and that on inner surface 55% by taking the thickness decrease of 8% into account[6], and these estimation was consistent with those observed in Figure 1. The value of the change in thickness was obtained from our previous paper.

By applying bending modulus of 10 kP and the strain on the gel surface, the stress produced between the inner and outer surfaces is estimated to be 5.8 kP. This value is a little bigger than 3.6 kP, suggesting that the presence of PVA and its network structure plays critical roll for the effective huge deformation of the gel.

Crawling Motion: As PVA-DMSO gel was shown to be electrically actuated by the charge-injected solvent drag, a different type of motion can be expected to be induced by changing electrode alignment. We put a PVA-DMSO gel on an electrode array as shown in Figure 4. The gel crawled by applying an electric field. The motion was explained as follows: The quick swelling on the cathode edge raised the gel upward, and this motion results in the crawling to the anode side since the gravity center of the gel remains on the anode side in this particular case. This is a novel type of motion of the gel, and suggests the new type application.

Orientation of DMSO: As DMSO has a dipole moment in is chemical structure, we can expect the effect of its orientation on the deformation process[7]. We investigated the effect of the electric field on the orientation of DMSO by using laser Raman spectroscopy. It turned out that the orientation induced at rather low electric field like 25 V/mm, and reaches maximum at around 60 V/mm. Over this value of the electric field, the orientation is depressed. The decrease of the DMSO orientation seems to be accompanied by the vigorous motion induced by the liquid flow. The deformation of the gel becomes remarkable over this value of the electric field. Therefore, we have to conclude the orientation can not be the origin of the deformation.

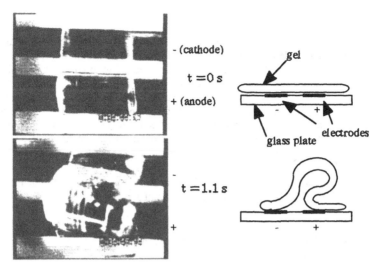

Figure 4. Crawling motion of PVA-DMSO gel induced by placing it on an array of electrodes fixed on a glass plate. Width of electrode is 3 mm. Distance between the electrodes is 2 mm. Gel length is 10 mm. Applied field was 250 V/mm, and the current observed was 60 µA.

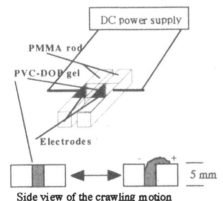

Side view of the crawling motion

Figure 5. Crawling motion of PVC-DOP gel. Thickness of the gel is 2 mm. Width of the electrodes are 10 mm.

PVC-DOP gel

PVC plasticized with large amount of DOP could also show an interesting motion as shown in Figure 5. The DOP content of the gel shown in this paper is 90 wt%. The gel was placed between the aluminum electrodes which are tightly fixed on the poly(methyl methacylate) rod. When the field was applied the gel crawls out onto the anode surface like a tongue. Once the field is off, the "tongue" restores its original position instantly. The moving rate is much slower than that observed in PVA-DMSO gel. The flow of DOP in the gel was difficult to detect. The crawling phenomenon is somehow similar to that observed in PVA-DMSO solution in Figure 3(c).

Although we will show the detailed results on PVC-DOP gel elsewhere, we would like to point out that the phenomena observed on PVA-DMSO gel and the discussion made on it is applicable to other conventional polymers and their gels.

CONCLUSIONS

PVA gel swollen with DMSO was found to be actuated very efficiently by applying a dc electric field. Bending angle of ca. 90° was shown to be attained within 60 ms. The constant deformation was sustained as far as the field was on. The gel restored its original shape as soon as the field was off. By using the electrode array, crawling motion could be induced. The orientation of DMSO did not induce the large strain treated in this paper. The mechanism of the motion was suggested to be charge-injected solvent drag. The injected charge was suggested to be transferred to PVA and caused the effective deformation. The concept could be applied to PVC, and the PVC-DOP gel was shown to crawl by an electric field.

ACKNOWLEGMENTS

This work was partially supported by a Grant-in-Aid for COE Research (10CE2003) by the Ministry of Education, Science, Sports and Culture of Japan.

REFERENCES

1. D. DeRossi, K. Kajiwara, Y. Osada, and A. Yamauchi, Polymer Gels. Fundamentals and Biomedical Applications, Plenum(1991).

2. Y. Osada and M. Hasebe, Chem. Lett., p. 1285 (1985). Y. Osada, H. Okuzaki, and H. Hori, Nature, 355, 242 (1991).

3. T. Hirai, H. Nemoto, T. Suzuki, M. Hirai, and S. Hayashi, J. Appl. Polym. Sci., 53, p. 79 (1994).

4. M. Hirai, T. Hirai, A. Sukumoda, H. Nemoto, Y. Amemiya, K. Kobayashi, and T. Ueki, J. Chem. Soc., Faraday Trans. 91, 473 (1995).

5. O. M. Stuetzer, J. Appl. Phys. 30, p. 984 (1959).

6. T. Hirai, J. Zheng, and M. Watanabe, will be submitted.

7. T. Hirai, Jianming Zheng, and M. Watanabe, Proceedings of SPIE 3669, p. 209, SPIE's 6th Annual International Symposium on Smart Structures and Materials, Electro-active Polymer Actuators and Devices, Newport Beach, CA 1999.

Effect of the Pore Formation on the Solution Flow through Acrylamide Gel

H. TAMAGAWA, S. POPOVIC, M. TAYA
Department of Mechanical Engineering, University of Washington, Box 352600, Seattle, WA
98195-2600, U.S.A, tayam@u.washington.edu

ABSTRACT

A simple synthesis method of porous acrylamide gel is proposed. Pore formation in gel body
is dominated by the amount of polymerization initiator, accelerator and gelation temperature.
We investigated the influence of these factors on the number and the size of pores. Gelation
temperature control is the easiest and the most effective way to create a number of large pores in
a gel body. We also investigated the pore volume fraction dependence of solution flow through
the gels. Solution flow rate was found to be promoted with the increase of pore volume fraction.

INTRODUCTION

Gel actuator is one of the future practical gel products, which has fascinated the people in the
field of polymer science for long years. However, it is well known that volume change of gels is
so slow, since it is dominated by solution diffusion [1]. Such a property stagnates the realization
of gel actuators. Creation of tiny pores in gel body is one method to design faster responsive
gels [2, 3], since they will shorten the solution diffusion path. Kabra et.al synthesized porous
PVME gels by applying γ-ray to the pregel solution and extensively investigated its volume
change property [2]. Wu et.al synthesized macroporous polyNIPPAm gels by the use of HPC
(hydroxypropylcellulose) as a pore formation agent [3]. They mixed pregel solution with HPC
and the HPC was removed from the gel body after the polymerization. Neither method is simple. γ-ray radiation method needs γ-ray facility, which is not commonly accessible, while HPC
method implies a difficulty in removing the HPC from the polymerized gel. In this paper, we
propose a simple method of a porous acrylamide gel synthesis and investigate the controlling
parameter of pore formation and the solution flow rate through the porous gels.

EXPERIMENT

a. Porous gel synthesis and controllability of the pore size and the number

porous gel synthesis Porous acrylamide gels were prepared by the following method, where all
chemicals were purchased from Aldrich (Milwaukee, WI) and used without further purification.
The pregel solution which consists of acrylamide (monomer), N,N'-methylenebisacrylamide
(crosslinker), N,N,N',N'-tetramethylethylenediamine (accelerator) and deionized water (solvent)
is heated up for 30 minutes. Once ammonium persulfate (polymerization initiator) is added to it,
immediately the gelation and pore formation started simultaneously and they were completed
soon. This method is different from the conventional gel synthesis method that the pregel solution in the present method is heated up completely without adding polymerization initiator [4].
controllability of the pore size and number The number and size of pores are greatly influenced by the amount of polymerization initiator, accelerator and gelation temperature. We investigated these effects on the pore formation. All gel specimens described below were prepared
by following the porous gels synthesis procedure described above, and their composition and
gelation temperature are summarized in Table1.

Mat. Res. Soc. Symp. Proc. Vol. 600 © 2000 Materials Research Society

i) **initiator** Four different porous gels, **G-I^i**, **G-II^i**, **G-III^i** and **G-IV^i**, were prepared by varying the amount of initiator used, 0.08g, 0.16g, 0.24g and 0.60g, respectively.

ii) **accelerator** Four different porous gels, **G-I^a**, **G-II^a**, **G-III^a** and **G-IV^a**, were prepared by varying the amount of accelerator used, 0.03g, 0.06g, 0.09g and 0.30g, respectively. **G-I^a** is identical to **G-I^i**.

iii) **temperature** Four different porous gels, **G-I^T**, **G-II^T**, **G-III^T** and **G-IV^T**, were prepared by varying the gelation temperature, 50, 60, 70 and 80°C, respectively.

Table1 The synthesis conditions of porous gel specimens[†1]

	G-I^i	G-II^i	G-III^i	G-IV^i	G-I^{a}[†1]	G-II^a	G-III^a	G-IV^a	G-I^T	G-II^T	G-III^T	G-IV^T
I[†2]	0.08	0.16	0.24	0.60	0.08	0.08	0.08	0.08	0.16	0.16	0.16	0.16
A[†3]	0.03	0.03	0.03	0.03	0.03	0.06	0.09	0.30	0.06	0.06	0.06	0.06
T[†4]	70	70	70	70	70	70	70	70	50	60	70	80

[†1] The amount of monomer, crosslinker and solvent used for the synthesis of all the gels shown in this table are 11.56g, 0.154g and 100g, respectively. **G-I^a** is identical to **G-I^i**.

[†2] I : initiator / g

[†3] A : accelerator / g

[†4] T : gelation temperature / °C

b. Effect of pore volume fraction on solution diffusion through gel

We investigated the diffusion coefficient of iodine solution flowing through porous gels.

specimen preparation By varying the gelation temperature, gel sheets (thickness = 4.97mm) with different volume fraction of pores were prepared. To study the solvent diffusion through the gels, we made use of iodostarch reaction. Thus for the preparation of gel specimens, starch particles and saturated iodine solution were used in addition to the ingredients of gel. They were purchased from Aldrich (Milwaukee, WI) and Fisher Scientific (Pittsburgh, PA), respectively. Monomer (11.56g), crosslinker (0.154g), accelerator (0.09g) and starch particles (0.56g) were dissolved in deionized water (100g). The pregel solution was heated up at given gelation temperature for 30 minutes with a gel mold followed by addition of initiator (0.24g).

measurement of diffusion coefficient The method of measuring diffusion coefficient, D, is explained here. Immediately after the synthesis of porous gel sheet, it was cut into a 3cm-length × 1cm-width × 4.97mm-thickness strip. Its weight, W_{gel}, is measured to calculate the pore volume fraction later. And it was sandwiched between the acrylic plates with silicone rubber sealants (Fig.1). The edge of the gel is dipped in the iodine solution (Fig.2) where this moment is defined as t = 0 (t : time). Due to the iodostarch reaction, the gel strip changes its color from transparent to blue. The diffusion distance of the iodine solution, L, was measured as a function of time at 20°C (Fig.2). If the pore volume fraction is quite large, the gel volume decreases immediately after the synthesis due to the gas release from the pores, namely pores flatten. Therefore in such a case, the gel was immersed in the deionized water until it was fully swollen in order to open those flatten pores completely again, then followed by the experiment described above.

Next, the method of measuring the pore volume fraction is explained. First, it is necessary to define what pores are. Even if no pore formation reaction is induced in acrylamide gel body, initially it may consist of voids. Namely, water swollen acrylamide gel consists of small volume of molecular chains and large volume of voids filled with the deionized water. The gaps among molecular chains are considered to be all voids (Fig.3(a)), and hereafter such a type of void is called "micro pore". The pores created by the pore formation reaction described earlier are different type of voids. Their sizes are large compared with micro pores (Fig.3(b)). Hereafter this type of voids created by the present process is called "macro pore".

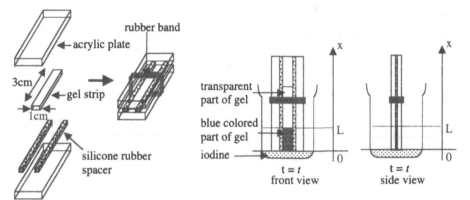

Fig.1 Illustration of specimen preparation   Fig.2 Illustration of experimental setup

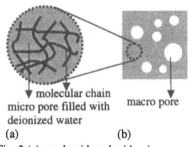

(a)   (b)

Fig. 3 (a) acrylamide gel with micro pores
(b) acrylamide gel with micro and macro pores

The method to calculate the volume fractions of micro and macro pore, which are designated as Vf_{micro} and Vf_{macro}, respectively, is explained, where the total pore volume fraction is given by $Vf_{total} = Vf_{micro} + Vf_{macro}$. The total gel weight is the same as total weight of gel ingredients, 112.604g, and its volume can be measured experimentally, 108cm³, as long as there exist no macro pores. Therefore the density of gel without macro pores, ρ, is 112.604g ÷ 108cm³ = 1.0426g cm⁻³. If there is no macro pores in gel specimen, its volume, V_{gel}^{o}, should be 3cm × 1cm × 4.97mm = 1.491cm³.

However, if there exist macro pores in the gel specimen, its volume without the macro pores, V_{gel}, should be given by eq(1).

$$V_{gel} = W_{gel} \div \rho = \frac{W_{gel}}{1.0426}$$ (1)

Therefore Vf_{macro} is given by eq(2).

$$Vf_{macro} = \frac{\left(V_{gel}^{o} - V_{gel}\right)}{V_{gel}^{o}} = \left(1.491 - \frac{W_{gel}}{1.0426}\right) \bigg/ 1.491$$ (2)

Since all micro pores are filled with deionized water, the volume ratio of micro pores to the gel specimen except for macro pores should be same as the volume ratio of total deionized water (100cm³) used for the total gel volume prepared (108cm³), which is 0.9259. Therefore Vf_{micro} is given by eq(3).

$$Vf_{micro} = 0.9259\left(1 - Vf_{macro}\right)$$ (3)

RESULTS and DISCUSSION

a. Porous gel synthesis and controllability of the pore size and the number

   (a)   (b)

Fig.4 (a) pregel solution   (b) pore formation

porous gel synthesis Fig.4 is photos of the pore formation and the gelation process of acrylamide gel, where the white bar on the bottom of the beaker is a stir bar. This gel consists of 11.56g of monomer, 0.154g of crosslinker, 0.75g of accelerator, 100g of solvent and 0.08g of initiator. Gelation temperature was 75°C. Although the pregel solution is transparent, Fig. 4(a), the polymerized gel color becomes opaque due to the pore formation and growth, Fig. 4(b).

controllability of the pore size and number

    i) **initiator** Fig.5 is a photograph of a typical shape of pores created in **G-I^i**. The shapes of pores created in **G-II^i**, **G-III^i** and **G-IV^i** are the same as that of **G-I^i**, i.e. spherical. It was difficult to obtain the pore size distribution, since some pores diameter are too small to be measured from the optical microscope image, and we could not obtain clear image of **G-IV^i** due to the creation of too many pores in a small region, either. However mainly the pores of 20 ~30 μm diameter were observed in **G-I^i**, **G-II^i** and **G-III^i**, and the number of pores appears to increase in the order of **G-I^i**, **G-II^i**, **G-III^i**, **G-IV^i**. The number of pores is not proportional to the actual amount of initiator used. **G-III^i** and **G-IV^i** contain far larger number of pores than **G-I^i** and **G-II^i**. This might be due to the slow gelation speed of **G-I^i** and **G-II^i** for trapping the created bubbles in them due to small amount of initiator used, while the amount of initiator used for **G-III^i** and **G-IV^i** is large enough for gas bubbles trapping. In fact, the gelation speed becomes faster with increase of initiator amount. The initiator does not appear to influence the pore growth so significantly, but the number of pores certainly becomes larger with higher amount of initiator. Therefore the initiator must have worked as a pore nucleus. And the creation of small gas bubbles, which grow into pores, is observed, as soon as the initiator is added to the pregel solution. This evidence also supports the conclusion that the initiator works as a pore nucleus.

Fig.5 A photo of a typical pore shape created in **G-I^i**

    ii) **accelerator** After addition of initiator to pregel solution, gas bubbles were created. They grew gradually and the gel became opaquer with time. Due to the same reason described above, it is not possible to obtain the size distribution of pores, but **G-I^a**, **G-II^a** and **G-III^a** appear to contain mainly the pores of 20 ~30 μm diameter, whereas **G-IV^a** appears to contain mainly the pores of 50 ~ 60 μm diameter. This evidence suggests that the pore diameter increases with the accelerator.

In order to assure this conclusion, another gel, **G-V^a**, was prepared by using larger amount of accelerator (0.90g). Fig.6 shows the photographs of typical pores created in **G-V^a**, indicating larger pore size than the pores created in other gels. This evidence strongly supports the conclusion that the accelerator promotes the pore growth. It is noted in Fig.6(a) and (c) that there exist unshaped (non-spherical) pores. They were created due to the continuation of pore growth even after the completion of gelation. After the completion of gelation, pores could not grow into pure spherical shape, instead they grow in the direction along which the gel body strength is the weakest, namely less polymerized region. The creation of unshaped pores is also attributed to the coalescence of pores with one another. The higher amount of accelerator does not seem to increase the number of pores. It is concluded that the accelerator works as a pore growth agent.

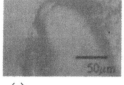

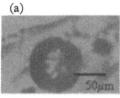

(a)

(b)

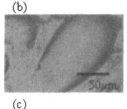

(c)

Fig.6 Photos of typical pores created in **G-V^a** (a) and (c) are the typical unshaped pores, while (b) is spherical in shape.

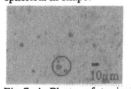

Fig.7 A Photo of typical pores created in **G -I^T** The circle diameter is 30μm.

iii) temperature We could not obtain the size distribution of the pores, and we could not obtain the clear image of **G-IV^i** due to the creation of too many pores in a small region. However, it was clear that the pore size increased with the temperature. G-IT contains mainly the pores less of 10 μm in diameter. **G-II^T** and **G- III^T** appear to have mainly the pores of 20 ~ 30 μm in diameter. But **G-IV^T** contains quite large pores of around 100 microns. This is due to the pore expansion, often resulting in the rupture of gel body of **G-IV^T**. We also observed many unshaped pores in **G- IV^T**. This is attributed to the same reason associated with **G-V^a**. For the synthesis of **G-IV^T**, the amount of initiator and accelerator used were only 0.16 and 0.06g, respectively, but there exist many huge sized pores. Among **G-II^T, III^T** and **IV^T**, there is a trend that the number of pores per unit volume increases with temperature. However, only **G-I^T** does not follow this trend. The number of pores in **G-I^T** is larger than that in **G-II^T** and **G-III^T**. This must be due to the extra smallness of pores created in **G-I^T**. Fig.7 is the photograph of pores created in **G-I^T**. Many smaller pores exist in the narrow region. If the temperature is raised, each pore will grow and merge together. For example, three pores circled in Fig.7 are located in a narrow region where the circle diameter is only 30μm, which is almost equivalent to the size of pores created in other gels. Once they grow, three pores presumably would coalescence to form one unshaped pore.

Generally speaking, heating up the pregel solution, especially at and above 80°C to induce gel expansion, is very effective way to create a number of large pores, even though the amount of initiator and accelerator is small.

b. Effect of pore volume fraction on solution diffusion through gel Fig.8 shows the gelation temperature dependence of Vf_{macro}. It is clear from Fig.8 that the dependence of Vf_{macro} on gelation temperature is so critical above 80°C. Without a mold, Vf_{macro} should become larger, since it prevent the pores from growing largely.

The diffusion distance of iodine tracer, L, is plotted as a function of (time)$^{1/2}$, $t^{1/2}$, in Fig.9. All data exhibit approximately linear relationship between L and $t^{1/2}$, which suggests one-dimensional diffusion. It follows $x = \sqrt{2Dt}$. Thus by the least square analysis on Fig.9, absolute value of diffusion coefficient is calculated and summarized in Table2. Obviously, D increases with Vf_{macro}.

Vf_{total}, Vf_{micro} Vf_{macro} and D are summarized in Table2. The data shown in this table are represented as a diagram in Fig.10. It follows from Fig.10 that diffusion coefficient drastically increases with the increase of Vf_{macro} as described above, despite that Vf_{total} remains constant and

Vf_{micro} decreases. This evidence suggests that the creation of larger pores is quite effective to enhance the solution flow rate in gel. Otherwise, the solution is strongly trapped by gel network.

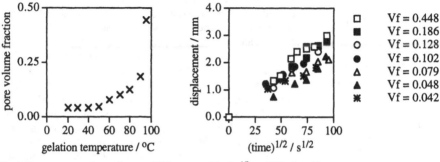

Fig.8 Temperature dependence of Vf_{macro} Fig.9 $t^{1/2}$ vs diffusion distance

Table2 The volume fraction of the micro, macro and total pores and the diffusion coefficient

Vf_{micro}	0.37	0.76	0.85	0.88	0.92	0.93	0.93
Vf_{macro}	0.60	0.18	0.09	0.05	0.01	0.00	0.00
Vf_{total}	0.97	0.94	0.93	0.93	0.93	0.93	0.93
$D^{\dagger}$	5.45	4.81	4.21	3.92	2.21	2.42	2.21

$\dagger \times 10^{6} cm^{2}s^{-1}$

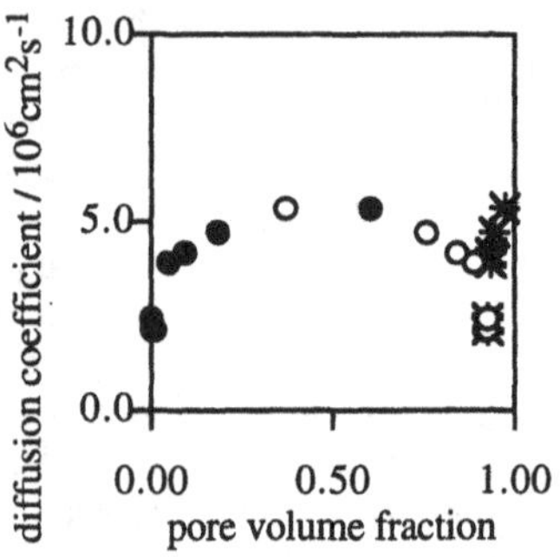

CONCLUSION

It was found from the present study that heating up the pregel solution to higher temperature is the most effective way to create a number of large pores.

The most effective way to promote the solution flow through gel is creation of larger size pores as well as increasing the volume fraction of pores.

Fig.10 Pore volume fraction dependence of diffusion coefficient ● : Vf_{macro} vs D, ○ : Vf_{micro} vs D, ✳ : Vf_{total} vs D

ACKNOWLDGMENTS

This work was conducted under the supports from National Sceince Foundation (EEC 9700705) and NEDO/RIMCOF, Japan.

We would like to express our gratitude to Dr.Tsutomu Mori (Department of Mechanical Engineering, University of Washington) for his valuable suggestions and enduring teaching.

REFERENCES

1. T. Tanaka and D. J. Fillmore, J. Chem. Phys. **70**, p. 1214-1218 (1979).
2. B. G. Kabra, M. K. Akhtar and S. H. Gehrke, Polymer **33**, p. 990-995 (1992).
3. X. S. Wu, A. S. Hoffman and P. J. Yager, Polym. Sci. Part A Polym. Chem. **30**, p. 2121-2129 (1992).
4. T. Tanaka, I. Nishio, S-T. Sun, S. Ueno-Nishio, Science, **218**, p. 467-469 (1982).

Composites and Others

HIGH DIELECTRIC CONSTANT POLYMER CERAMIC COMPOSITES

Y. BAI, V. BHARTI, Z.-Y. CHENG, H. S. XU, Q. M. ZHANG
Materials Research Laboratory, Pennsylvania State University, University Park. PA 16802,

ABSTRACT

A new polymer-ceramic composite, using the newly developed relaxor ferroelectric polymer that has a high room temperature dielectric constant as the matrix, is reported. Different kinds of ceramic powders were studied and homogeneous composite thin films (20μm) were fabricated. It was observed that the increase of the dielectric constant of the composites with the ceramic content could be described quite well by the expression developed by Yamada *et al.*, when the ceramic content is below 60% by volume. The experimental data shows that the relative dielectric constant of composites using PMN-PT powders can reach more than 250 with weak temperature dependence (i.e., the dielectric constant changes little in a broad temperature range). In addition to high permittivity, the composite prepared in clean environment also has high breakdown field strength (120MV/m), which yields an energy storage density more than 14J/cm^3. The dielectric behavior of the composite at various frequencies was also studied and the results show that the material is promising for high frequency applications.

INTRODUCTION

Ferroelectric ceramics, such as barium titanate ($BaTiO_3$), with very high dielectric constants, are widely used in the applications of capacitors and high-density energy storage devices. However, the fragility, inflexibility and poor processibility, which are inherent to ceramics, limit their applications. Using polymers to replace ceramics has been the focus of many studies in the past several decades, since they are flexible, easy to process and possess higher breakdown strength. But their dielectric constants are very low. Currently, the polymeric capacitors, which account for nearly half of the market, are made from polystyrene, polypropylene, polyester and polycarbonate. None of them has a dielectric constant higher than 10. For applications of capacitors and energy storage devices, it is highly desirable to develop polymer-based materials with high dielectric constant to achieve high volume efficiency and energy storage density. In past twenty years, there has been a great interest in the development of polymer ceramic composites with high dielectric constant [1,2]. However, due to the low dielectric constant of polymer matrix, the relative dielectric constants of the ceramic powder-polymer composites (0-3 composites) developed up to date can only reach about 60 [3,4,5]. It is clear that the dielectric constant of a composite is determined by the dielectric constants of its polymer and ceramic components and their volume percentages [6,7] and past experimental results indicate that the main limitation to the high dielectric constant composite is the low dielectric constant of the polymer matrix. A recent research on P(VDF-TrFE) copolymer found that electron irradiation with proper dosage can increase its dielectric constant up to 50 with a broad peak [8]. In addition, it is well known that relaxor ferroelectric material PMN-PT has much higher dielectric constant compared with traditional ferroelectric materials such as PZT and BT. These results offer new opportunities to prepare 0-3 composite with high dielectric constant.

EXPERIMENT

To prepare 0-3 composite, P(VDF/TrFE) 50/50 mol% copolymer (supplied by Solvay and Cie, Belgium) was dissolved in methyl-ethyl ketone (MEK), and a proper amount of PMN-PT ceramic powders (PMN-85, TRS Ceramics, Inc.) with average particle diameter of 0.5 μm was added into the solution. The suspension was stirred for 12 hours at room temperature. Then it was put in an ultrasonic bath for 30 minutes before poured onto a glass plate placed in a clean bench. The plate was kept in the clean bench and maintained at room temperature for 1 hour. It is sufficient for the evaporation of the solvent and a composite film was left on the plate. The film was heated in a vacuum oven at 70°C for 12 hours in order to remove any remaining traces of the solvent. Then it was folded to an average area of 3×2 inches and melt pressed at 170°C, 15,000lbs. The thickness of the pressed thin film was about 20 micron. Finally, the sample was annealed at 140°C in vacuum oven for 12 hours and allowed to slowly cool down to room temperature. The volume percentage of the ceramic varied from 10-60%. Those films with optimal perpetuities were irradiated at 120°C and the energy of the electron source was 2.55MeV and various dosages (40, 60 and 80MRad) were used to compare the effects of irradiation on composite. The film was cut to small pieces of 5×5mm for measurement.

In order to characterize the dielectric property of the composite, circular gold electrodes with 3mm radius were sputtered in the center on both sides of each sample. The permittivity and dissipation factor of the composite versus temperature were measured at 100, 1k, 10k and 100kHz using a dielectric analyzer (DEA 2970, TA Instrument). The frequency dependence of dielectric constant and dissipation factor at a constant temperature were measured by means of an impedance analyzer (Model HP4194A, Hewlett-Packard) in the range from 1k to 100MHz. The dielectric constant was calculated as a parallel plate capacitor. In order to evaluate the breakdown strength, the sample was put in silicon oil and applied a DC voltage using a high voltage supply (Model 610D, Trek).

RESULTS

Figure 1 shows the variation of relative dielectric constant with volume percentage of ceramic PMN-PT for 0-3 composite using P(VDF-TrFE) 50/50 copolymer as matrix (un-irradiated). As expected, the dielectric constant increased with volume fraction of ceramic filler. There have been many efforts in developing theoretical model for 0-3 composites [9,10,11]. In Figure 1, the theoretical curve calculated from the expression developed by Yamada et $al.$ [9], is also presented. This model relates the ε value of the composite with the relative dielectric constant of its two components as:

$$\varepsilon = \varepsilon_p \left[1 + \frac{nq(\varepsilon_c - \varepsilon_p)}{n\varepsilon_p + (\varepsilon_c - \varepsilon_p)(1+q)} \right] \qquad (1)$$

where ε_p and ε_c are the relative permittivity of the polymer matrix and the ceramic particles, respectively, q is the volume fraction of ceramic and n is a parameter related to the geometry of ceramic particles [10].

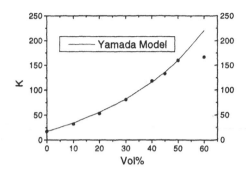

Figure 1 Variation of ε (f=100Hz) with the volume percentage of the ceramic at 25°C

The ε_p is 17 at 100Hz at room temperature, which is measured by a dielectric analyzer (DEA 2970, TA Instrument). The values of ε_c and n, which best fit the theoretical values calculated form Equation 1 with those experimental results, are 1400 and 10.6, respectively. After irradiated with 40 Mrad, ε_p changes to 46, while n keeps same since it is only related with the geometry of ceramic powders. The value of ε_c back calculated from the dielectric constant of irradiated sample is 1360. This result indicates that the irradiation does not have significant effect on the dielectric properties of the ceramic filler, which is consistent with previous study [12].

From the Figure 1, we can see that the experiment results fit very well with the equation. But when volume percent increases up to 60%, the measured dielectric constant of composite becomes much lower than the theoretical value. This can probably be explained by two ways. First the increase of volume percent of ceramic will lead to agglomeration of ceramic powders, which will make the composite not homogeneous, so in some part of composite, the polymer matrix will break and no longer continuous. In the other word, the composite does not have 0-3 type configurations any more. Secondly, the high volume fraction of ceramic filler in composite may result in a sharp rise in porosity of the material, which will reduce the relative dielectric constant of the composite. Therefore, in this study, only the samples with 50% volume fraction of loading were chosen for further study.

Frequency dependence of dielectric constant and tanδ of the composite irradiated with 40Mrad at 120°C is shown in Figure 2 (a). It is evident that the dielectric absorption of the composite with a maximum at 1MHz is a simple relaxation process, as shown in Figure 2 (b), the Cole-Cole plot is symmetric with respect to a line parallel to the ε″ axis.

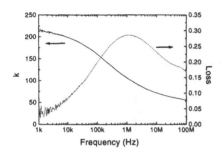 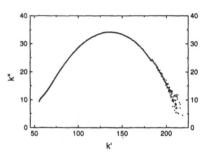

| (a) Frequency dependences of dielectric constant and tanδ | (b) Cole-Cole representation of complex permittivity |

Figure 2 Dielectric behavior of PMN-PT/P(VDF-TrFE) composite at high frequency (40Mrad)

The curve of dielectric constant fits Davidson-Cole empirical dependence of ε on frequency:

$$\varepsilon = \varepsilon_\infty - \frac{\Delta\varepsilon}{1 + (i\omega\tau)^\alpha} \qquad (2)$$

where ε_∞=50.832, $\Delta\varepsilon$=173.2, α=0.484, τ=61.69μs. The composite has a dielectric constant of 50 at 100MHz and this value is comparable to those of current materials used in microwave applications. Moreover, with increased temperatures, the relaxation frequency $1/\tau$ moves progressively to higher frequency as shown in Figure 3. The data in Figure 3(a) also show that the relative dielectric constant of the composite can be higher than 150 at 10MHz, when measured at 90°C and above.

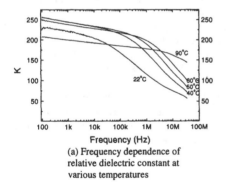 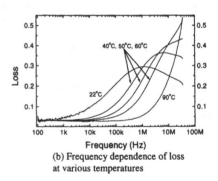

| (a) Frequency dependence of relative dielectric constant at various temperatures | (b) Frequency dependence of loss at various temperatures |

Figure 3 Dielectric behavior of PMN-PT/P(VDF-TrFE) composite at various temperatures (1kHz)

In addition to relatively high dielectric constant, the irradiated composite also exhibits relatively weak temperature dependence. That is, the dielectric constant remains high over a broad temperature range as shown in Figure 4. The data in Figure 4 present the temperature dependence of the dielectric of 50vol% PMN-PT irradiated P(VDF-TrFE) composite. Further more, by adjusting the dosage. The level of the dielectric constant and the flatness of temperature dependence can also be varied. These can be observed in Figure 4, where the dielectric constant for composites irradiated at different dosages is shown.

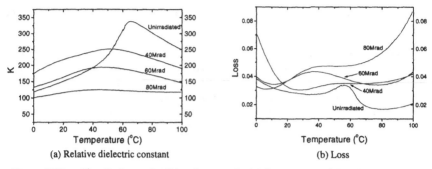

(a) Relative dielectric constant   (b) Loss

Figure 4 Effect of irradiation on the dielectric properties (1kHz) of PMN-PT/P(VDF-TrFE) composite

As expected from the irradiation effect on P(VDF-TrFE) copolymer, by varying the irradiation dosage, the peak of dielectric constant is moved from 65°C to 35°C. The dielectric peak becomes broader while the dielectric constant decreases with increased irradiation dosage.

We also characterized the irradiated composite for possible application of electric energy storage devices. In that application, the maximum stored energy density is an important parameter. For a linear dielectric, the maximum stored energy is

$$U = \frac{1}{2} K \varepsilon_0 E_{max}^2 \qquad (3)$$

where K is the relative dielectric constant and E_{max} is the breakdown field of the material. The breakdown fields of the composites prepared in previous work are lower than 80 MV/m [13], which is low compared with pure polymer. The low breakdown field may be due to several reasons. One of the reasons could be due to inhomogeneous distribution of the ceramic powder, which could result in a high local electrical field. The situation would be even worse when some ceramic powders agglomerated together. In addition, the residue solvent and dust introduced during the processing will also dramatically reduce the breakdown field. In order to reduce these effects, several steps were taken in the processing to improve the quality of the composites. Suspension was put into ultrasonic bath before the mixture was poured on glass substrate to improve the dispersion of ceramic powders. The composite films were prepared in a clean bench (class 1000). The solvent was allowed to evaporate in the clean bench for a longer time. After these modifications, the breakdown field of the irradiated composite thin film (40 Mrad), with the thickness of 20 microns, can be improved to 120MV/m, while the dielectric constant remains same. Using Equation 3 to calculate the maximum stored energy of the composite, where K equals 220 at room temperature, we get the value of 14J/cm³. It is higher than those of ceramic multilayer capacitors.

CONCLUSIONS

In the present study, a relaxor-relaxor composite (both PMN-PT ceramic and irradiated copolymer are relaxor ferroelectric materials) was studied. Making use of the high dielectric constant polymer matrix in a recently developed irradiated P(VDF-TrFE) copolymer, a high dielectric constant 0-3 composite has been developed. It was also observed that the properties of the composite can be changed by varying the irradiation conditions. The dielectric constant of the composite can be varied from 120 to 350 and the transition temperature can be shifted from 65 to 35°C. A relatively flat dielectric response from 0 to 100°C can be achieved. The composite also has high dielectric constant (>120) at 10MHz and 90°C, which gives the material an opportunity to be used in high frequency applications. The high dielectric strength of the material prepared in clean environment leads to a high-energy storage density of 14J/cm^3.

REFERENCES

1. C. J. Dias and D. K. Das-Gupta, in *Ferroelectric Polymer and Ceramic-Polymer Composites*, pp. 217-248, edited by D. K. Das-Gupta, Trans-Tech Publications (1994).
2. C. J. Dias and D. K. Das-Gupta, *IEEE Trans. Electr. Insul.*, Vol.3(5), pp. 706-734 (1996).
3. K. A. Hanner, A. Safari, R. E. Newnham and J. Runt, *Ferroelectrics*, 100, pp. 255-260 (1989).
4. R. Gregorio Jr., M. Cestari and F. E. Bernardino, *J. Mater. Sci.*, 31, p2925-2930 (1996).
5. H. L. W. Chan, W. K. Chan, Y. Zhang and C. L. Choy, *IEEE Trans. Electr. Insul.*, Vol.5(4), pp. 505-512 (1998).
6. R. E. Newnham, *Ann. Rev. Mater. Sci.*, 16, p. 47-68 (1986).
7. J. Wolak, *IEEE Trans. Electr. Insul.*, Vol.28(1), pp. 116-121 (1993).
8. Q. M. Zhang, V. Bharti and X. Zhao, *Science*, 280, p. 2101-2104 (1998).
9. T. Yamada, T. Ueda and T. Kitayama, *J. Appl. Phys.*, 53(4), pp. 4328-4332 (1982).
10. D. K. Das-Gupta, *Ferroelectrics*, 118, pp. 165-189 (1991).
11. T. Furukawa, K. Ishida and E. Fukada, *J. Appl. Phys.*, 50(7), pp. 4904-4912 (1979).
12. J. Gao, L. Zheng, J. Zeng and C. Lin, *Jpn. J. Appl. Phys.*, 37, pp. 5126-5127 (1998).
13. D. K. Das-Gupta and R. Zhang, *Ferroelectrics*, 134, pp. 71-79 (1992).

THE VOLTAGE AND COMPOSITION DEPENDENCE OF SWITCHING IN A POLYMER CURRENT LIMITER DEVICE

ANIL R. DUGGAL
General Electric Corporate Research and Development, Niskayuna, NY 12309

ABSTRACT

The high power electrical switching properties of a polymer current limiter device are studied as a function of applied voltage. It is shown that a dramatic change in switching behavior occurs at a characteristic voltage. Below this voltage, the device switches to a high resistance state whereas at higher applied voltages it does not. It is shown that the high voltage, low resistance state has similar electrical characteristics to an arc discharge. Material variation experiments are also described which demonstrate that the changeover depends sensitively on the contact resistance between the filler particles of the composite material.

INTRODUCTION

Recently it was demonstrated that conductor-filled polymer composites can be fashioned into "polymer current limiter" devices that rapidly and reversibly switch from a low to high resistance state when high current densities flow through them.[1-3] A typical device consists simply of two metal electrodes pressure contacted to the polymer composite material. Such a device holds great promise for the power distribution industry because it is fast-acting, reusable, and relatively inexpensive.

Figure 1A shows typical current and voltage waveforms when a polymer current limiter device is placed in a high power circuit. The device utilizes a nickel-filled epoxy composite material and the circuit consists of a high power amplifier system attempting to apply a 300V voltage pulse. With the application of the pulse, the current initially increases, but then, within 200 microseconds, decreases to a low level as the device switches to a high resistance state. The switching event is accompanied by a loud noise, visible light generation, and ablation of material at the material/electrode interfaces. In spite of this, when the voltage pulse is terminated, the device regains its low initial resistance and, under low power conditions, exhibits ohmic behavior.

Previous studies have brought understanding to some aspects of the high power switching phenomenon. It has been determined that the phenomenon is not a bulk effect and is not based on the positive temperature coefficient of resistance (PTCR) effect which is exhibited by many conductor-filled polymer composite materials.[3] With pressure-contacted electrodes, the switching is localized at the electrode/material interfaces and it has been shown that the time required to initiate switching can be explained quantitatively with an adiabatic contact heating model.[4] The current qualitative model for device operation is based on polymer ablation and dielectric breakdown or arcing. Specifically, it is postulated that with a high power input, polymer ablation occurs at the material/electrode interfaces due to the excess constriction resistance heating that occurs in these regions. The ablation causes separation of filler-electrode and filler-filler contacts which tends to increase the device resistance. At the same time, transient arcs form between separated contacts and this tends to mitigate the increase in resistance caused by the contact separations. The overall high device resistance state is then postulated as a dynamic equilibrium that includes contacts continually separating and arcs continually forming and extinguishing between these separating contacts. With the cessation of the high power input, the ablation-induced separation of contacts ceases and, since the electrodes are under pressure, the contacts reform and the device regains its low resistance.

In this work, switching experiments as a function of voltage and material parameters aimed at better characterizing the switched state of the device are described.

RESULTS

Switching experiments as a function of voltage were performed on nickel-epoxy based devices with contact resistances configured such that, under high power input, switching only occurred on one of the two material/electrode interfaces. High power pulses were applied by means of an amplifier system described elsewhere.[3] This system has a maximum current output capability of 200A and, for the short pulses utilized in this work, can sustain a voltage of 300V while operating at maximum current. For all voltages, switching occurred and was accompanied by a loud noise, light generation, and material ablation from the high resistance contact. However, a marked change in the current and voltage waveforms was observed as the applied voltage was increased above ~200V. This is illustrated in figure 1 which shows the current and voltage waveforms when 100V and 300V pulses are applied to a device. In both experiments, the initial device resistance was 0.2 ohm. The 100V pulse data shown in figure 1A exhibits typical switching behavior as described in the introduction. By contrast, figure 1B shows that when a 300V pulse is applied, the current maintains an amplifier-limited value of 200A and the voltage across the device rises to and then maintains a value ranging from 160-200V. This implies a device resistance of ~1 ohm which is larger than the initial device resistance but clearly much less than that attained by the device subject to the 100V voltage pulse.

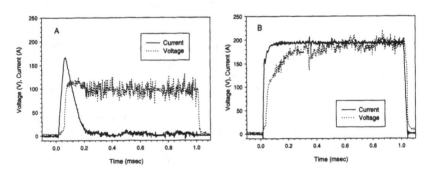

Figure 1: Current through and voltage across a single switching-contact polymer current limiter device with an applied voltage of 100V (A) and 300V (B).

Experiments were also performed with longer applied pulse length and with varying values of series resistance added to the test circuit. In all tests with applied voltage below 200V, the device switched to a high resistance and the data exhibited the same features as those depicted in figure 1A. At higher applied voltages, the devices did not switch to a high resistance and the data exhibited the features depicted in figure 1B. However, at the higher applied voltages, the value of resistance attained exhibited a strong dependence on the added series resistance value. This is illustrated in figure 2 which shows typical results when a 400V, 6 msec pulse is applied to a device with different series resistors added to the test circuit. Figure 2A depicts the case with no added series resistance. One can see that the data exhibit the same behavior as the analogous shorter time data of figure 1B except that the voltage across the device increases to a higher value of ~230V over the longer time period. Figures 2B and 2C show the current through and voltage across a device when 1.8 ohm and 5 ohm resistors are added to the circuit. With increasing series resistance, the magnitude of both the current and voltage in the switched state decrease. However, as the current changes from ~190A in figure 2A to ~40A in figure 2C, the voltage only changes from ~230V to ~175V. Thus, an almost 500% change in current through the device is accommodated by only a 30% change in voltage. This implies a nonlinear relationship between current and voltage so that it does not make sense to characterize this state in terms of a resistance value.

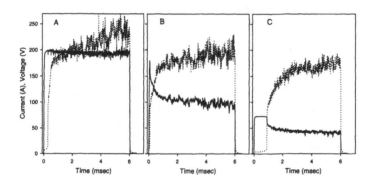

Figure 2: Current through (solid curves) and voltage across (dotted curves) a device with an added series resistance of 0 ohm (A), 1.8 ohm (B), and 5 ohm (C).

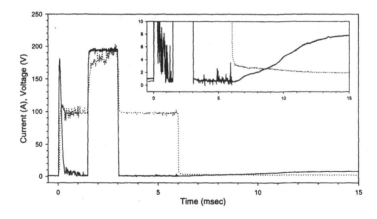

Figure 3: Current through (solid curves) and voltage across (dotted curves) a device during the application of a multistep voltage pulse. Inset shows an expanded view of the same data.

The existence of a change in switching behavior above a characteristic voltage can be further illustrated by applying a multistep voltage pulse to a device. Figure 3 shows the current and voltage data from such an experiment where the applied pulse consists of 100V for 1.5 msec followed by 400V for 1.5 msec followed by 100V for 3 msec followed by 2V for 9 msec. The figure inset displays the same data with expanded current and voltage scales. Note that during the first 1.5 msec, the device switches to a high resistance state as in figure 1A but then when the applied voltage is increased to 400V for the next 1.5 msec, the voltage across the device reaches a value of 190V and the current reaches the amplifier limited value of 200A. This is analogous to the data in figure 1B. When the applied voltage is then decreased again to 100V the device again reverts to the high resistance behavior depicted in figure 1A. Note also that in the high resistance state, fluctuations in the current, presumably due to arcs repeatedly forming and quenching, are apparent. During the final 2V voltage step, these fluctuations cease and the current increases. This current increase marks the recovery of a low device resistance as the device is no longer in the high power switching regime.

Experiments were also performed to probe the dependence of the characteristic "changeover" voltage on the properties of the polymer composite material. Here, we describe experiments in which the effect of one small but controlled variation of the nickel-epoxy material is investigated. The specific variation examined is the degree of fusion between the conducting filler particles of the composite. The method of nickel-epoxy sample fabrication utilized in this study consisted of infiltrating the epoxy through a ~4 mm thick layer of nickel powder (2.5 micron particle size, 0.74 g/cm3 apparent density) that had been allowed to settle to its natural tap density. Through scanning electron microscopy analysis, it was found that, prior to epoxy infiltration, the nickel powder particles could be partially fused together without greatly decreasing their natural tap density by means of low temperature, short-time (15 minutes) sintering in a hydrogen atmosphere. Increasing degrees of fusion could be accomplished by increasing the sintering temperature. This method was employed to fabricate samples with varying degrees of particle fusion. For each sample, the volume percent of nickel loading was determined through density measurements and the resistivity was determined through 4-point probe measurements.

Table 1: Properties of fabricated nickel-epoxy composite materials.

Sinter Temperature (C)	Ni Loading (V%)	Resistivity (mohm-cm)
25	17	21.5 +/- 1.1
300	16	1.7 +/- 0.2
350	16	0.94 +/- 0.05
400	18	.59 +/- 0.02

The results are given in Table I along with the sinter temperature. One can see that even with the lowest temperature sintering, the resistivity of the composite decreases dramatically. This indicates that a large portion of the composite material resistivity for the non-sintered samples is due to contact resistance between nickel particles. With increasing sintering temperature, the resistivity decreases due, presumably, to further reduction in particle-particle contact resistance.

Devices were made with each sample and tested with high power pulses as a function of voltage. For all samples, the change in behavior depicted in figure 1 occurred at a characteristic voltage. When voltages above this characteristic value were applied, the voltage across the device remained fixed at the characteristic value and the current rose to the maximum that could be supported by the test circuit. Interestingly, it was found that the actual characteristic voltage value is sensitive to the sinter temperature. This is illustrated in figure 4 which shows the current/voltage data for each device during the application of a 400V, 1 msec pulse using our amplifier system. Note that the highest characteristic voltage value occurs with the unsintered sample and that, with increased sintering temperature, the characteristic voltage value clearly decreases. It was also observed that the amount of material ablation during each switching event, as measured by sample weight loss, decreased with increasing sintering temperature.

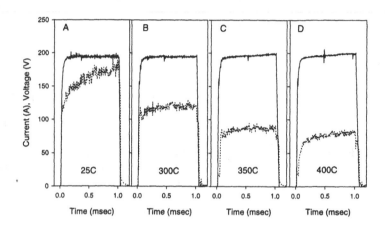

Figure 4: Current (solid) and voltage (dotted) data for devices utilizing nickel-epoxy with the nickel filler sintered at 25C (A), 300C (B), 350C (C), and 400C (D).

DISCUSSION

The results of this study clearly indicate that a change in high power switching behavior occurs as the voltage is increased above a characteristic voltage. When the applied voltage is below this characteristic voltage, the device limits current effectively by switching to a high resistance state whereas at higher voltages it does not. The electrical properties of the higher voltage, non-current-limiting state are similar to those of an arc discharge in that, it is only observed above a characteristic voltage and it exhibits a nonlinear current/voltage relationship. Interpreting this state as some form of arc discharge is consistent with the idea that the high resistance switched state of a device is governed by an interplay between ablation-induced contact separations and transient arcs between those separating contacts. In this scenario, the high resistance attained is essentially controlled by the reduced number of contacts between the electrode and the bulk of the composite material since all arcs that form are rapidly quenched. One would expect that, with increasing voltage, a point would be reached where the electrical fields are high enough that any arcs forming between separated contacts would no longer quench. In such a case, one would expect one or more high current arcs to connect the electrode with the bulk of the material so that now the switched state would be determined by the electrical properties of the arc. We postulate that this changeover in behavior is what we are observing.

Our results show that the voltage at which this changeover from transient arcs to one or more sustained arcs occurs depends sensitively on the contact resistance between the conducting filler particles of the composite. In particular, a lower interparticle contact resistance is associated with a lower changeover voltage and a smaller amount of material ablation. It is well known that the presence of material ablation in an arc increases the voltage required to sustain the arc.[5-6] Thus a possible explanation for these results is that materials with higher interparticle contact resistance exhibit higher rates of material ablation through joule heating and that this higher rate of ablation results in a higher required voltage for a sustained arc. This is the simplest explanation, but it should be noted that other more complicated explanations involving arcs in series and/or arc/electrode interactions are also possible. Further work is clearly necessary to better understand the influence of this and other material composition/fabrication influences on device behavior.

CONCLUSIONS

High power polymer current limiter devices represent a promising technology for the circuit protection industry as they provide a new, nonmechanical way to rapidly limit short-circuit currents. In order for such devices to operate in significant real-world applications, it is desirable to increase the upper range of operating voltage of a device to as high a value as possible. This study shows that there is a definite "changeover voltage" above which current limiting does not occur and that this voltage depends sensitively on composite material fabrication conditions. Further work aimed at understanding and optimizing the relationship between composite material characteristics and switching performance will be critical in enabling the widespread application of high power polymer current limiter technology.

ACKNOWLEDGEMENTS

The author thanks Hemant Mody and Sylvain Coulombe for insightful discussions and Kevin Mcevoy, Lauraine Denault, and Harold Patchen for expert technical assistance.

REFERENCES

1 "PROLIM Current Limiter", *Asea Brown Boveri Control*, Vasteras, Sweden.
2 R. Strumpler, G. Maidorn, and J. Rhyner, *J. Appl. Phys.*, **81**, 6786 (1997).
3 A. R. Duggal and L. M. Levinson, *J. Appl. Phys.*, **82**, 5532 (1997).
4 A. R. Duggal and F. G. Sun, *J. Appl. Phys.*, **83**, 2046 (1998).
5 J. Slepian and C. L. Denault, *AIEE Trans.*, **51**, 157 (1932).
6 P. F. Hettwer, *IEEE Trans. Power Apparatus and Systems*, **PAS-101**, 1689 (1982).

SMART POLYMER COMPOSITE THERMISTOR

RALF STRÜMPLER, JOACHIM GLATZ-REICHENBACH,
ABB Corporate Research, CH-5405 Baden-Dättwil, Switzerland, ralf.struempler@ch.abb.com

ABSTRACT

An industrially important class of passively smart materials is electrically nonlinear polymer composites. The transition of conducting composites from low to high resistivity can be utilized for current limitation. Due to Joule losses the material is heated by a fault or short-circuit current. With increasing temperature the polymer matrix expands and the current paths over the conducting filler particles are interrupted. Within milliseconds, the material responds to the fault current by an increase in resistivity up to eight orders of magnitude. Due to the strong nonlinear resistivity - temperature relation, a narrow hot-zone is formed even for long samples. The length of the hot-zone limits the maximum switching voltage. By adding a second filler material of varistor-type, however, the maximum voltage can be considerably increased. When a hot spot is formed in one of the current paths over the conducting particles, a small voltage increase allows already a commutation of the current to neighboring varistor particles. Consequently, the current can still flow to a certain degree and allows to heat also the rest of the material around its path. This leads finally to a very broad hot area, which can resist much higher voltages. By the development of a smart material with two strong non-linearities, a dramatic improvement has been achieved for the application of thermistor composites in current limitation.

INTRODUCTION

Smart materials are designed to perform both sensing and actuation. They are capable to detect a change in the environment and to respond to it in a technical suitable way. A number of intelligent or smart materials with such response have already been proposed[1-4].

The positive temperature coefficient (PTC) of resistance of conductive composites[5] is used for self-limited heating systems and fault current limiters in circuit protection. When the composite is heated by the passage of electrical current, the volume expansion of the polymer disrupts the conductive filler network, causing the resistance of the composite to increase tremendously. The resistance increase is usually six to eight orders of magnitude within a temperature range of only 20°C [6]. The composite thus shows two distinct, current dependent behaviors. At low current, it conducts electricity, but it becomes a current limiting device when it is heated by a fault current. The thermal expansion between room temperature (RT) and the transition temperature is in the range of 1 %. For external heating of the PTC material, the relationship of resistivity and thermal expansion can be proved by dilatometry[7, 8]. Even for the dynamic case, i.e., the limitation of a short circuit current, the fast expansion of the PTC material can be measured by using a laser distance sensor and correlated to the resistivity increase[9].

Commercial current limiting devices are available for electronic circuit protection[10] for protection of computers[11] or batteries[12,13]. Applications at higher voltage are limited by the intrinsic material properties. Due to the very strong non-linearity of the PTC effect one thin layer, perpendicular to the direction of current flow, heats up fastest. Across this high resistive, hot layer the material has to resist the entire system voltage. It has been shown that thermistor material consisting of 50 vol.% TiB_2 particles in polyethylene can limit repetitively short-circuit currents at 400 V in a capacitor discharge[6] or up to 220 V a.c.[14]. The hot layer has for such loads a typical thickness of 3 - 5 mm. For industrial applications, however, it is necessary to protect 400 to 690 V a.c. in low voltage applications or even 12 to 24 kV in medium voltage networks.

CONCEPT OF SMART THERMISTORS

In order to use PTC thermistors as fault current limiter for higher voltages, they can be connected in series, but the voltage across the limiting elements must be controlled. This can be achieved by using varistors, i.e. voltage dependent resistors, which are connected in parallel to the PTC resistors. Recently it has been shown that polymer composite current limiters can be operated at voltage levels far beyond their intrinsic voltage resistance by using such an arrangement[15]. The varistors (60 elements for 15 kV) serve for a very homogeneous voltage distribution although the PTC resistors switch at different times. The current limiter can be used at least for three times, which is about the maximum number of faults during the equipment life time in most of the medium voltage networks. However, such an external voltage control is rather expensive and works well only for rather large fault currents.

In order to reduce the electrical stress inside the PTC material, the electrical field between adjacent conducting particles has to be controlled across the PTC element on a microscopic level. A second microscopic filler can realize this with an electrical conductivity, which depends strongly on the electrical field. The second filler particles have to be in intimate electrical contact with the conducting ones. They can consist of a varistor material as doped SiC or doped ZnO. Compared to the conducting particles they should be larger in size. Then one can achieve a relative high conductivity of the composite due to the formation of a core-shell structure[16].

Figure 1 illustrates a polymer composite containing conducting and varistor particles. If the filler content of the conducting particles is high enough, they form percolating paths through the material. When the current exceeds a critical value, first the conducting particles are heated. With some time delay the matrix material around them heats up as well and expands[6]. The voltage across the created gaps between the particles increases until the breakdown voltage of adjacent varistor particles is reached. This means a rapid decrease in resistivity due to their strong non-linear j(E)-relationship. Then the resistance through the by-pass via the varistor particles is much lower and the current is commutated. As a consequence, the current flow is more or less not reduced and able to heat other parts of the conducting chain. Hence, the material has the chance to be heated at another place in current flow direction. This hot-spot will then again be by-passed. The process continues until such a long part of the PTC material has become tripped that the driving voltage is not sufficient anymore to sustain the breakdown at every involved varistor particle contact. Then the resistance increases strongly due to the j(E)-characteristic. Finally even the varistor particles might be separated by the thermal expansion of the polymer and the whole composite becomes insulating.

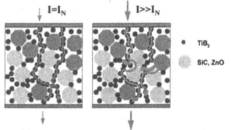

Figure 1 Current paths through PTC/varistor composites at nominal current I_N and fault current $I>I_N$ The shaded areas (\\\\\\ ) indicate the hot spots, which are formed if a fault current is applied.

EXPERIMENT

The investigated thermistor material consists of 30 vol.% conducting TiB_2 particles and either doped 20 vol.% ZnO varistor powder or 20 vol.% Al-doped SiC powder in high-density polyethylene (HDPE). The TiB_2 powder has a particle size of <45 μm with an electrical conductivity of 5 - 15 μΩ cm. The ZnO varistor powder (D70, Merck AG, Germany) has a particle size distribution of 60 - 160 µm and is made especially for low breakdown fields. The green powder was sintered at 1100 °C for 8 hours to get ceramic particles[17]. The varistor effect in ZnO is a grain boundary effect and depends therefore on the internal structure of adjacent varistor grains. Consequently E_B depends on the doping and on the preparation of the varistor material.

The intrinsic resistivity of pure SiC is with a value of about 10^{12} Ωcm much too high for this application. SiC becomes conducting, however, by doping with Al (p-type) or N (n-type). The used powder (ESK Kempten, Germany, SiC Extra) had size distributions of 45 - 75 µm, 90 - 125 µm and 150 – 212 µm. Pressed at 9.38 MPa the resistivities of the SiC powder were 18.2 Ωcm (at an applied field of 3164 V/cm), 13.5 Ωcm (at 2292 V/cm), and 11.0 Ωcm (at 1888 V/cm), respectively.

In order to achieve a homogeneous dispersion, filler and matrix material are compounded for several minutes using a Brabender Plasticorder at temperatures around 180°C. After compounding the composites are hot-pressed at 145°C to form several mm thick plates. The dependence of the electrical resistivity of the composite material on the temperature is determined by measuring test samples of $20 \times 10 \times 2$ mm in four-probe geometry using a HP 4247 A impedance analyzer.

Figure 2 shows the j(E)-characteristics of all powders. The ZnO varistor powder exhibits compared to the SiC powders a stronger non-linearity with a breakdown field $\approx$ 220 V/mm (at 10^{-4} A/cm^2). The critical field of SiC decreases as the particle size increases. The explanation is that the nonlinear effect occurs for SiC at the interface of the particles. With increasing particle-particle contact density, i.e. decreasing particle size, the breakdown field becomes higher. The ceramic varistor particles, however, contain many grain boundaries at which the non-linearity occurs. The inter-particle contacts play a minor role in this case. As much more grain boundaries exist per unit length, the breakdown field is larger for ZnO than for SiC.

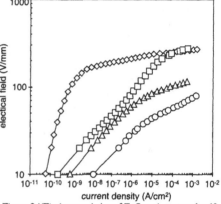

Figure 2 j(E) characteristics of ZnO varistor powder 60 - 160µm (◊), SiC 150 - 212µm (▢), SiC 90 - 125 µm (Δ) and SiC 45 - 75 µm powder (◯).

Short-circuit tests were performed using a capacitor battery at prospective currents of I_{pros} = 12 kA (scaled to 50 Hz) and loading voltages between 300 and 1200 V. The capacitance and inductance of the circuit were 14.56 mF and 54 µH for 300 V, 3.64 mF and 153 µH for 850 V, and 2.8 mF and 153 µH for 1200 V. In all tests a circuit breaker is making the short-circuit. The generated short-circuit current is limited and interrupted by the current limiting thermistor within parts of a millisecond. Then, 100 ms after making the short-circuit, the switch opens again and interrupts the circuit. The thermistor is protected against the occurring overvoltage by ceramic varistors (ABB MRV0.14ZS) connected in parallel. One varistor for 300 V, three varistors in series for 850 V, and four for 1200 V. Test samples were 1.5 mm thick plates of 3 cm length and 4 cm width contacted by pressed aluminum bars. In order to avoid tripping close to the contacts due to current constriction, the cross-section of the PTC center part is reduced by holes, either two holes of 5 mm diameter or one long-hole of 1 cm width and 2 cm length. During the experiments thermal images of the sample surface are recorded at video frequency (i.e. 30 pictures/s) by an infrared (IR) camera (Nippon Avionics Co. Ltd., TVS-2200ST). In this way the surface temperature of the PTC sample can be monitored online to provide additional information to estimate size and shape of the region where the thermistor trips.

RESULTS AND DISCUSSION

Figure 3 shows the specific resistivity ρ of a HDPE/ 30vol%TiB$_2$/ 20vol%SiC composite as a function of temperature. The resistivity increases from 0.1 Ω cm at RT to about 10^6 Ω cm at 130 - 140 °C. After cooling to RT, ρ is slightly increased to 0.3 Ω cm. Compared to the $\rho(T)$ characteristic of a pure PTC composite (HDPE/vol.50 % TiB$_2$), which is shown in Figure 3 as well, the RT resistivity is about one order of magnitude higher. With increasing temperature the resistivity starts earlier to increase and reaches the maximum value at a slightly higher temperature.

The electrical response during a short-circuit test at a loading voltage of 1200 V is shown in Figure 4. The PTC resistor reacts fast while the current rises. The maximum current is limited to 1290 A after 250 μs and is suppressed to about zero after further 400 μs. As the resistance of the PTC resistor increases, the current is commutated to the external varistor, which absorbs the energy stored in the inductance of the circuit. The transient recovery voltage is limited by the varistor to a peak value of 1400 V. The recovery voltage of 1000 V, which is well below the breakdown voltage of the used varistor, remains over the thermistor for 100 ms after switching. During this period no further current can be detected within a resolution of about 1A. The current through the PTC/SiC composite looks very similar to the related current through a pure PTC material in a 300 V test [14].

Before suppression to zero, however, the current shows a tail indicating a slow vanishing current contribution. We assume that the current, which is passed internally through the SiC particles, causes this. This becomes also apparent in a comparison of TiB$_2$/SiC to TiB$_2$/ZnO material, as shown in Figure 5. This short-circuit test has been performed at 850 V on three samples (with two 5 mm holes) in parallel, which would allow about the same current rating as a single element of HDPE/TiB$_2$. As a consequence of the larger cross-section, the

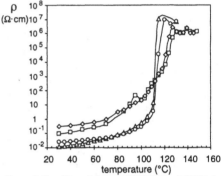

Figure 3 Specific resistivity of HDPE/30 vol.%TiB$_2$/ 20vol% SiC (90-125 μm): (□) heating up, (◊) cooling, and HDPE /50 vol.% TiB$_2$ composites: (△) heating up, (○) cooling, as a function of temperature.

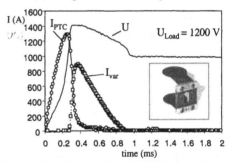

Figure 4 Short circuit testing at 1200 V loading voltage of polyethylene / 30 vol.% TiB$_2$ (<45 μm)/ 20vol% SiC (45-75 μm) composite; insert: sample with long hole; (○) PTC current, (□) varistor current, (—) voltage

current is limited at a higher value of 2000 A, both for ZnO and SiC. That is more or less the same as for the HDPE/TiB$_2$[14]. However, with ZnO varistor as second filler, the current drops earlier to zero. For the time period from 0.6 ms to 1.2 ms there is an additional current contribution only in the case of SiC filler. Therefore we assume that the ZnO varistor powder is inactive due to an E$_B$ level being too high. This assumption is supported by further tests using different particle size distributions. Table I shows the materials which survived tests up to

Table I Hot-zone lengths of different thermistor materials at 300, 850, and 1200 V.

	Voltage		
Material	300 V	850 V	1200 V
PE/TiB$_2$	3 mm	-	/·
PE/TiB$_2$/ZnO	3 mm	-	/·
PE/TiB$_2$/SiC 150-212 μm	6 mm	18 mm	/·
PE/TiB$_2$/SiC 90-125 μm	4 mm	18 mm	20 mm
PE/TiB$_2$/SiC 45-75 μm	/·	/·	20 mm

1200 V and the lengths of the hot-zones as observed by the IR camera. Composites with SiC 150-212 μm can resist at least 850 V, those with SiC 90-125 μm or SiC 45-75 μm can resist 1200 V. A reason for the enhanced voltage resistance is the increased length of the hot-zone. The tripping area becomes 20 mm long at 1200 V. Although the hot-zone lengths can not be measured precisely, they seem to be more or less proportional to the applied voltage. This is illustrated in Figure 6, which shows IR pictures taken during the short-circuit limitation of PE/TiB$_2$/SiC 90-125 μm samples (with long holes) at 300 and 850 V. At 300 V the hot-zone has a length of about 4 mm, at 850 V, however, it is 18 mm long. Due to the length of the sample the hot-zone can establish free from any geometric constriction.

Therefore, these tests are a clear indication that the concept of current commutation via the SiC particles really serves for a longer hot-zone and hence a higher voltage resistance. Compared to the pure PTC and the PTC/ZnO composite the energy density which is dissipated in the hot-zone is lower for all PTC/SiC composites as listed in Table II. This helps to survive switching off at higher voltages. If the commutation from the conducting to the varistor particles does not become effective, the material overheats and is damaged.

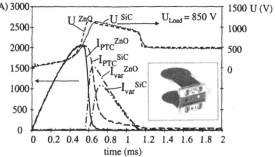

Figure 5 Short circuit testing at 850 V loading voltage of HDPE/30 vol.% TiB$_2$/20vol% ZnO varistor and HDPE/30 vol.% TiB$_2$/20vol% SiC (90-125 μm); sample with two 5 mm holes, three in parallel.

CONCLUSION

Our experiments have shown that smart polymer composite thermistors have a high potential for industrial applications in current limitation. The smartness of the thermistor is essentially the combination of two nonlinear functionalities: the PTC effect and the varistor effect. The use of the PTC effect is intrinsically limited to lower voltages due to the occurrence of hot-spots or hot-zones. The added varistor function can, however, serve for an increase of the length of the hot-

Table II Energy densities dissipated in the hot-zone of different thermistor materials at 300, 850, and 1200 V. "-" indicates that the samples were damaged, " /· " that no test was performed.

	Voltage		
Material	300 V	850 V	1200 V
PE/TiB$_2$	890 J/cm^3	-	/·
PE/TiB$_2$/ZnO	520 J/cm^3	-	/·
PE/TiB$_2$/SiC 150-212 μm	250 J/cm^3	216 J/cm^3	/
PE/TiB$_2$/SiC 90-125 μm	420 J/cm^3	185 J/cm^3	203 J/cm^3
PE/TiB$_2$/SiC 45-75 μm	/·	/·	230 J/cm^3

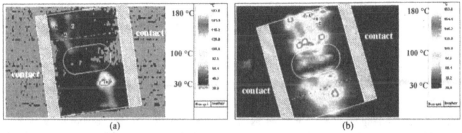

(a)   (b)

Figure 6 Infrared pictures of the hot-zone of HDPE/TiB$_2$/SiC (90-125 μm) during current limitation at (a) 300 V and (b) 850V charging voltage; samples with long hole; the shaded areas indicate the electrical contacts.

zone according to the applied voltage. As a consequence the voltage resistance is tremendously increased, from 300 V for the pure PTC material to at least 1200 V for the smart thermistor. For the control of the electrical field strength on a microscopic scale a fairly low breakdown field E_B is necessary. This is achieved by the use of SiC as second filler material. It has a much lower breakdown field than ZnO varistor powder, even if this is made in particular for low voltage applications. With the development of the smart thermistor material a variety of power applications now become possible.

ACKNOWLEDGEMENT

The authors thank F. Greuter, C. Schüler, J. Skindhoj and A. Steffens for many helpful discussions and R. Loitzl, B. Meyer, R. Kessler, D. Minichiello, and P. Huser for technical assistance.

REFERENCES

[1] R. E. Newnham and G.R. Ruschau, J. Am. Ceram. Soc. **74**, pp. 436-480 (1991)
[2] R. E. Newnham, G.R: Ruschau, J. Intelligent Mater. Syst. and Struct. **4**, pp. 289-294 (1993)
[3] R. Strümpler, P. Kluge-Weiss, F. Greuter, in *Advances in Science and Technology,* edited by P. Vincenzi (Techna S.r.I., Faenza-Ra, Italy, 1995) vol 10, pp. 15 - 22
[4] R. Strümpler, J. Rhyner, F. Greuter, P. Kluge-Weiss, J. of Smart Mater. & Struct. **4**, pp. 215-222 (1995)
[5] F. Bueche, J. Appl. Phys. **44**, pp. 532 (1973)
[6] R. Strümpler, G. Maidorn, J. Rhyner, J. Appl. Phys. **81**, pp. 6786 - 94 (1997)
[7] R. Strümpler, J. Glatz-Reichenbach, and F. Greuter, Mater. Soc. Symp. Proc. **411**, pp. 393-398 (1996)
[8] M. B. Heaney, Appl. Phys. Lett., **69**, pp. 2602-2604 (1996)
[9] J. Glatz-Reichenbach, J. Skindhøj, R. Strümpler, Proc. of 11th Int. Conf. on Comp. Mater., Gold Coast, Australia, July 14-18, 1997, ed.: M. L. Scott, Woodhead Publ., vol. 5, (1997) pp. 749-758
[10] F. A. Doljack, IEEE Trans. on Comp., Hybrids and Manufact. Techn. CHMT-4, p.372 (1981)
[11] T. Fang, St Morris, Elektron, January 97, pp.103-104 (1997)
[12] M. Stoessl, Power Control in Motion, June 93, pp. 50 - 55 (1993)
[13] T. Kobayashi, H. Endo, NEC Research and Development 86, pp. 81-90 (1987)
[14] J. Skindhøj, J. Glatz-Reichenbach, R. Strümpler, IEEE Trans. on Power Delivery 13, No.2, pp. 489-494 (1998)
[15] R. Strümpler, J. Skindhøj, J. Glatz-Reichenbach, J. H. W. Kuhlefelt, F. Perdoncin, IEEE Trans. on Power Delivery 14, No. 2, pp. 425 – 436. (1999)
[16] R. Strümpler, J. Glatz-Reichenbach, J. Electroceramics 3, No. 3, pp. 329-46 (1999)
[17] J. Glatz-Reichenbach, B. Meyer, R. Strümpler, P. Kluge-Weiss, and F. Greuter, J. Mater. Sci. 31, pp. 5941-5944 (1996)

FAST PROTON CONDUCTORS FROM INORGANIC-ORGANIC COMPOSITES : I. AMORPHOUS PHOSPHATE-NAFION AND SILICOPHOSPHATE-PMA/PWA HYBRIDS

YONG-IL PARK, JAE-DONG KIM, MASAYUKI NAGAI
Advanced Research Center for Energy and Environment, Musashi Institute of Technology,
1-28-1 Tamazutsumi, Setagaya-ku, Tokyo 158-8557, JAPAN

ABSTRACT

A drastic increase of electrical conductivity was observed in the composite of amorphous phosphate and ion-exchange resins (Nafion) as phosphorus concentration increased. Incorporation of amorphous phosphate into Nafion caused a large increase of conductivity to about 4×10^{-1} S/cm at 23°C. However, the fabricated composite showed very low chemical stability.

A high proton conductivity was also observed in a new inorganic-organic hybrids through incorporating PMA(molibdo-phosphoric acid)/PWA(tungsto-phosphoric acid) as a proton source in amorphous silicophosphate gel structure. Obtained gels were homogeneous and chemically stable. Resulting proton conductivity is very high (up to 5.5×10^{-3} S/cm) compared to those of silicophosphate gels.

INTRODUCTION

Nowadays, environmental conservation and new energy development have become serious issues for human beings. As the most practical fuel cell candidate, solid polymer electrolyte fuel cells (PEFCs) are receiving attention.

Non-production of CO_2 and suitability for electric vehicles are advantages of PEFCs. The polymer electrolyte membrane (PEM) in PEFC is an electronic insulator but an excellent conductor of hydrogen ions. The PEM is used in the form of 50μm-200μm-thick membrane and works as a gas separator and also as an electrolyte. The electrolyte and electrodes are pressed together to produce a single membrane-electrode assembly (MEA). Therefore, electrolyte material in PEFC needs to have : (1) high specific conductivity, (2) good mechanical, chemical stability, and (3) low cost (<$200/m²) [1, 2]. The membranes developed by DuPont (Nafion series) and Dow are most widely used. Both these membranes consist of perfluorinated copolymers with sulfonic-acid functionalized side chains. These materials are expensive and have complicated synthetic procedures.

In this work, we investigated the inorganic-organic hybrids for PEM applications. The inorganic-organic hybrids can be one of the most promising materials, because they represent a new class of materials that possess both inorganic and organic functionality. These materials have been synthesized by utilizing the low temperature reaction of the sol-gel process. Some kind of these materials fabricated by co-polymerization of metal alkoxides and organic-substituted silicon alkoxides or organic-based oligomers have been reported as `ORMOSILs` [3] and `CERAMER` [4]. Especially, the hybrid composite from tetraethoxysilane (TEOS) and polydimethysiloxane (PDMS) have been intensively investigated. Incorporating of other inorganic species can be also used to control the chemical and physical properties of hybrid composites. In spite of many advantages, very few attempts to develop these materials for proton-conducting electrolyte have been reported.

Here, we report fabrication process and properties of fast proton-conducting inorganic-organic hybrid composites of :
(1) amorphous phosphate incorporated in Nafion membranes (P-Nafion) : Nafion membrane as modified by the incorporation of an amorphous phosphate phase via *in situ* sol-gel reaction for di-isopropyl phosphate.
(2) Silicophosphate-PWA/PMA hybrid gels (SiP-PMA/PWA) : PMA($H_3[Mo_{12}PO_{40}]\cdot29H_2O$) and PWA($H_3[W_{12}PO_{40}]\cdot29H_2O$), which are representative solid acids having high proton conductivity about 2×10^{-1} S/cm at 25°C [5], are incorporated as a proton source in amorphous silicophosphate gel structure.

299

EXPERIMENTAL PROCEDURE

Sample preparation

Perfluorinated ion-exchange resin (Nafion, 5wt%) and di-isopropyl phosphate($HPO(OC_3H_7)_2$) were used as starting materials for fabrication of P-Nafion composites. Through hydrolysis reaction between the mixed starting materials and water at room temperature, clear precursor solution was acquired. P-Nafion composite membranes were obtained after drying and heat-treatment at 25°C - 150°C for several times. The sample composition was controlled in weight ratio of di-isopropyl phosphite/Nafion (P/Nafion) from 0.0 to 8.0. (Fig.1, (a))

SiP-PMA/PWA hybrid gels were prepared using tetraethoxylsilane (TEOS, $Si(OC_2H_5)_4$, 1mol), di-isopropyl phosphate (0.5mol) and ethanol as starting materials. After 30minutes of stirring, 0.005mol - 0.08mol of tungsto-phosphoric acid (PWA, $H_3[W_{12}PO_{40}]\cdot29H_2O$)) or molibdo-phosphoric acid (PMA, $H_3[Mo_{12}PO_{40}]\cdot29H_2O$) was added to the starting solution. And then 1mol of di-methylformamide (DMF) was added as a drying condition control agent. 1.5mol of distilled water diluted with ethanol was added to the mixed solution for hydrolysis reaction. The obtained clear precursor solutions were allowed to gelate for 3days at 23°C, 13%RH. (Fig.1, (b))

Conductivity Measurement

The conductivity of all the fabricated composites was obtained from Cole-Cole plots by an ac method using an impedance analyzer (IM6 impedance measurement system, Zhaner Electric Co.) using platinum plate electrodes fixed on both sides of the samples. The conductivity measurements were carried out at 23°C, and all the samples were immersed in distilled water for 10 minutes before the measurement.

RESULTS AND DISCUSSION

P-Nafion Composites

The obtained P-Nafion membranes are dense and soft, and also show high water absorption due to the amorphous phosphate network formed between Nafion polymers.

In Fig.2, XRD patterns of P-Nafion membranes prepared with various P/Nafion ratio are shown. There is not any distinguishable difference in the peak patterns of all the composition. This indicates the formation of an amorphous phosphate phase, not a crystalline phase.

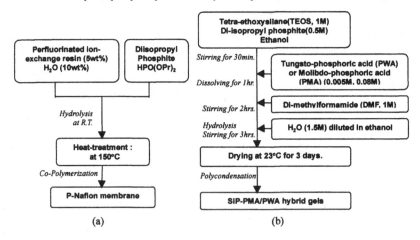

(a)   (b)

Fig.1. Flow diagram of the processing for : (a) P-Nafion composite and (b) SiP-PMA/PWA gels.

FT-IR spectra for P-Nafion and Nafion membranes are illustrated in Fig. 3. Scholze [6] found that there are three main IR absorption bands due to hydroxyl groups, i.e., band-1 in the range of 3640 cm⁻¹ ~ 3390cm⁻¹, band-2 around 2900cm⁻¹ and band-3 at 2340cm⁻¹. However, the existence of the band-3 is not certain yet and various absorption bands due to network formers are overlapped. Band-1 is due to hydrogen bonding-free hydroxyl groups and band-2 is to strongly hydrogen-bonded hydroxyl groups. It has been reported that protons in band-2 are much more mobile than those in band-1. These three bands are observed more strongly in the spectrum of P-Nafion membrane (Fig.3, (a)) compared to the spectrum of Nafion (Fig.3, (b)). Fig. 4 shows SEM images of the surfaces of P-Nafion membranes : (a) dried at 70°C for 24hrs before heat-treatment and (b) directly heat-treated at 150°C. Compared to the flat and homogeneous surface of the sample surface in Fig. 4, (a), directly heat-treated sample shows many open pores on rough surface. Therefore, all the samples used for conductivity measurement are prepared through intermediate drying procedure. The measured conductivity of P-Nafion membranes are illustrated

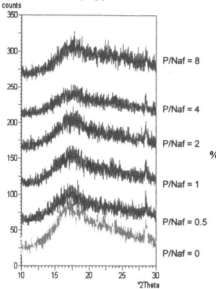

Fig.2. XRD patterns of P-Nafion membranes prepared at different heat-treatment temperature..

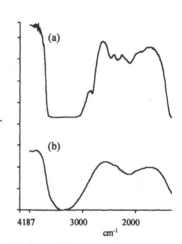

Fig.3. FT-IR spectra of (a) P-Nafion and (b) Nafion membranes.

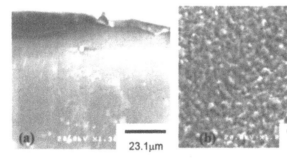

Fig.4. SEM images of the surfaces of the P-Nafion membranes : (a) dried at 70°C for 24hrs. before heat-treatment and (b) directly heat-treated. (P/Nafion=8)

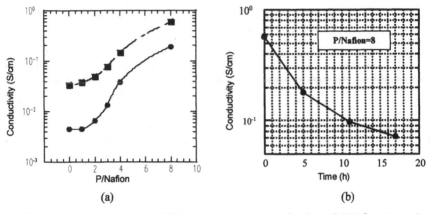

(a)   (b)

Fig.5.   Conductivity change of P-Nafion membrane as a function of P/Nafion (a), and
conductivity change in distilled water as a function of time (b). (■ : after immersed in water for
30min., ● : 15RH%, P/Nafion represents di-isopropyl phosphite/Nafion weight ratio in (a)).

in Fig.5, (a). With increasing P/Nafion ratio, the proton conductivity of the samples drastically increased.
This is due to an increase of the proton carrier density due to formation of amorphous phosphate network
and also an increase of water content in P-Nafion membranes.

However, the chemical stability in water of the P-Nafion membranes is very low as shown in Fig. 5,
(b). A drastic decrease of conductivity is observed, and the pH value of the water in which the sample is
treated also drops down implying the hydrolysis and dissolution of phosphate.

Silicophosphate-PMA/PWA Hybrids (SiP-PMA/PWA)

Fig. 6 shows photographs of the obtained SiP-PMA/PWA hybrid gels. The samples prepared with
small amount of PMA/PWA are translucent and rigid, and those prepared with large amount of
PMA/PWA are opaque and brittle. It should be noted that all the samples are very stable in water in spite
of the high solubility of PMA and PWA in water. SiP-PMA has a yellowish green color, and the SiP-PWA
is translucent at 0.005mol and becomes white at 0.08mol of PWA. Some cracks are grown on the surface
of the samples during drying procedure, and results in breaking of the samples to several parts. Fig. 7
shows FT-IR patterns of SiP-PMA and SiP-PWA hybrids which are immersed in distilled water for 1h
before characterization. The three IR bands($3640cm^{-1} \sim 3390cm^{-1}$, $\approx 2900cm^{-1}$, $2340cm^{-1}$) from the

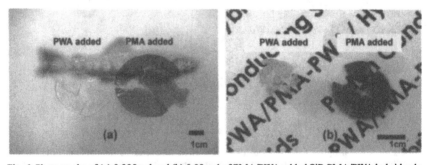

Fig. 6. Photographs of (a) 0.005mol and (b) 0.08mol of PMA/PWA added SiP-PMA/PWA hybrid gels.

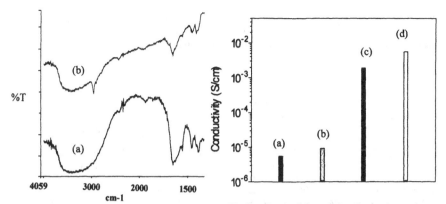

Fig.7. FT-IR spectra of (a) SiP-PMA (PMA:0.08mol) and (b) SiP-PWA (PWA:0.08mol).

Fig.8. Conductivity of the obtained samples; (a) SiP-PWA (PWA:0.005mol), (b) SiP-PMA (PMA:0.005mol), (c) SiP-PWA (PWA:0.08mol), (d) SiP-PMA (PMA:0.08mol).

absorbed hydroxyl groups discussed in Fig. 3, and absorption peaks by molecular water around 1600cm^{-1} are observed in the IR patterns of both samples.

The measured conductivities of SiP-PMA and SiP-PWA hybrids prepared at different composition are illustrated in Fig. 8. High proton conductivities of 1.9×10^{-3}S/cm for SiP-PWA and 5.5×10^{-3}S/cm for SiP-PMA at the composition of PMA(PWA)/Si=0.08 are obtained. With increasing of PMA(PWA)/Si molar ratio from 0.005 (Fig.8, (a) and (b)) to 0.08 (Fig.8, (c) and (d)), conductivity increased about 10^3 times. Fig. 9 shows a proposed proton conduction mechanism in the SiP-PMA/PWA hybrid gels. It is assumed that the SiP-PMA/PWA hybrid gels contain both the trapped ionized-PMA(PWA) as a proton source, and hydroxyl groups bonded to phosphorus atoms fixed by P-O-Si bonds which work as a proton conduction path. Moreover, the molecular water which is generally contained in gel structure provides high proton mobility. According to Abe et al. [7], the activation energies for the dissociation or ionization of protons in hydroxyl groups are assumed to be essentially the same when the values for υ_{OH} (peak wavenumber of the infrared absorption band due to fundamental O-H stretching vibration) are the same,

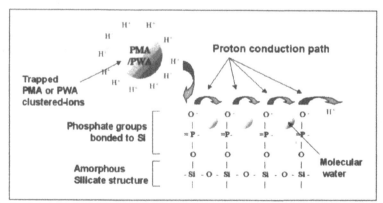

Fig. 9. Schematic diagram of a proposed proton conduction mechanism in the SiP-PMA/PWA hybrids.

but the activation energies for transport (proton hopping) are different. That is, the activation energy for hopping through molecular water is much less than those for hopping through non-bridging oxygen. This may be another advantage of sol-gel derived proton conducting materials.

SUMMARY

Fast proton-conducting inorganic-organic hybrids of amorphous phosphate-Nafion (P-Nafion) membranes and silicophosphate-PMA/PWA (SiP-PMA/PWA) hybrid gels were successfully fabricated. All the samples showed appropriate mechanical strength enough for measuring their conductivity.

In the P-Nafion composite, a drastic increase of proton conductivity up to 4×10^{-1}S/cm was measured at 23°C as phosphate concentration increases to 8.0 of P/Nafion weight ratio. However, the fabricated membranes were very unstable in water, and hydrolysis and dissolution of phosphate occurred.

Obtained SiP-PMA/PWA hybrid gels were brittle and have some cracks. However, the proton conductivity of these materials was high, and increased about 10^3 times with increasing PMA(PWA)/Si molar ratio from 0.005 to 0.08. The measured conductivities were 1.9×10^{-3}S/cm for SiP-PWA and 5.5×10^{-3}S/cm for SiP-PMA at PMA(PWA)/Si=0.08, respectively. These high proton conductivities may be due to the gel structure which contains both the trapped ionized PMA(PWA) as a proton source and hydroxyl groups bonded to P-O-Si oxide bonds which work as a proton conduction path. Moreover, the molecular water generally contained in gel structure also provides high proton mobility resulting in an increase of proton conductivity.

Study on polydimethylsiloxane(PDMS)-based hybrid system to give a flexibility to SiP-PMA/PWA gels is now in progress.

REFERENCES

1. S. J. Sondheimer, N. J. Bunce, C. A. Fyfe, and J. Macromol, Sci. Rev. in Macromol Chem. Phys., **C26**, p. 353 (1986).

2. S. Holmberg, T. Lehtinen, J. Nasman, D. Ostrovskii, M. Paronen, R. Serimaa, F. Sundholm, G. Sundholm, L.Torell, M. Tokkeli, J. Mater. Chem., **6**, p. 1309 (1996).

3. G.Phillipp and H. K. Schmidt, J. Non-Cryst. Solids, **63**, p. 283 (1984).

4. G. L. Wilkes, B. Orler, and H. Hwang, Polym. Prep., **26**, p. 300 (1985).

5. T. Maruyama, Y. Saito, Y. Matsumoto, and Y. Yano, Solid-State Ionics, **17**, p. 181 (1985).

6. H. Scholze, Glastech. Ber. (a) **32**, p. 81 (1959) ; (b) **32**. P. 142 (1959) ; (c) **32**. P. 314 (1959).

7. Y. Abe, G. Li, M. Nogami, T. Kasuga and L. L. Hench, J. Electrochem. Soc., 143, **1**, p. 144 (1996).

FAST PROTON CONDUCTORS FROM INORGANIC-ORGANIC COMPOSITES : II. AMORPHOUS PHOSPHATE-PTFE AND ZrP-PTFE COMPOSITES

YONG-IL PARK, JAE-DONG KIM, MASAYUKI NAGAI
Advanced Research Center for Energy and Environment, Musashi Institute of Technology,
1-28-1 Tamazutsumi, Setagaya-ku, Tokyo 158-8557, JAPAN

ABSTRACT

A high proton-conductivity was observed in the composite of amorphous phosphate and poly-tetafluoroethylene (PTFE). Incorporation of amorphous phosphate into PTFE emulsion caused a large increase of conductivity to about 4×10^{-2}S/cm at 23°C. However, the conductivity decreased with increasing heat-treatment temperature, and the fabricated composite showed very low chemical stability.

As a chemically stable composite, PTFE-based composite was also synthesized from α- or γ-zirconium phosphate crystalline powders dispersed in partially polymerized PTFE particles. By addition of zirconium phosphate powders, the proton conductivity jumped up to 2.2×10^{-3}S/cm from 10^{-13}S/cm of PTFE.

INTRODUCTION

Nowadays, environmental conservation and new energy development have become serious issues for human beings. As the most practical fuel cell candidate, solid polymer electrolyte fuel cells (PEFCs) are receiving attention. Non-production of CO_2 and suitability for electric vehicles are advantages of PEFCs. The polymer membrane in PEFC works as both the separator and the electrolyte. Therefore, electrolyte material in PEFC needs both high proton conductivity and high mechanical strength [1, 2].

For PEFC application, the inorganic-organic hybrids composites can be one of the most promising electrolyte materials because inorganic-organic hybrids represent a new class of materials that possess both inorganic and organic functionality. These materials have been synthesized by utilizing the low temperature reaction of the sol-gel process. Some kind of these materials fabricated by co-polymerization of metal alkoxides and organic-substituted silicon alkoxides or organic-based oligomers have been reported. Incorporating of other inorganic species can be also used to control the electrochemical and physical properties of hybrid composites. However, almost no attempt to use these inorganic-organic hybrids as a proton-conducting electrolyte has been reported. Here, we report fabrication procedure and properties of proton-conducting inorganic-organic composites of :

(1) amorphous phosphate-PTFE composite (P-PTFE) : PTFE membrane in which an amorphous phosphate phase incorporated via *in situ* sol-gel reaction for di-isopropyl phosphate.

(2) α- or γ-zirconium phosphate-PTFE composite (ZrP-PTFE) : PTFE-based hybrid incorporating layer-structured α- or γ-zirconium phosphate crystalline (α-ZrP or γ-ZrP) in partially polymerized PTFE grain boundaries.

EXPERIMENTAL PROCEDURE

Sample preparation

Poly-tetrafluoroethlyene (PTFE) emulsion (60wt%, grain size : 0.2μm-0.4μm) and di-isopropyl phosphite (HPO(OC$_3$H$_7$)$_2$) were used as starting materials for fabrication of P-PTFE composite. By mixing the starting materials and adding water at room temperature, fast agglomeration and co-polymerization were occurred resulting in a flexible bulk composite. After drying at room temperature for 4 hrs, the obtained samples were heat-treated at 25°C ~ 300°C for 1hr.(Fig.1, (a)).

α-ZrP-PTFE and γ-ZrP-PTFE composites were prepared using the PTFE emulsion, guaranteed-grade commercial α-ZrP (Zr(HPO$_4$)$_2$·H$_2$O) and γ-ZrP (Zr(HPO$_4$)$_2$·2H$_2$O) crystalline powders (Daiichi Kigenso Kaguku Kogyo Co. Ltd) as starting materials. Fast agglomeration and polymerization of PTFE were also occurred during mixing the starting materials. All the sample composition was fixed as 50wt% of α-ZrP or γ–ZrP. After drying at 80°C for 3hrs, the obtained composites were pressed at 0.1MPa and 10MPa. Bulk composites were obtained through drying at 80°C and heat-treatment at 150°C for 1hr. (Fig.1, (b)).

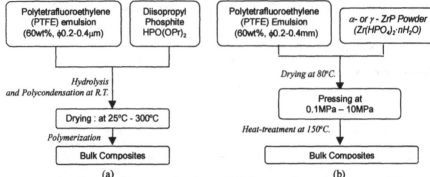

Fig. 1. Flow diagram of the processing for : (a)P-PTFE composites and (b) α- or γ - ZrP-PTFE composites.

Conductivity Measurement

The conductivity of all the fabricated composites was obtained from Cole-Cole plots by an ac method using an impedance analyzer (IM6 impedance measurement system, Zhaner Electric Co.) using platinum plate electrodes fixed on both sides of the samples. The conductivity measurements were carried out at 20°C, and all the samples were immersed in distilled water for 10 minutes before the measurement.

RESULTS AND DISCUSSION

P-PTFE Composites

The fabricated P-PTFE composites are shown in Fig. 2. Since P-PTFE composite has high plasticity before heat-treatment, various shapes could be easily prepared. After drying, the composites become to white and dense solids. However, high humidity absorption is also observed, and this is due to the amorphous phosphate network formed between partially polymerized PTFE particles. In Fig. 3, XRD patterns of P-PTFE composites heat-treated at 70°C ~ 280°C are shown. For comparison, the XRD pattern of commercial PTFE is also presented. In the temperature range of 70°C ~ 180°C, any distinguishable difference in the XRD patterns is not detected. This indicates the formation of an amorphous phosphate phase, not a crystalline phase. However, in XRD patterns of the samples heat-treated over 180°C, a new-phase assignable to crystalline P_2O_5 is observed. The P_2O_5 phase grows gradually from 180°C to 250°C. However, the peak intensity of crystalline P_2O_5 phase decreases at 280°C.

The volatilization of phosphorus observed over 260°C during heat-treatment can be a reason for the decreased P_2O_5 peak intensity at 280°C although the phosphorus peaks could not be detected from the XRD patterns. FT-IR spectra for PTFE and P-PTFE composites are illustrated in Fig.4. Scholze [3] found that

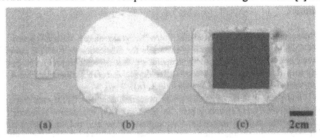

Fig. 2. Photographs of the fabricated P-PTFE composites. : (a) bulk (thickness=3mm), (b) film (thickness=50μm) and (c) membrane (thickness=450μm) with Pt/C-Nafion electrodes.

306

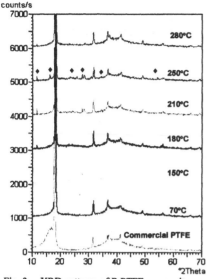

Fig. 3. XRD patterns of P-PTFE composites prepared at different heat-treatment temperature. (♦ :P₂O₅)

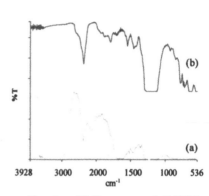

Fig. 4. FT-IR spectra of P-PTFE composite (a) and commercial PTFE film (b).

there are three main IR absorption bands due to hydroxyl groups, i.e., band-1 in the range of 3640 cm⁻¹~ 3900cm⁻¹, band-2 around 2900cm⁻¹ and band-3 at 2340cm⁻¹. However, the existence of the band-3 is not certain yet and various absorption bands due to network formers are overlapped. Band-1 is due to hydrogen bonding-free hydroxyl groups and band-2 is to strongly hydrogen-bonded hydroxyl groups. It has been reported that protons in band-2 are much more mobile than those in band-1. The three bands are clearly observed in the spectrum of P-PTFE composite (Fig.4, (a)) compared to the spectrum of commercial PTFE membrane (Fig.4, (b)) which shows almost no hydroxyl group absorption. A strong absorption peak around 1600cm⁻¹ by absorbed water is also observed in P-PTFE spectrum. Broad absorption in the low wavenumber region from 530cm⁻¹ - 1000cm⁻¹ is due to the P-O bond formation in P-PTFE composite. Fig.5 shows FT-IR spectra of P-PTFE composite heat-treated at 23°C - 280°C. The characterization was carried out using the samples maintained in the atmosphere of 23°C, 15%RH for 5hrs. As heat-treatment temperature increases, the absorption peaks due to X-OH groups and absorbed water are decreased. That is, the ability of the composite to absorb hydroxyl group or water is decreased with increasing heat-treatment temperature. This result can be explained as: (1) breaking of the amorphous phosphate networks, which have been origin of water absorption property, (2) volatilization of phosphorus at high heat-treatment temperature. Fig. 6 shows SEM images of P-PTFE composites. As heat-treatment temperature increases, the thread-like parts observed in cross-section of P-PTFE composite increase. Melting of surfaces of PTFE particles occurs and polymerization of PTFE progresses with elevating heat-treatment temperature. The large open pores observed in the surface of the sample heat-treated at 280°C seem to be due to volatilization of phosphorus. The energy dispersive X-ray spectrometer (EDX) analysis result of P-PTFE composite is shown in Fig. 7. Fluorine from PTFE and phosphorus from amorphous phosphate network show uniform distribution in P-PTFE composite. Increase of magnification could not be done because of low X-ray intensity of phosphorus. The measured conductivity of P-PTFE composite is shown in Fig.8, (a). With increasing heat-treatment temperature, the proton conductivity of the samples drastically decreased. The breaking of amorphous phosphate network by the progress of polymerization of PTFE and volatilization of phosphorus may be the reason for the decrease of conductivity. The chemical stability in water of the P-PTFE composite heat-treated at 150°C is shown in Fig. 8, (b). The sample was immersed in distilled water and vibrated using a ultrasonic bath. A drastic decrease of conductivity is observed and the pH value of the water in which the sample is treated also drops down implying hydrolysis and dissolution of phosphate. However, the P-PTFE composite with Pt-carbon-Nafion electrodes shows increased chemical stability, although Pt-carbon-Nafion electrode

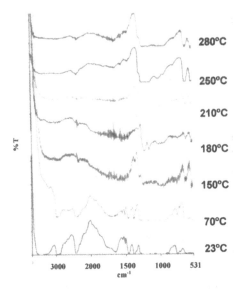

Fig.5. FT-IR spectra of P-PTFE composites prepared at different temperature.

Fig.6. FE-SEM images of P-PTFE composites heat-treated at the presented temperatures.

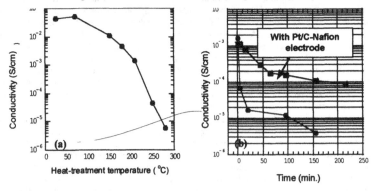

Fig.7. EDX analysis images FE-SEM images of P-PTFE composite heat-treated at 150°C ; (a)SEM image, (b) fluorine distribution and (c) phosphorus distribution.

Fig.8. Conductivity of P-PTFE composite at different heat-treatment temperature (a) and conductivity change in distilled water (ultrasonic vibration is engaged.) (b).

itself is porous. Therefore, dense proton-conducting polymer (as like Nafion) coating will be effective to prevent phosphate dissolution of P-PTFE composites.

α-ZrP-PTFE and γ-ZrP-PTFE composites

Zirconium phosphate (ZrP) is a well-known two dimensional layered-structure compound. The intercalation property and relatively high proton-conductivity makes this compound more attractive. Reported proton conductivity for α-ZrP($Zr(HPO_4)\cdot H_2O$) is about ~ 10^{-5}S/cm [4]. However, Ablerti et.al. [5] found that the mobility of surface ion in α-ZrP is 10^4 times larger than those of interlayer ions, and that the low activation energy (11~13kJ/mol) for ionic conduction was due to the transport of surface ions. Therefore, the surface ions of layer-structured zirconium phosphate compound make an important contribution to the total conduction, although the fraction of the carriers present on the surface is very small. And it should be noted that α-ZrP and γ-ZrP ($Zr(HPO_4)\cdot 2H_2O$) are very stable in water. For the application of the high surface ion mobility of ZrP in the field of PEFC, it is required to fix fine ZrP particles in the electrolyte material. In this study, partially polymerized PTFE particles are examined as a fixing matrix. Fig. 9 shows the XRD patterns of commercial PTFE film, α-ZrP-PTFE composite and γ-ZrP-PTFE composite. The samples in Fig.9 are fabricated by pressing dried composites with a pressure of 0.1MPa. Both α-ZrP-PTFE (Fig. 9, (b)) and γ-ZrP-PTFE (Fig. 9, (c)) composites show simply mixed phases of α-ZrP and PTFE, and γ-ZrP and PTFE, respectively. Fig. 10 shows FT-IR patterns of α-ZrP-PTFE, γ-ZrP-PTFE and commercial PTFE film which are immersed in distilled water for 1h before the characterization. The three IR bands(3640cm $^{-1}$~3390cm^{-1}, ≈2900cm^{-1}, 2340cm^{-1}) from the absorbed hydroxyl groups discussed in Fig. 4, and the peak due to absorbed water around 1600cm^{-1} are observed. The measured conductivities of ZrP-PTFE composites prepared at different pressure are illustrated in Fig.11. By addition of ZrP, the proton conductivity jumps up to over 10^{-4}S/cm from 10^{-13}S/cm of PTFE (Fig. 11, (a)). This high proton conductivity over those of ZrP single crystal may be due to the high ion mobility on the surface of ZrP particles. The ZrP-PTFE composites show conductivity change as the measuring orientation is changed. The conductivity measured with the electrodes arranged vertically to the pressure engaged direction (⊥) (Fig.11, (c) and (e)) shows higher value than those with parallel electrodes (Fig.11, (b) and (d)). The proton-conductivity change with orientation is due to the anisotropy of ZrP crystal structure, and the electrical anisotropy is also its own nature as a two-dimensional structured compound. However, the γ-ZrP-PTFE composite prepared at 0.1MPa, with parallel electrodes (Fig.11, (f)) shows higher conductivity (2.2×10^{-3}S/cm) than those at 10MPa, with vertical electrodes (Fig.11, (e), (1.2×10^{-3}S/cm)). The high conductivity of the sample of Fig.11, (f) may be due to low density resulting in the increase of water accommodation. Therefore, pressure is also an important factor to control the electrical properties in ZrP-PTFE composites.

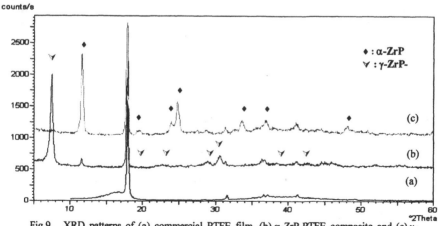

Fig.9.   XRD patterns of (a) commercial PTFE film, (b) α-ZrP-PTFE composite and (c) γ-ZrP-PTFE composite.

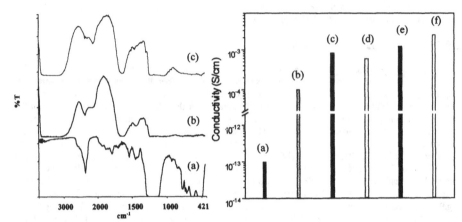

Fig.10. FT-IR spectra of (a) commercial PTFE film, (b) α-ZrP-PTFE composite and (c) γ-ZrP-PTFE composite.

Fig.11. Conductivity of the obtained ZrP-PTFE composites; (a) commercial PTFE film, (b) a-ZrP-PTFE (=, 10MPa), (c) α-ZrP-PTFE (⊥, 10MPa), (d) γ-ZrP-PTFE (=, 10MPa), (e) γ-ZrP-PTFE (⊥, 10MPa), (f) γ-ZrP-PTFE (=, 0.1MPa).
(⊥ and = represent vertical and parallel direction of electrode to pressure engaged direction, respectively.)

SUMMARY

The fast proton-conducting composites of amorphous phosphate-PTFE and α-/γ-ZrP-PTFE systems were successfully fabricated. All the composites showed high mechanical strength by PTFE polymer network formation.

In the P-PTFE composite, incorporation of amorphous phosphate into PTFE emulsion caused a large increase of proton conductivity to about 4×10^{-2} S/cm at 23°C. However, the conductivity decreased with increasing heat-treatment temperature, and the fabricated composite showed very low chemical stability. In the ZrP-PTFE composites, the measured conductivities of α- and γ- ZrP-PTFE composites jumped up to over 10^{-3} S/cm from 10^{-13} S/cm of PTFE. This high proton conductivity over those of ZrP single crystal may be due to the high ion mobility associated with large surface area of ZrP particles. The obtained ZrP-PTFE composites were very stable in water and showed high water absorption.

REFERENCES

1. S. J. Sondheimer, N. J. Bunce, C. A. Fyfe, and J. Macromol, Sci. Rev. in Macromol Chem. Phys., **C26**, p. 353 (1986).

2. S. Holmberg, T. Lehtinen, J. Nasman, D. Ostrovskii, M. Paronen, R. Serimaa, F. Sundholm, G. Sundholm, L.Torell, M. Tokkeli, J. Mater. Chem., **6**, p. 1309 (1996).

3. H. Scholze, Glastech. Ber. (a) **32**, p. 81 (1959) ; (b) **32**. P. 142 (1959) ; (c) **32**. P. 314 (1959).

4. S. Yamanaka and M. Hattori in *Inorganic Phosphate Materials*, edited by T. Kanazawa, Kodansha, Tokyo, 1989, pp. 131-156.

5. G. Alberti, M. Casiola, U. Costantino, G. Levi, and G. Ricciard, J. Inorg. Nucle. Chem., **40**, p. 533 (1978).

DIELECTRIC RELAXATION SPECTROSCOPIC MEASUREMENTS ON A NOVEL ELECTROACTIVE POLYIMIDE

SAADI ABDUL JAWAD[1], ABDALLA ALNAJJAR[2], MAMOUN M. BADER[3]
[1] Department of Physics, The Hashemite University, P. O. Box 1504591, Zarqa, Jordan
[2] Department of Physics, Sharja University, Sharja, UAE
[3] Department of Chemistry, Pennsylvania State University, Hazleton, PA 18201

ABSTRACT

AC electrical behavior of a novel aromatic electro-optic polyimide was investigated in the temperature range 25 °C to 300 °C and a frequency range from 1 Hz to 10^6 Hz. Three electrical quantities: impedance, permittivity and electric modulus are reported. The dependence of imaginary and real components of these quantities on temperature and frequency are discussed. The experimental results show that the polymer has high thermal stability below 200 °C, where the resistivity, dielectric constant and permittivity are nearly temperature-independent indicating highly rigid structure. Above this temperature, however, a well-defined broad peak corresponding to a relaxation process was observed for which the activation energy was calculated to be 8.5 Kcal/mole. This relaxation is associated with a restricted local rotational motion of the side chain chromophore.

INTRODUCTION

Significant progress in synthesis and processing of polymers has led to the development of new materials with favorable mechanical, thermal, optical and electrical properties. These developments, along with theoretical advances have led to increasing knowledge of structure-property relations. Development of polymers based on aromatic structure, led high performance polymers with high thermal and electrical stability. An intensive research work was performed on this class of polymers in order to relate the structure to physical and mechanical properties. The development of polymers as structural materials focused attention on the consideration of change of polymer physical properties with temperature and frequency. A new class of highly stable aromatic polyimide with electro-optic and photo-refractive properties has been reported. An intensive work has been published concerning the synthesis[1] and properties[2] of these materials. The interest behind this is the potential applications of these materials in three-dimensional holography, light processing, phase conjugation and the handling of large quantities of information in real time [3,4]. Other dielectric relaxation spectroscopic measurements were performed on reactive polymers[5] and to investigate the changes in molecular dynamics during bulk polymerization of an epoxide -amine system[6]. In addition dielectric relaxation was also used to study relaxation in nonlinear optical materials[7-9]. In this paper, the ac electrical behavior of a novel electro-optic polyimide will be reported as a function of frequency and temperature.

Mat. Res. Soc. Symp. Proc. Vol. 600 © 2000 Materials Research Society

EXPERIMENTAL

The polymer used in this study was prepared according to a procedure similar to that described by Yu et.al [10]. The structure is shown below

The polymer film was cast onto an ITO glass with thickness of 6.16 μ. The number average molecular weight (Mn) = 27,000 and the weight average molecular weight (Mw) = 42,000. The ac impedance measurements were carried out in the temperature range 25-300°C, and covering a frequency range 1Hz to 10^6 Hz, using a Solarton - 1260 Impedance/Gain Phase Analyzer (Shlumberger Instrument). The instrument is controlled by Z-60 and Z-View package, which maximizes the performance and data handling of the system. An aluminum electrode was evaporated on the surface of the film. The sample setup was then kept in a shielded cavity to improve low frequency measurements. Best signal generator amplitude and dc bias, were selected after performing a series of amplitude and dc bias sweeping tests. Then for the measurements 0.5 V for amplitude and zero dc bias were chosen. Using this setup, ac complex impedance Z* and the phase angle θ were measured. From the measurements of Z* and θ, the real and imaginary components of ac-impedance (Z*), permittivity(ε*) and electric modulus (M*) were determined as function of frequency at different temperatures.

RESULTS AND DISCUSSION

In alternating field, different regions in a sample will oscillate with different relaxation times. These regions may be characterized by resistance and capacitor in parallel, and each will have an associated frequency where maximum loss occurs, given by,

$$\omega_{max} RC = 1$$

In ac electrical measurement the complex quantities of ac impedance (Z^*) permittivity (ε^*) and admittance (Y^*) have been in use for many years. The fourth member of this group is the complex electric modulus ($M^* = \varepsilon^{*-1}$) was introduced by Macedo et.al , who were first to exploit the modulus and used it for analyzing conductivity relaxations in glasses and in concentrated aqueous solutions.[11] The extension of the modulus treatment to solid electrodes was done by Hodge et.al.[12] More recently Starkweather et.al[13] found that the study of relaxation transitions of polymers using the electric modulus has two advantages over the loss factor ε'', first the maximum in M'' will occur at a lower temperature than the maxima in ε''. Second , since ε' appears in the denominator of the M'' equation to the second power, its tendency to overwhelm the loss function is minimized. He was able to resolve the high temperature relaxation in Nylon 6 to two well defined relaxations. Nevertheless, using ac impedance analysis, it is possible to identify the various regions of the polycrystalline samples, where an equivalent circuit consisting from resistive (R) and capacitive (C) elements represent the grain boundary, the bulk and electrode regions. These regions might appear as a series of semi-circles in the plot of the imaginary (Z'') and real (Z') and the plot of real (M') and imaginary (M'') components of the complex ac impedance and electric modulus respectively. The analysis of ac impedance is represented in references.[14]

Measurements of ac resistivity, dielectric constant and electric modulus as a function of temperature and a frequency range 10^3 Hz and 10^6 Hz, show that below 200 ^{0}C these quantities are nearly temperature-independent as shown in Figure 1. However, above this temperature the three quantities increase with increasing temperature. Variation of ac impedance and phase angle are similar to that observed in RC network in parallel, where the impedance is frequency independent at low frequencies range and then becomes proportional to the inverse of frequency (Figure 2). The phase angle at low frequencies is nearly zero and decreases with increasing frequency to attain a value of -90.0. Therefore the bulk material at temperatures above 200 ^{0}C can be considered to consist of capacitive and resistive components in parallel with ohmic behavior dominating at high temperatures and low frequencies. Using ac impedance analysis at temperatures above 200 ^{0}C, where the material can be represented as RC network in parallel, the equivalent bulk resistance and capacitance can be determined. The bulk resistance at 200 ^{0}C, is in the order of 2.9 x 10^8 ohm and decreases to 2.34 x 10^6 ohm , where the capacitance component remains nearly constant in the temperature range 220 ^{0}C to 300 ^{0}C as given in Table 1.

Table 1: Equivalent bulk Resistance and Capacitance

Temperature, ^{0}C	R(ohm)	C (farad)
225	3.29 X 10^8	1.97 X 10^{-10}
250	4.29 X 10^7	5.58 X 10^{-11}
275	3.76 X 10^6	5.5 X 10^{-11}
300	2.43 X 10^6	4.54 X 10^{-11}

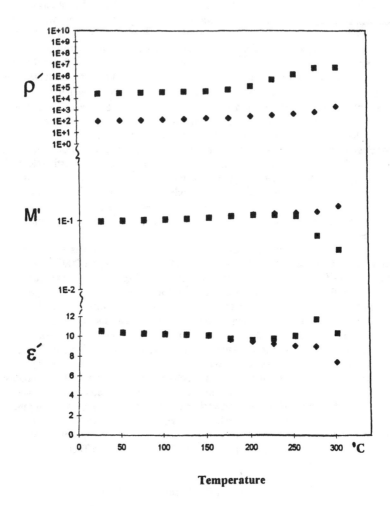

Figure 1: ac resitivity (Ω.cm), electric modulus (M`) and dielectric constant (ε`) at
■10³ Hz and ◆ 10 ⁶ Hz Vs. temperatuure for a novel electro optic polyimide.

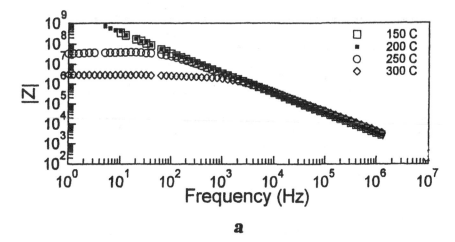

Figure 2: (a) ac impedance (Z) (Ω) Vs. frequency at different temperatures for a novel electro optic polyimide

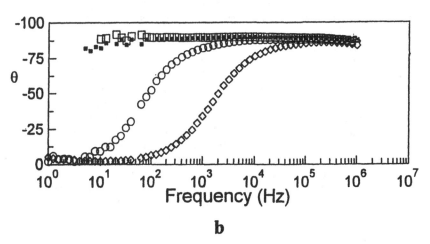

Figure 2: (b) Phase angle (θ) Vs. frequency at different temperatures for a novel electo optic polyimide.

No plateau region was observed below 200 °C was observed in the ac resistivity dependence on frequency at different temperatures, however, above this temperature a plateau region starts to appear (Figure 3). The frequency range of this plateau increases with increasing temperature and extends from 1 Hz to about 1 KHz at 300 °C. In the plot of Z" or M`` versus frequency no relaxation peak was observed below 200 °C, however above this temperature a relaxation peak starts to appear and became a well defined peak at 225 °C indicating that the chains below 200 °C are nearly frozen or rigid. The peak moves to a higher frequency with increasing temperature as can be seen in Figures 4 and 5. This peak is observed when $\omega\tau_{max} = 1$, where τ is the apparent relaxation time. This relaxation time decreases with increasing temperature and therefore, it is expected to be satisfied at a higher frequency. Nevertheless, the dependence of the relaxation time is generally assumed to take the Arrhenius form,

$$\tau = \tau_0 \exp{^{-E/RT}}$$

where, E is the apparent activation energy, τ_0 is the relaxation time at infinite temperature, T is the absolute temperature and R is the gas constant. The relaxation time is taken to be the reciprocal of the experimental frequency at which the dielectric loss factor or the mechanical loss factor exhibit a maximum. The apparent activation energy was obtained from the plot of log f versus 1/T, where f is the frequency at maximum Z" or M" where, the maximum frequencies of the relaxation peaks coincide on each other. For this polyimide, the activation energy was calculated to be about 8.5 Kcal /mole which is less than the activation energy of low temperature relaxations in polymers associated with crank-shaft mechanism and much less than the activation energy assigned to full rotational polar groups in liquid crystalline copolymers which is in the order of 30 Kcal/mole [15-18]. Therefore, we believe that the observed relaxation process in our sample is more likely due to highly restricted rotational local motion of the side chain chromophore. It is also worth pointing out that the relaxation peaks in Z" and M" are broad peaks, which can be assigned to a summation of relaxations occurring within the bulk material. This has been verified by plotting Z" versus Z' (Figure 6). At temperatures 275 °C and 300 °C an almost perfect semi-circle with a tail is observed. It might be possible to construct another smaller circle in the low frequency region. The first semi circle can be assigned to the local motion from the rotation of the side chain chromophore as described above. The smaller circle can be associated with some sort of phase separation, similar to the effect of grain boundaries in polycrystalline materials with rather small activation energy. Finally we like to point out that the real components of permittivity and electric modulus are nearly independent of frequency at temperatures below 200 °C. Increasing the temperature results in sharp increase in the real component of permittivity at low frequencies and a sharp drop in the real component of electric modulus (Figure 7). The strong frequency dependence of ε` and M ' at low frequencies and high temperatures indicate that the interfacial polarization contributes significantly to the polarization in the sample at low frequencies. However, interfacial polarization arises wherever phases with different conductivities are present. This may results from the formation of different micro-regions with varying conductivities, therefore if this is indeed the case this might be the

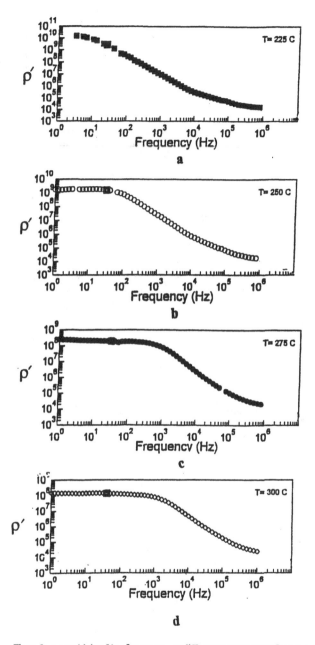

Figur 3: ac resitivity Vs. frequency at different temperatures for a novel electro optic polimide

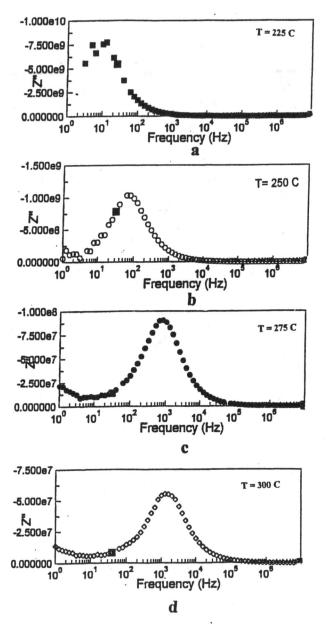

Figure 4: Imaginary component of ac impedance (Z``) Vs. frequency at different temperatures for a novel electro optic polyimide

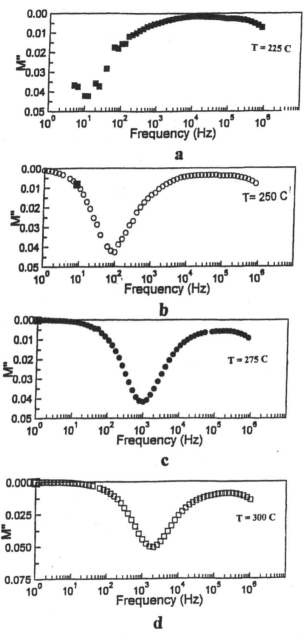

Figure 5: Imaginary component of electric modulus (M'') Vs. Frequency at different temperatures for a novel electro optic polyimide polyimide

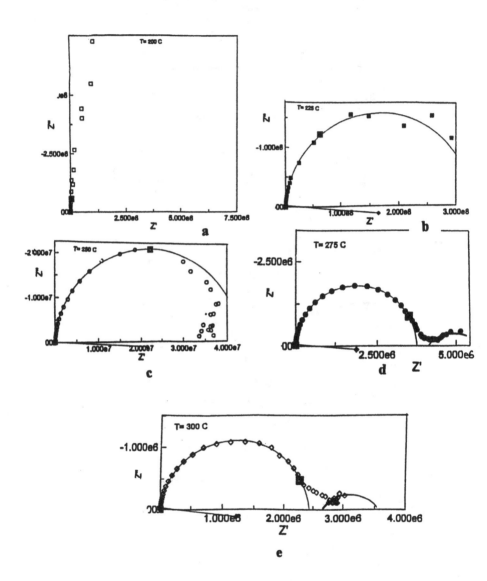

Figure 6: Imaginary component (Z'') Vs. reral component (Z') of ac impedance at different temperatures for a novel electro optic polyimide

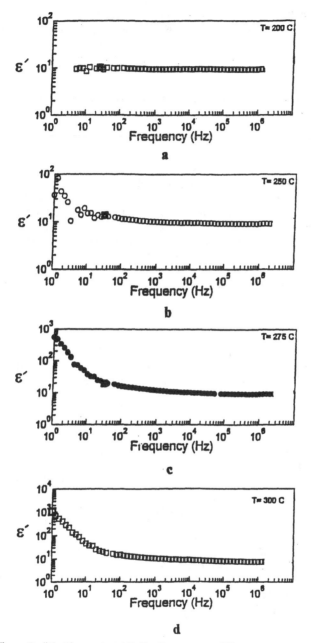

Figure 7: dielectric constant (ε') Vs. Frequency at different temperatures for a novel electro optic polyimide

reason for observing a distorted semicircle at low frequencies in the plot of Z`` versus Z`. The constancy in ac resisitivity, dielectric constant and electric modulus up to 200 ^{0}C indicates that this polyimide exhibits very high stability in dipole configuration. This may suggests that the mobility of the side chain is highly restricted to a limited localized motion. At 200 ^{0}C which represents a threshold temperature for this polymer, the mobility of the side chain chromophore starts to increase in a way to affect the electrical behavior of the material. These findings concerning thermal stability of electrical behavior of this polymer makes it a favorable candidate for electro-optical applications.

CONCLUSIONS

The experimental results reveal the following conclusions.
1. The polyimide shows high thermal stability below 200 ^{0}C due to its highly rigid structure. This result clearly demonstrates and strengthens the notion that the suitability of such polymeric materials and related structures for electro optical applications.
2. The main relaxation process in this system has an activation energy of 8.5 Kcal/mole associated with a restricted local rotational motion of side chain chromophore.
3. The mobility of the polymer chains increases above 200 ^{0}C associated with a dramatic change in electrical behavior of the polymer with a sharp increase in dielectric constant at low frequencies.

ACKNOWLEDGEMENTS

The authors would like to thank Professor L.Yu for providing the polyimide samples used in this study. M.B. would like to thank the DAA at PSU-Hazleton and the PDF program at the College of Science at PSU for travel support to attend the MRS meeting to present this work.

REFERENCES

1. Yu, L.; Chan,W. K.; Peng, Z.; Gharavi, A., *Acc. Chem. Res.* 1996, 29,13.
2. Moerner,W. E.; Silence, S. M., *Chem. Rev.* 1994, 94,129.
3. Gunter, P.; Huignard, J. P. *Photorefractive Materials and their Applications;* Springer-Verlag, Berlin, Vols.1 and 2 , 1988.
4. Yarvin, A.*Optical Electronics;* 4th ed., Har Court Brace, Orlando, FL , 1991.
5. Peng, Z. H.; Bao, Z. N.; Chen.Y. M.; Yu, L. P., *J. Am. Chem. Soc.* 1994, 116, 6003, Yu, L. P.; Peng, Z. H.; *Pol. Mater. Sci. Eng.* , 1994,71,441., Chan,W. K; Chen,Y. M; Peng, Z. H.; Yu, L. P., *J. Am. Chem. Soc.* , 1993, 155 ,11735., Sasone, M. J.; Teng,C. C; East, A. J.; Kwiatek, M. S., *Opt. Lett,*1993,18,1400.
6. Bellucci, F.; Maio,V.; Monetta, T.; Nicodemo, L.; Mijovic, J., *J.Poly. Sci., Part B: Polymer Phys.,*1996, 34,1280.
7. Fournier,J. ;Williams,G.; Duch;C.: Aldredge,G. A., *Macromolecules,* 1996, 29, 7097.

8. Koehler,W.; Robello,D. R.; Dao, P.T.; Willand,C.,S.; Williams, D.J., *Macromolecules*, **1992**, 24, 5589.

9. Teroka, I.; Jungbauer, D.; Reck, B.; Yoon, D.Y.; Twieg, R.; Wilson,C. G. *J. Appl. Phys.*, **1991**, 69, 2568.

10. Yu, D.; Gharavi, A.; Yu , L., *J. Amer. Chem Soc.*, **1995**,117,11680.

11. Macedo, P. B.; Bose, R.; Provenzano,V.; Litovitz, T. A. *Amorphous Materials:* Douglas, R.W. and Ellis, B. (Eds.), Interscience, London , **1972**.

12. Hodge, I. M; Ingram, M. D; West, A. R, *J. Electroanal. Chem.*, **1975**, 58, 429.

13. Starkweather, H. W.; Avakian, P. *J. Poly. Sci. Part B: Poly. Phys.*, **1992**, 30, 637.

14. See for example: Handbook of Conductive Polymers , Skotheim , T.A (ed.) , Marcel-Dekker Inc. Basel, 1989., Impedance Spectroscopy , Macdonald, R (ed.) , Wiley , New-York , 1987.

15. Issa, M. A, J. Mat. Sci, 1992, 27, 3685.

16. Boyd, R. H.; G. H. Porter, J. Poly.Sci., A-2, 1972, 10, 647.

17. Abdul-Jawad,S; Al-Haj-Mohammad, M.S, Mat. Let, 1992, 13, 312.

18. Green,D.I; Davies,G.R; Ward, I. M; Al-Haj-Mohammad,M.H; Abdul-Jawad,S, Pol. Adv Tech.,1990,1,41.

COMPUTER CONTROLLED PULSED PECVD REACTOR FOR LABORATORY SCALE DEPOSITION OF PLASMA POLYMERIZED THIN FILMS

P.D. PEDROW**, L.V. SHEPSIS*, R. MAHALINGAM*, AND M.A. OSMAN**
*ChE Department, Washington State University, Pullman, WA 99164
**EECS Department, Washington State University, Pullman, WA 99164

ABSTRACT

A pulsed PECVD reactor has been successfully constructed for laboratory scale studies of plasma polymerized thin films. A computer control system based on National Instrument's LABVIEW software controls power supply sequence, feed injection, and introduction of RF energy. An optical fiber and a photo diode allow the user to monitor the emitted light for each pulse. A fast ionization gauge is used to characterize the pressure evolution over time, subsequent to acetylene gas injection. Substrates with diameter as large as 10 cm can be accommodated within the reactor. Both aniline liquid and acetylene gas have been used as reactor feed. The deposited plasma-polymerized films were characterized using AFM and SEM. Electrical conductivity of plasma polymerized acetylene film was also measured

INTRODUCTION

As device sizes shrink and new applications are being found for thin films, plasma enhanced chemical vapor deposition (PECVD) is becoming more and more popular as a thin film deposition technique. In this technique plasma is used to decompose feed chemicals into reactive species which then deposit onto the substrate. Plasma processing offers such advantages as low temperature processing, film smoothness and uniformity, ease of deposition, and the ability to polymerize a wide variety of organic compounds. At the present time, continuous plasma discharges are almost universally used in industry. Pulsed plasma deposition, however, offers significant advantages in terms of better control of the process and film structure. Additional advantages include possible reductions in raw material usage, and process emissions.

Pulsed plasma reactors have been used in the past to grow SiO_2 and silicon nitride films [1,2], and to deposit hydrocarbon [3,4], halogenated hydrocarbon [5,6], and amorphous silicon films [7]. In this paper a laboratory scale pulsed plasma reactor is described. In our studies the reactor has been used for deposition of acetylene (gas) and aniline (liquid) polymer films. The films were characterized using AFM and SEM. Electrical properties of the films were also measured. Several diagnostic tools were employed to characterize plasma and pre-plasma conditions in the reactor. Fast ionization gauge (FIG) was used for measurement of transient pressures in the reactor. Plasma light emission was measured at different reactor locations and pressures using a photodiode detector. Data obtained from the above measurements are presented and discussed here.

EXPERIMENTAL SETUP

The schematic of the reactor is shown in Figure 1. Gas or liquid monomer is pulse-fed into the reaction chamber using a conventional automotive gasoline fuel injector. The setup for gas and liquid injection was different in the way the monomer was supplied to the injector. Liquid is delivered to the injector from a syringe storing a small (1-10 mL) volume of liquid monomer. Gaseous monomer is delivered from a compressed gas cylinder through a regulator and a valve. The amount of monomer delivered to the system is set by the duration of voltage pulse delivered to the injector by a computer control unit. Plasma is generated using a single turn inductively coupled copper coil. Each plasma pulse was produced by discharging a 1.86 μF capacitor with an initial voltage equal to 23 kV.

National Instrument's Labview was used to interface with the user through controlling the capacitor voltage, activating the feed injector, and injecting the RF energy. The computer controlled injector pulse duration and an auxiliary time delay unit (analog oscilloscope) controlled the exact firing time for RF energy injection. Use of the oscilloscope for timing the delay between monomer injection

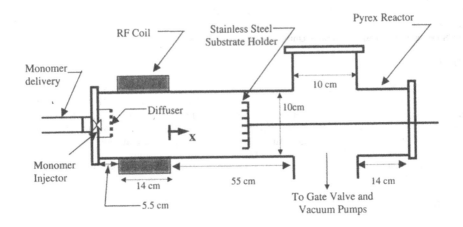

Figure 1. Schematic of the pulsed plasma reactor

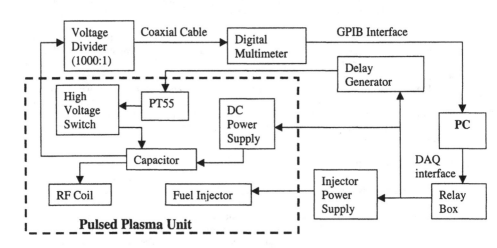

Figure 2: Computer control system block diagram

and RF coil activation facilitated a more accurate timing since the Windows operating system does not provide for real time control. Figure 2 shows a diagram of the computer control system.

A reactor base pressure of 5×10^{-5} torr is maintained by a diffusion pump (Consolidated Vacuum Corporation) in series with a roughing pump (Welch Vacuum Pump Model 1402). Pump-down can be halted at any time by closing a gate valve isolating the reactor chamber from the vacuum pumps. Silicon wafers of various sizes (up to 10 cm in diameter) and gold foil have been previously used as substrates for plasma polymer deposition. The substrates can either be mounted on the wall (parallel to flow) or on a seven-fingered metal stage (perpendicular to flow). The metal stage can either be biased or insulated from ground so that its electrostatic potential tracks the floating potential of the plasma.

Steady-state pressure readings were obtained with a Pirani gauge (BOC Edwards PRL10) mounted on top of the reactor. Pirani gauge pressure correction factors for gases other than air were obtained from the manufacturer. Fast ionization gauge (FIG) was used to measure pressure transients in the reactor resulting from gas injection. FIG is a useful tool for making low pressure measurements when fast response time is required [8,9]. FIG consists of three electrodes: electron emitter, electron collector and ion collector. The pressure is related to ion current as follows [9]:

$$i_+ = \sigma \lambda P i_. = S P i_.$$ (1)

where λ = MFP, $i_.$ = electron current, i_+ = ion current, P = pressure, S = gauge sensitivity, and σ represents the probability that any given electron will create an ion that will be collected on the ion collector. As described elsewhere [10], a 6AU6A vacuum tube with 40% of the plate removed was used here. The sensitivity factor, S, was obtained from the literature as described in Ref. [10]. It can be seen from Equation (1) that knowing electron current is important for obtaining accurate pressure results.

A photodiode detector was used to measure plasma light emission. This is a technique similar to plasma emission spectroscopy, commonly used as a plasma diagnostic tool [11-14]. The photodiode used here can not resolve the emission from individual plasma species but rather it detects a light emission signal produced by light at several different wavelengths. The light emitted by the plasma propagated through an optical fiber into the photodiode, where it is converted into a voltage signal (see Ref. 14 for photodiode circuit details).

Films produced in the reactor were inspected under the Scanning Electron Microscope (Hitachi S-570). Gold was sputter deposited onto the polymer surface prior to SEM analysis to prevent charging. Film thickness was measured using an AFM (Digital Instruments Model MMAFM-2 AFM). For this analysis, films were deposited on Si wafers coated with approximately 100 nm SiO_2 layer. A scratch in the polymer film was next made using a razor. The scratch region was then scanned with the AFM. This technique for measuring film thickness has been successfully used before [5]. Figure 3 shows a typical depth profile of a scratch region obtained using the AFM.

RESULTS AND DISCUSSION

Fast ionization gauge was used to characterize the pressure evolution in the reactor immediately after injection of the acetylene gas. Figure 4 shows sample FIG data for a 25 ms pulse of acetylene gas. The pressure evolution over time is shown in part (c) of Figure 4. The initial delay of about 3 ms is due to injector dead time. As can be seen from Figure 4c the pressure rises rapidly at first but after about 2 ms the slope of the curve decreases. This is due to the presence of a small injector plenum which stores gas for immediate injection. Once this gas is exhausted, a quasi steady state is established between the gas injection from the monomer delivery system and the pump-down.

Light measurements were made at different reactor conditions and locations. Figure 5 shows how light emission varies with acetylene pressure in the reactor. Diamonds represent runs with gate valve open, with the gas being pumped out as it is injected, and stars represent runs with gate valve closed. It can be observed from Figure 5 that light emission increases at low pressures and then decreases as pressures rises above approximately 40 mtorr. It is known from kinetic theory that pressure has a significant influence on mean free paths in the reactor. It is believed that, at very low pressures, light emission is low because there are fewer molecules available for excitational collisions. At the other

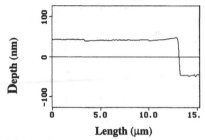

Figure 3: AFM cross sectional view of the film scratch (Vertical distance between the two arrow heads was 88 nm)

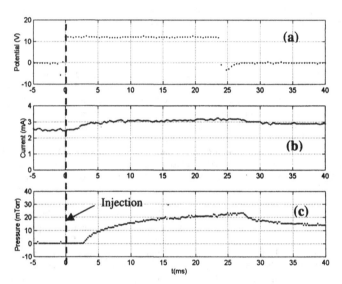

Figure 4. FIG data showing pressure transients after acetylene injection. (a) Injector voltage, showing duration of pulse; (b) Electron emission current; (c) Pressure obtained from Equation (1)

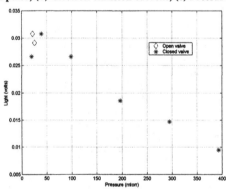

Figure 5. Variation of plasma light emission with initial reactor pressure –Acetylene studies

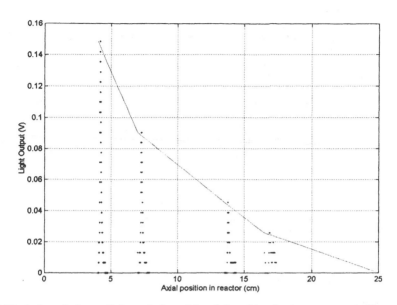

Figure 6. Variation of plasma light emission with axial position in the reactor - Aniline studies. Each series of data points is from an oscillogram of photodiode voltage. Duration of these pulses is about 100 μs.

Figure 7. SEM micrograph of acetylene film (approx. 0.3 μm thick) where a portion of the film is peeled off.

extreme of very high pressures electrons are not efficient at gaining energy from the electric field which also leads to fewer excitational collisions. This phenomenon can easily be related to Paschen's curve. Experiments are under way for obtaining a correlation between light emission and deposition rate data, which would potentially allow the in-situ estimation of deposition rates using a photodiode.

Figure 6 shows variation of plasma light emission with axial position in the reactor, for the case of aniline discharge. It is observed that light emission decreases as one moves away from the RF coil. This is most likely due to loss of electrons and electron energy as the plasma travels down the reactor tube. Figure 6 should be useful in predicting energy decay as a function of distance from the coil. It can also be used to measure the velocity of the light emitting portion of the plasma.

Figure 7 shows an SEM micrograph of an acetylene film deposited on gold foil mounted perpendicular to gas flow at a location x = 30 cm. The acetylene film shown in Figure 7 was deposited onto a gold foil substrate with 100 plasma pulses each at an initial reactor pressure of 26 mtorr. The film in Figure 7 is seen to be smooth, uniform and cohesive. As more research is performed using the pulsed plasma reactor, SEM will be used to study how film characteristics change as a function of pressure, power input, and position in the reactor.

Conductivity of acetylene films has been measured by constructing a polymer resistor as described elsewhere [15]. A conductivity of 1.9×10^{-10} S/cm was measured immediately after film deposition and was observed to decrease slightly over time. This value appears to be significantly higher than a value of 4×10^{-18} S/cm previously reported for plasma polyacetylene [16] and values of between 2×10^{-13} S/cm and 8×10^{-17} S/cm for other plasma polymers [17]. This illustrates that plasma deposition conditions can have a significant effect on polymer film conductivities and the relationship between the two should be further investigated.

CONCLUSIONS

A pulsed PECVD reactor has been constructed for laboratory scale deposition of plasma polymerized films. Acetylene and aniline films produced in the reactor have been characterized using SEM and AFM. Conductivity of the polyacetylene film was also measured. SEM analysis demonstrated that smooth films can easily be produced at certain conditions. A conductivity of 1.9×10^{-10} S/cm for acetylene films was measured. Conditions during plasma discharge were also studied. Fast ionization gauge was used for measurement of transient reactor pressures and a photodiode was used for measuring plasma light emission.

ACKNOWLEDGEMENTS

This work was partially supported by the Department of Chemical Engineering and the School of Electrical Engineering and Computer Science at Washington State University in Pullman, Washington. The computer control system was designed and assembled by Wing Lo, Keith Fahlenkamp and Mike Collins.

REFERENCES

1. A. Dollet, L. Layeillon, J.P. Couderc, and B. Despax, Plasma Sources Sci. Technol. 4 (1995) 459-473
2. G. Scarsbrook, I.P. Llewellyn, S.M. Ojha, and R.A. Heinecke, Vacuum, 38 (1988) 627-631
3. H. Yasuda and T. Hsu, J. Polym. Sci.: Polym. Chem., 15 (1987) 81-97
4. K. Ebihara and S. Maeda, Mat. Res. Soc. Symp. Proc., 98 (1987) 249-255
5. C. B. Labelle and K.K. Gleason, J. Vac. Sci. Technol. A 17 (1999) 445-452
6. V. Panchalingam, X. Chen, C.R. Savage, R.B. Timmons, Polym. Prepr. 34 (1993) 681-682
7. K. Ebihara and S. Maeda, J. Appl. Phys., 57 (1985) 2482-2485
8. J.H. Leck, *Pressure measurement in vacuum systems*, 2nd ed., London: Chapman and Hall LTD, 1964.
9. G.L. Weissler, R.W. Carlson, *Methods of Experimental Physics*, Vacuum Physics and Technology, vol. 14, New York: Academic Press, 1979, p.60.
10. P. D. Pedrow, A. M. Nasiruddin, and R. Mahalingam, J. Appl. Phys., 67 (1990) 6109-6113.
11. D.M. Gruen, C.D. Zuiker, A.R. Krauss, and X. Pan, J. Vac. Sci. Technol. A 13 (1995) 1628-1632.
12. K.S. Mogensen, S.S. Eskildsen, C. Mathiasen, J. Bottiger, Surf. Coat. Technol. 102 (1998) 41-49

13. M. Chen, T.C. Yang, Z.G. Ma, IEEE Trans. Plasma Sci. 23 (1995) 151-155.
14. A.M. Nasiruddin, "Pulsed RF plasma source for materials processing", Electrical Engineering PhD Thesis, Washington State University, May 1991.
15. K.O. Goyal, "Studies on preparation and characterization of thin acetylene plasma polymer films in a pulsed PECVD reactor", Chemical Engineering Master's Thesis, Washington State University 1995.
16. T. Nakano, S. Koike, and Y. Ohki, J. Phys. D: Appl. Phys. 23 (1990) 711-718
17. H. Yasuda, *Plasma Polymerization*, Orlando:Academic Press, 1985, p.396.

AUTHOR INDEX

SUBJECT INDEX

Printed in the United States
By Bookmasters